AF828609

Reinforcement Learning and Optimal Control

by

Dimitri P. Bertsekas

Massachusetts Institute of Technology

WWW site for book information and orders

http://www.athenasc.com

Athena Scientific, Belmont, Massachusetts

Athena Scientific
Post Office Box 805
Nashua, NH 03060
U.S.A.

Email: info@athenasc.com
WWW: http://www.athenasc.com

Cover photography: Dimitri Bertsekas

© 2019 Dimitri P. Bertsekas
All rights reserved. No part of this book may be reproduced in any form by any electronic or mechanical means (including photocopying, recording, or information storage and retrieval) without permission in writing from the publisher.

Publisher's Cataloging-in-Publication Data

Bertsekas, Dimitri P.
Reinforcement Learning and Optimal Control
Includes Bibliography and Index
1. Mathematical Optimization. 2. Dynamic Programming. I. Title.
QA402.5 .B465 2019 519.703 00-91281

ISBN-10: 1-886529-39-6, ISBN-13: 978-1-886529-39-7

2nd Printing (includes editorial revisions)

ABOUT THE AUTHOR

Dimitri Bertsekas studied Mechanical and Electrical Engineering at the National Technical University of Athens, Greece, and obtained his Ph.D. in system science from the Massachusetts Institute of Technology. He has held faculty positions with the Engineering-Economic Systems Department, Stanford University, and the Electrical Engineering Department of the University of Illinois, Urbana. Since 1979 he has been teaching at the Electrical Engineering and Computer Science Department of the Massachusetts Institute of Technology (M.I.T.), where he is McAfee Professor of Engineering. In 2019, he joined the School of Computing, Informatics, and Decision Systems Engineering at the Arizona State University, Tempe, AZ, as Fulton Professor of Computational Decision Making.

Professor Bertsekas' teaching and research have spanned several fields, including deterministic optimization, dynamic programming and stochastic control, large-scale and distributed computation, and data communication networks. He has authored or coauthored numerous research papers and seventeen books, several of which are currently used as textbooks in MIT classes, including "Dynamic Programming and Optimal Control," "Data Networks," "Introduction to Probability," "Convex Optimization Algorithms," and "Nonlinear Programming."

Professor Bertsekas was awarded the INFORMS 1997 Prize for Research Excellence in the Interface Between Operations Research and Computer Science for his book "Neuro-Dynamic Programming" (co-authored with John Tsitsiklis), the 2001 AACC John R. Ragazzini Education Award, the 2009 INFORMS Expository Writing Award, the 2014 AACC Richard Bellman Heritage Award, the 2014 INFORMS Khachiyan Prize for Life-Time Accomplishments in Optimization, the 2015 MOS/SIAM George B. Dantzig Prize, and (jointly with John Tsitsiklis) the 2018 INFORMS John von Neumann Theory Prize for the contributions of the research monographs "Parallel and Distributed Computation" and "Neuro-Dynamic Programming." Professor Bertsekas was elected in 2001 to the United States National Academy of Engineering for "pioneering contributions to fundamental research, practice and education of optimization/control theory, and especially its application to data communication networks."

ATHENA SCIENTIFIC

OPTIMIZATION AND COMPUTATION SERIES

1. Reinforcement Learning and Optimal Control, by Dimitri P. Bertsekas, 2019, ISBN 978-1-886529-39-7, 388 pages
2. Abstract Dynamic Programming, 2nd Edition, by Dimitri P. Bertsekas, 2018, ISBN 978-1-886529-46-5, 360 pages
3. Dynamic Programming and Optimal Control, Two-Volume Set, by Dimitri P. Bertsekas, 2017, ISBN 1-886529-08-6, 1270 pages
4. Nonlinear Programming, 3rd Edition, by Dimitri P. Bertsekas, 2016, ISBN 1-886529-05-1, 880 pages
5. Convex Optimization Algorithms, by Dimitri P. Bertsekas, 2015, ISBN 978-1-886529-28-1, 576 pages
6. Convex Optimization Theory, by Dimitri P. Bertsekas, 2009, ISBN 978-1-886529-31-1, 256 pages
7. Introduction to Probability, 2nd Edition, by Dimitri P. Bertsekas and John N. Tsitsiklis, 2008, ISBN 978-1-886529-23-6, 544 pages
8. Convex Analysis and Optimization, by Dimitri P. Bertsekas, Angelia Nedić, and Asuman E. Ozdaglar, 2003, ISBN 1-886529-45-0, 560 pages
9. Network Optimization: Continuous and Discrete Models, by Dimitri P. Bertsekas, 1998, ISBN 1-886529-02-7, 608 pages
10. Network Flows and Monotropic Optimization, by R. Tyrrell Rockafellar, 1998, ISBN 1-886529-06-X, 634 pages
11. Introduction to Linear Optimization, by Dimitris Bertsimas and John N. Tsitsiklis, 1997, ISBN 1-886529-19-1, 608 pages
12. Parallel and Distributed Computation: Numerical Methods, by Dimitri P. Bertsekas and John N. Tsitsiklis, 1997, ISBN 1-886529-01-9, 718 pages
13. Neuro-Dynamic Programming, by Dimitri P. Bertsekas and John N. Tsitsiklis, 1996, ISBN 1-886529-10-8, 512 pages
14. Constrained Optimization and Lagrange Multiplier Methods, by Dimitri P. Bertsekas, 1996, ISBN 1-886529-04-3, 410 pages
15. Stochastic Optimal Control: The Discrete-Time Case, by Dimitri P. Bertsekas and Steven E. Shreve, 1996, ISBN 1-886529-03-5, 330 pages

Contents

Preface

Turning to the succor of modern computing machines, let us renounce all analytic tools.

Richard Bellman [Bel57]

From a teleological point of view the particular numerical solution of any particular set of equations is of far less importance than the understanding of the nature of the solution.

Richard Bellman [Bel57]

In this book we consider large and challenging multistage decision problems, which can be solved in principle by dynamic programming (DP for short), but their exact solution is computationally intractable. We discuss solution methods that rely on approximations to produce suboptimal policies with adequate performance. These methods are collectively known by several essentially equivalent names: *reinforcement learning*, *approximate dynamic programming*, and *neuro-dynamic programming*. We will use primarily the most popular name: reinforcement learning.

Our subject has benefited greatly from the interplay of ideas from optimal control and from artificial intelligence. One of the aims of the book is to explore the common boundary between these two fields and to form a bridge that is accessible by workers with background in either field. Another aim is to organize coherently the broad mosaic of methods that have proved successful in practice while having a solid theoretical and/or logical foundation. This may help researchers and practitioners to find their way through the maze of competing ideas that constitute the current state of the art.

There are two general approaches for DP-based suboptimal control. The first is *approximation in value space*, where we approximate in some way the optimal cost-to-go function with some other function. The major alternative to approximation in value space is *approximation in policy space*, whereby we select the policy by using optimization over a suitably restricted class of policies, usually a parametric family of some form. In some schemes these two types of approximation may be combined, aiming to capitalize on the advantages of both. Generally, approximation in value space is tied more closely to the central DP ideas of value and policy iteration than approximation in policy space, which relies on gradient-like descent, a more broadly applicable optimization mechanism.

While we provide a substantial treatment of approximation in policy space, most of the book is focused on approximation in value space. Here, the control at each state is obtained by optimization of the cost over a limited horizon, plus an approximation of the optimal future cost. The latter cost, which we generally denote by $\tilde{J}$, is a function of the state where we may be. It may be computed by a variety of methods, possibly involving simulation and/or some given or separately derived heuristic/suboptimal policy. The use of simulation often allows for implementations that do not require a mathematical model, a major idea that has allowed the use of DP beyond its classical boundaries.

We discuss selectively four types of methods for obtaining $\tilde{J}$:

(a) *Problem approximation*: Here $\tilde{J}$ is the optimal cost function of a related simpler problem, which is solved by exact DP. Certainty equivalent control and enforced decomposition schemes are discussed in some detail.

(b) *Rollout and model predictive control*: Here $\tilde{J}$ is the cost function of some known heuristic policy. The needed cost values to implement a rollout policy are often calculated by simulation. While this method applies to stochastic problems, the reliance on simulation favors deterministic problems, including challenging combinatorial problems for which heuristics may be readily implemented. Rollout may also be combined with adaptive simulation and Monte Carlo tree search, which have proved very effective in the context of games such as backgammon, chess, Go, and others.

Model predictive control was originally developed for continuous-space optimal control problems that involve some goal state, e.g., the origin in a classical control context. It can be viewed as a specialized rollout method that is based on a suboptimal optimization for reaching a goal state.

(c) *Parametric cost approximation*: Here $\tilde{J}$ is chosen from within a parametric class of functions, including neural networks, with the parameters "optimized" or "trained" by using state-cost sample pairs and some type of incremental least squares/regression algorithm. Approximate policy iteration and its variants are covered in some detail, including several actor and critic schemes. These involve policy evaluation with simulation-based training methods, and policy improvement that may rely on approximation in policy space.

(d) *Aggregation*: Here $\tilde{J}$ is the optimal cost function of some approximation to the original problem, called aggregate problem, which has fewer states. The aggregate problem can be formulated in a variety of ways, and may be solved by using exact DP techniques. Its optimal cost function is then used as $\tilde{J}$ in a limited horizon optimization scheme. Aggregation may also be used to provide local improvements to parametric approximation schemes that involve neural networks or linear feature-based architectures.

We have adopted a gradual expository approach, which proceeds along four directions:

(1) *From exact DP to approximate DP*: We first discuss exact DP algorithms, explain why they may be difficult to implement, and then use them as the basis for approximations.

(2) *From finite horizon to infinite horizon problems*: We first discuss finite horizon exact and approximate DP methodologies, which are intuitive and mathematically simple in Chapters 1-3. We then progress to infinite horizon problems in Chapters 4-6.

(3) *From deterministic to stochastic models*: We often discuss separately deterministic and stochastic problems. The reason is that deterministic problems are simpler and offer special advantages for some of our methods.

(4) *From model-based to model-free implementations*: Reinforcement learning methods offer a major potential benefit over classical DP approaches, which were practiced exclusively up to the early 90s: they can be implemented by using a simulator/computer model rather than a mathematical model. In our presentation, we first discuss model-based implementations, and then we identify schemes that can be appropriately modified to work with a simulator.

After the first chapter, each new class of methods is introduced as a more sophisticated or generalized version of a simpler method introduced earlier. Moreover, we illustrate some of the methods by means of examples, which should be helpful in providing insight into their use, but may also be skipped selectively and without loss of continuity.

The mathematical style of this book is somewhat different from the one of the author's DP books [Ber12], [Ber17], [Ber18a], and the 1996 neuro-dynamic programming (NDP) research monograph, written jointly with John Tsitsiklis [BeT96]. While we provide a rigorous, albeit short, mathematical account of the theory of finite and infinite horizon DP, and some fundamental approximation methods, we rely more on intuitive explanations and less on proof-based insights. Moreover, our mathematical requirements are quite modest: calculus, a minimal use of matrix-vector algebra, and elementary probability (mathematically complicated arguments involving laws of large numbers and stochastic convergence are bypassed in favor of intuitive explanations). In this way, we hope that the book will be accessible and useful to a broader cross section of readers.

Still in our use of a more intuitive but less proof-oriented expository style, we have followed a few basic principles. The most important of these is to maintain rigor in the use of natural language. The reason is that with fewer mathematical arguments and proofs, precise language is essential to maintain a logically consistent exposition. In particular, we have aimed to define terms unambiguously, and to avoid using multiple terms with essentially identical meaning. Moreover, when circumstances permitted, we have tried to provide enough explanation/intuition so that a mathematician can find the development believable and even construct the missing rigorous proofs.

We note that several of the methods that we present are often successful in practice, but have less than solid performance properties. This is a reflection of the state of the art in the field: there are no methods that are guaranteed to work for all or even most problems, but there are enough methods to try on a given challenging problem with a reasonable chance of success in the end.† To aid in this process, we place primary emphasis on developing intuition into the inner workings of each type of method. Still, however, it is important to have a foundational understanding of the analytical principles of the field and of the mechanisms underlying the central computational methods. To quote a statement from the preface of the NDP monograph [BeT96]: "It is primarily through an understanding of the mathematical structure of the NDP methodology that we will be able to identify promising or solid algorithms from the bewildering array of speculative proposals and claims that can be found in the literature."

Another statement from a recent NY Times article [Str18], in connection with DeepMind's remarkable AlphaZero chess program, is also worth quoting: "What is frustrating about machine learning, however, is that the algorithms can't articulate what they're thinking. We don't know why they work, so we don't know if they can be trusted. AlphaZero gives every appearance of having discovered some important principles about chess,

---

† While reinforcement learning rests on the mathematical principles of DP as well as machine learning, it also relies on multiple interacting approximations whose effects are hard to predict and quantify in practice. One may hope that with further theoretical and applications research, the state of the subject will improve and clarify, however it can be said that in its current form, reinforcement learning is an exploding field, which is complicated, unclean, and in many ways confusing (something that, incidentally, the front cover image of the book also tries to convey). Reinforcement learning is not unique in this. One may think of other important algorithmic optimization areas where a similar state of the art has prevailed for a long time.

but it can't share that understanding with us. Not yet, at least. As human beings, we want more than answers. We want insight. This is going to be a source of tension in our interactions with computers from now on."† To this we may add that human insight can only develop within some structure of human thought, and it appears that mathematical reasoning with algorithmic models is the most suitable structure for this purpose.

I would like to express my appreciation to the many students and colleagues that contributed directly or indirectly to the book. A special thanks is due to my principal collaborators on the subject, over the last 25 years, particularly John Tsitsiklis, Janey (Huizhen) Yu, and Mengdi Wang. Moreover, sharing insights with Ben Van Roy over the years has been important in shaping my thinking. Interactions with Ben Recht regarding policy gradient methods were also very helpful. The projects that my students worked on as part of DP courses I taught at MIT inspired many ideas that indirectly found their way into the book. I want to express my thanks to the many readers, who proofread parts of the book. In this respect I would like to single out Yuchao Li who made many helpful comments, and Thomas Stahlbuhk, who went through the entire book with great care, and offered numerous insightful suggestions.

The book took shape while teaching a course on the subject at the Arizona State University (ASU) during a two-month period starting in January 2019. Videolectures and slides from this class, as well as related research material, are available from my website

http://web.mit.edu/dimitrib/www/RLbook.html

and provide a good supplement and companion resource to the book. The hospitable and stimulating environment at ASU contributed much to my productivity during this period, and for this I am very thankful to Stephanie Gil, as well as other colleagues from ASU, including Heni Ben

---

† The two 1957 Bellman quotations at the beginning of this preface also express this tension, although the first of these, while striking and widely cited, is admittedly taken a little out of context (throughout his work on practical applications, Bellman remained a mathematical analyst at heart). Bellman's fascinating autobiography [Bel84] contains a lot of information on the origins of DP (and approximate DP as well!); selected quotations from this autobiography have been compiled by his collaborator Dreyfus [Dre02]. Among others, Bellman states that "In order to make any progress, it is necessary to think of approximate techniques, and above all, of numerical algorithms. Finally, having devoted a great deal of time and effort, mostly fruitless, to the analysis of many varieties of simple models, I was prepared to face up to the challenge of using dynamic programming as an effective tool for obtaining numerical answers to numerical questions." He goes on to attribute his motivation to work on numerical DP to the emergence of the (then primitive) digital computer, which he calls "Sorcerer's Apprentice."

Amor, Esma Gel, Subbarao (Rao) Kambhampati, Angelia Nedic, Giulia Pedrielli, Jennie Si, and Petr Sulc. Moreover, Stephanie together with her students, Sushmita Bhattacharya and Thomas Wheeler, collaborated with me on research and implementation of various methods, contributed many insights, and tested out several variations.

Dimitri P. Bertsekas
June 2019

1

Exact Dynamic Programming

Contents

In this chapter, we provide some background on exact dynamic programming (DP for short), with a view towards the suboptimal solution methods that are the main subject of this book. These methods are known by several essentially equivalent names: *reinforcement learning*, *approximate dynamic programming*, and *neuro-dynamic programming*. In this book, we will use primarily the most popular name: reinforcement learning (RL for short).

We first consider finite horizon problems, which involve a finite sequence of successive decisions, and are thus conceptually and analytically simpler. We defer the discussion of the more intricate infinite horizon problems to Chapters 4-6. We also discuss separately deterministic and stochastic problems (Sections 1.1 and 1.2, respectively). The reason is that deterministic problems are simpler and lend themselves better as an entry point to the optimal control methodology. Moreover, they have some favorable characteristics, which allow the application of a broader variety of methods. For example, simulation-based methods are greatly simplified and sometimes better understood in the context of deterministic optimal control.

Finally, in Section 1.3 we provide various examples of DP formulations, illustrating some of the concepts of Sections 1.1 and 1.2. The reader with substantial background in DP may wish to just scan Section 1.3 and skip to the next chapter, where we start the development of the approximate DP methodology.

1.1 DETERMINISTIC DYNAMIC PROGRAMMING

All DP problems involve a discrete-time dynamic system that generates a sequence of states under the influence of control. In finite horizon problems the system evolves over a finite number N of time steps (also called stages). The state and control at time k are denoted by x_k and u_k, respectively. In deterministic systems, x_{k+1} is generated nonrandomly, i.e., it is determined solely by x_k and u_k.

1.1.1 Deterministic Problems

A deterministic DP problem involves a discrete-time dynamic system of the form

$$x_{k+1} = f_k(x_k, u_k), \qquad k = 0, 1, \ldots, N-1, \tag{1.1}$$

where

k is the time index,

x_k is the state of the system, an element of some space,

u_k is the control or decision variable, to be selected at time k from some given set $U_k(x_k)$ that depends on x_k,

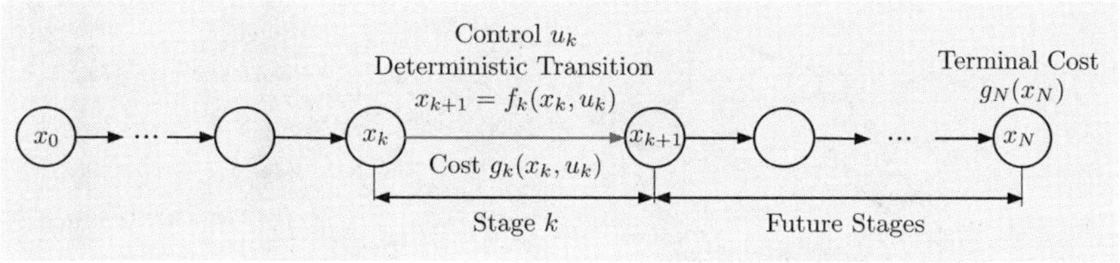

Figure 1.1.1 Illustration of a deterministic N-stage optimal control problem. Starting from state x_k, the next state under control u_k is generated nonrandomly, according to

$$x_{k+1} = f_k(x_k, u_k),$$

and a stage cost $g_k(x_k, u_k)$ is incurred.

f_k is a function of (x_k, u_k) that describes the mechanism by which the state is updated from time k to time $k + 1$.

N is the horizon or number of times control is applied.

The set of all possible x_k is called the *state space* at time k. It can be any set and can depend on k; *this generality is one of the great strengths of the DP methodology*. Similarly, the set of all possible u_k is called the *control space* at time k. Again it can be any set and can depend on k.

The problem also involves a cost function that is additive in the sense that the cost incurred at time k, denoted by $g_k(x_k, u_k)$, accumulates over time. Formally, g_k is a function of (x_k, u_k) that takes real number values, and may depend on k. For a given initial state x_0, the total cost of a control sequence $\{u_0, \ldots, u_{N-1}\}$ is

$$J(x_0; u_0, \ldots, u_{N-1}) = g_N(x_N) + \sum_{k=0}^{N-1} g_k(x_k, u_k), \tag{1.2}$$

where $g_N(x_N)$ is a terminal cost incurred at the end of the process. The total cost is a well-defined number, since the control sequence $\{u_0, \ldots, u_{N-1}\}$ together with x_0 determines exactly the state sequence $\{x_1, \ldots, x_N\}$ via the system equation (1.1). We want to minimize the cost (1.2) over all sequences $\{u_0, \ldots, u_{N-1}\}$ that satisfy the control constraints, thereby obtaining the optimal value†

$$J^*(x_0) = \min_{\substack{u_k \in U_k(x_k) \\ k=0,\ldots,N-1}} J(x_0; u_0, \ldots, u_{N-1}),$$

as a function of x_0. Figure 1.1.1 illustrates the main elements of the problem.

We will next illustrate deterministic problems with some examples.

† We use throughout "min" (in place of "inf") to indicate minimal value over a feasible set of controls, even when we are not sure that the minimum is attained by some feasible control.

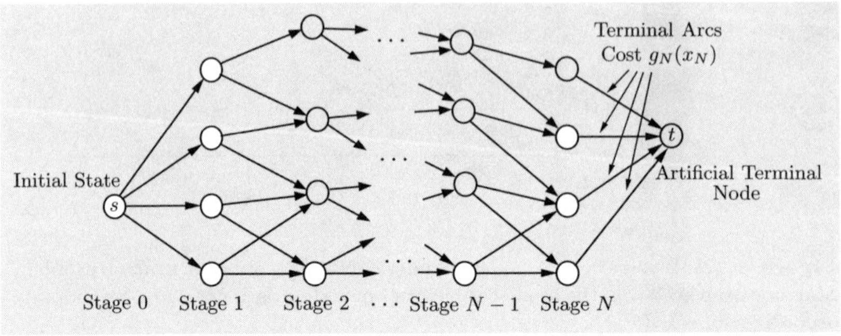

Figure 1.1.2 Transition graph for a deterministic finite-state system. Nodes correspond to states x_k. Arcs correspond to state-control pairs (x_k, u_k). An arc (x_k, u_k) has start and end nodes x_k and $x_{k+1} = f_k(x_k, u_k)$, respectively. We view the cost $g_k(x_k, u_k)$ of the transition as the length of this arc. The problem is equivalent to finding a shortest path from initial node s to terminal node t.

Discrete Optimal Control Problems

There are many situations where the state and control spaces are naturally discrete and consist of a finite number of elements. Such problems are often conveniently described with an acyclic graph specifying for each state x_k the possible transitions to next states x_{k+1}. The nodes of the graph correspond to states x_k and the arcs of the graph correspond to state-control pairs (x_k, u_k). Each arc with start node x_k corresponds to a choice of a single control $u_k \in U_k(x_k)$ and has as end node the next state $f_k(x_k, u_k)$. The cost of an arc (x_k, u_k) is defined as $g_k(x_k, u_k)$; see Fig. 1.1.2. To handle the final stage, an artificial terminal node t is added. Each state x_N at stage N is connected to the terminal node t with an arc having cost $g_N(x_N)$.

Note that control sequences $\{u_0, \ldots, u_{N-1}\}$ correspond to paths originating at the initial state (node s at stage 0) and terminating at one of the nodes corresponding to the final stage N. If we view the cost of an arc as its length, we see that *a deterministic finite-state finite-horizon problem is equivalent to finding a minimum-length (or shortest) path from the initial node s of the graph to the terminal node t.* Here, by the length of a path we mean the sum of the lengths of its arcs.†

Generally, combinatorial optimization problems can be formulated as deterministic finite-state finite-horizon optimal control problem. The following scheduling example illustrates the idea.

---

† It turns out also that any shortest path problem (with a possibly nonacyclic graph) can be reformulated as a finite-state deterministic optimal control problem, as we will show in Section 1.3.1. See [Ber17], Section 2.1, and [Ber91], [Ber98] for an extensive discussion of shortest path methods, which connects with our discussion here.

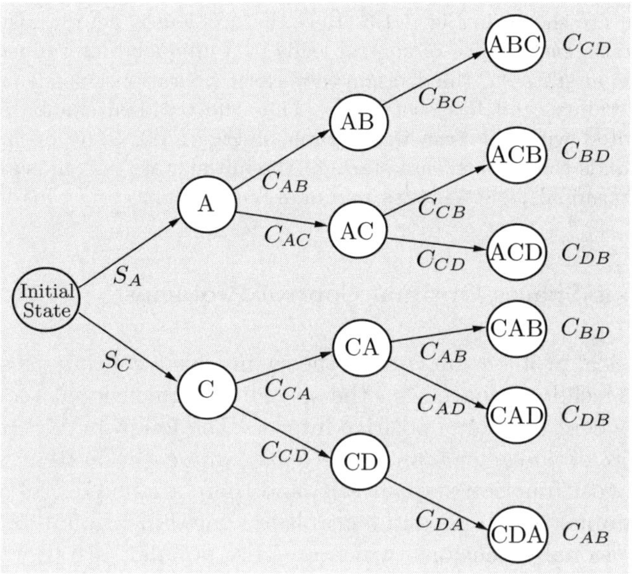

Figure 1.1.3 The transition graph of the deterministic scheduling problem of Example 1.1.1. Each arc of the graph corresponds to a decision leading from some state (the start node of the arc) to some other state (the end node of the arc). The corresponding cost is shown next to the arc. The cost of the last operation is shown as a terminal cost next to the terminal nodes of the graph.

Example 1.1.1 (A Deterministic Scheduling Problem)

Suppose that to produce a certain product, four operations must be performed on a certain machine. The operations are denoted by A, B, C, and D. We assume that operation B can be performed only after operation A has been performed, and operation D can be performed only after operation C has been performed. (Thus the sequence CDAB is allowable but the sequence CDBA is not.) The setup cost C_{mn} for passing from any operation m to any other operation n is given. There is also an initial startup cost S_A or S_C for starting with operation A or C, respectively (cf. Fig. 1.1.3). The cost of a sequence is the sum of the setup costs associated with it; for example, the operation sequence ACDB has cost

$$S_A + C_{AC} + C_{CD} + C_{DB}.$$

We can view this problem as a sequence of three decisions, namely the choice of the first three operations to be performed (the last operation is determined from the preceding three). It is appropriate to consider as state the set of operations already performed, the initial state being an artificial state corresponding to the beginning of the decision process. The possible state transitions corresponding to the possible states and decisions for this problem are shown in Fig. 1.1.3. Here the problem is deterministic, i.e., at a given state, each choice of control leads to a uniquely determined state. For example, at state AC the decision to perform operation D leads to state ACD with certainty, and has cost C_{CD}. Thus the problem can be conveniently represented with the transition graph of Fig. 1.1.3. The optimal solution corresponds to the path that starts at the initial state and ends at some state at the terminal time and has minimum sum of arc costs plus the terminal cost.

Continuous-Spaces Optimal Control Problems

Many classical problems in control theory involve a continuous state space, such as a Euclidean space, i.e., the space of n-dimensional vectors of real variables, where n is some positive integer. The following is representative of the class of *linear-quadratic problems*, where the system equation is linear, the cost function is quadratic, and there are no control constraints. In our example, the states and controls are one-dimensional, but there are multidimensional extensions, which are very popular (see [Ber17], Section 3.1).

Example 1.1.2 (A Linear-Quadratic Problem)

A certain material is passed through a sequence of N ovens (see Fig. 1.1.4). Denote

x_0: initial temperature of the material,

x_k, $k = 1, \ldots, N$: temperature of the material at the exit of oven k,

u_{k-1}, $k = 1, \ldots, N$: heat energy applied to the material in oven k. In practice there will be some constraints on u_k, such as nonnegativity. However, for analytical tractability one may also consider the case where u_k is unconstrained, and check later if the solution satisfies some natural restrictions in the problem at hand.

We assume a system equation of the form

$$x_{k+1} = (1 - a)x_k + au_k, \qquad k = 0, 1, \ldots, N - 1,$$

where a is a known scalar from the interval $(0, 1)$. The objective is to get the final temperature x_N close to a given target T, while expending relatively little energy. We express this with a cost function of the form

$$r(x_N - T)^2 + \sum_{k=0}^{N-1} u_k^2,$$

where $r > 0$ is a given scalar that trades off the error in attaining the final temperature with the expended energy.

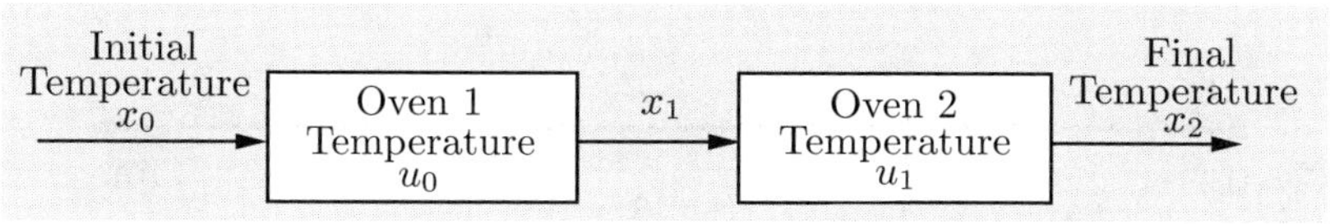

Figure 1.1.4 The linear-quadratic problem of Example 1.1.2 for $N = 2$. The temperature of the material evolves according to the system equation $x_{k+1} = (1-a)x_k + au_k$, where a is some scalar with $0 < a < 1$.

Linear-quadratic problems with no constraints on the state or the control admit a nice analytical solution, as we will see later in Section 1.3.7. In another frequently arising optimal control problem there are linear constraints on the state and/or the control. In the preceding example it would have been natural to require that $a_k \leq x_k \leq b_k$ and/or $c_k \leq u_k \leq d_k$, where a_k, b_k, c_k, d_k are given scalars. Then the problem would be solvable not only by DP but also by quadratic programming methods. Generally deterministic optimal control problems with continuous state and control spaces (in addition to DP) admit a solution by nonlinear programming methods, such as gradient, conjugate gradient, and Newton's method, which can be suitably adapted to their special structure.

1.1.2 The Dynamic Programming Algorithm

In this section we will state the DP algorithm and formally justify it. The algorithm rests on a simple idea, the *principle of optimality*, which roughly states the following; see Fig. 1.1.5.

Principle of Optimality

Let $\{u_0^*, \ldots, u_{N-1}^*\}$ be an optimal control sequence, which together with x_0 determines the corresponding state sequence $\{x_1^*, \ldots, x_N^*\}$ via the system equation (1.1). Consider the subproblem whereby we start at x_k^* at time k and wish to minimize the "cost-to-go" from time k to time N,

$$g_k(x_k^*, u_k) + \sum_{m=k+1}^{N-1} g_m(x_m, u_m) + g_N(x_N),$$

over $\{u_k, \ldots, u_{N-1}\}$ with $u_m \in U_m(x_m)$, $m = k, \ldots, N-1$. Then the truncated optimal control sequence $\{u_k^*, \ldots, u_{N-1}^*\}$ is optimal for this subproblem.

The subproblem referred to above is called the *tail subproblem* that starts at x_k^*. Stated succinctly, the principle of optimality says that *the tail of an optimal sequence is optimal for the tail subproblem.* Its intuitive

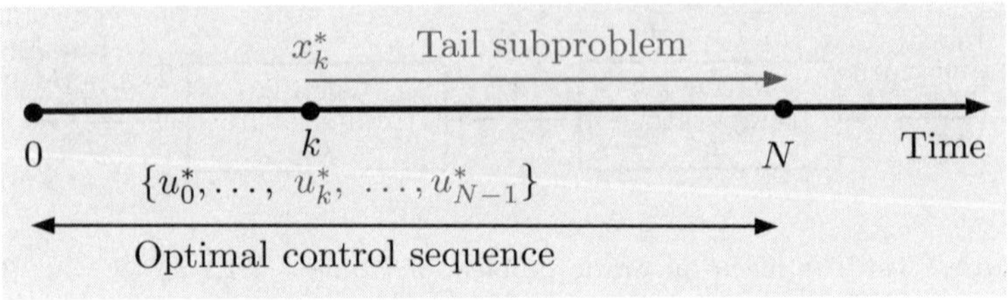

Figure 1.1.5 Illustration of the principle of optimality. The tail $\{u_k^*, \ldots, u_{N-1}^*\}$ of an optimal sequence $\{u_0^*, \ldots, u_{N-1}^*\}$ is optimal for the tail subproblem that starts at the state x_k^* of the optimal trajectory $\{x_1^*, \ldots, x_N^*\}$.

justification is simple. If the truncated control sequence $\{u_k^*, \ldots, u_{N-1}^*\}$ were not optimal as stated, we would be able to reduce the cost further by switching to an optimal sequence for the subproblem once we reach x_k^* (since the preceding choices of controls, $u_0^*, \ldots, u_{k-1}^*$, do not restrict our future choices). For an auto travel analogy, suppose that the fastest route from Los Angeles to Boston passes through Chicago. The principle of optimality translates to the obvious fact that the Chicago to Boston portion of the route is also the fastest route for a trip that starts from Chicago and ends in Boston.

The principle of optimality suggests that the optimal cost function can be constructed in piecemeal fashion going backwards: first compute the optimal cost function for the "tail subproblem" involving the last stage, then solve the "tail subproblem" involving the last two stages, and continue in this manner until the optimal cost function for the entire problem is constructed.

The DP algorithm is based on this idea: it proceeds sequentially, by *solving all the tail subproblems of a given time length, using the solution of the tail subproblems of shorter time length.* We illustrate the algorithm with the scheduling problem of Example 1.1.1. The calculations are simple but tedious, and may be skipped without loss of continuity. However, they may be worth going over by a reader that has no prior experience in the use of DP.

Example 1.1.1 (Scheduling Problem - Continued)

Let us consider the scheduling Example 1.1.1, and let us apply the principle of optimality to calculate the optimal schedule. We have to schedule optimally the four operations A, B, C, and D. There a cost for a transition between two operations, and the numerical values of the transition costs are shown in Fig. 1.1.6 next to the corresponding arcs.

According to the principle of optimality, the "tail" portion of an optimal schedule must be optimal. For example, suppose that the optimal schedule is CABD. Then, having scheduled first C and then A, it must be optimal to complete the schedule with BD rather than with DB. With this in mind, we solve all possible tail subproblems of length two, then all tail subproblems of length three, and finally the original problem that has length four (the subproblems of length one are of course trivial because there is only one operation that is as yet unscheduled). As we will see shortly, the tail subproblems of length $k + 1$ are easily solved once we have solved the tail subproblems of length k, and this is the essence of the DP technique.

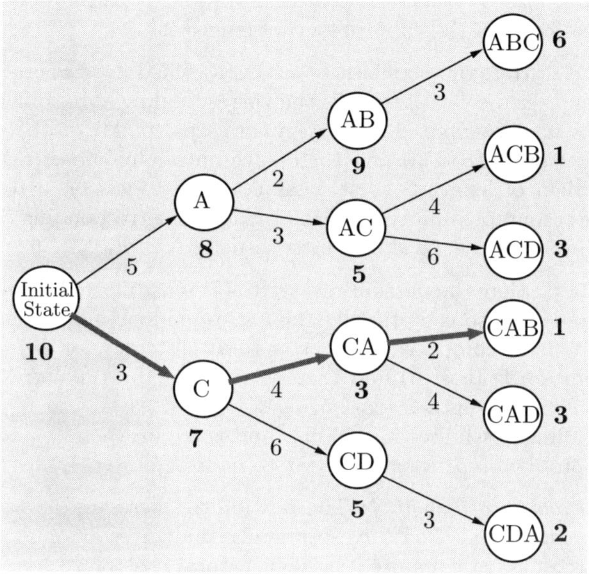

Figure 1.1.6 Transition graph of the deterministic scheduling problem, with the cost of each decision shown next to the corresponding arc. Next to each node/state we show the cost to optimally complete the schedule starting from that state. This is the optimal cost of the corresponding tail subproblem (cf. the principle of optimality). The optimal cost for the original problem is equal to 10, as shown next to the initial state. The optimal schedule corresponds to the thick-line arcs.

Tail Subproblems of Length 2: These subproblems are the ones that involve two unscheduled operations and correspond to the states AB, AC, CA, and CD (see Fig. 1.1.6).

State AB: Here it is only possible to schedule operation C as the next operation, so the optimal cost of this subproblem is 9 (the cost of scheduling C after B, which is 3, plus the cost of scheduling D after C, which is 6).

State AC: Here the possibilities are to (a) schedule operation B and then D, which has cost 5, or (b) schedule operation D and then B, which has cost 9. The first possibility is optimal, and the corresponding cost of the tail subproblem is 5, as shown next to node AC in Fig. 1.1.6.

State CA: Here the possibilities are to (a) schedule operation B and then D, which has cost 3, or (b) schedule operation D and then B, which has cost 7. The first possibility is optimal, and the corresponding cost of the tail subproblem is 3, as shown next to node CA in Fig. 1.1.6.

State CD: Here it is only possible to schedule operation A as the next operation, so the optimal cost of this subproblem is 5.

Tail Subproblems of Length 3: These subproblems can now be solved using the optimal costs of the subproblems of length 2.

State A: Here the possibilities are to (a) schedule next operation B (cost 2) and then solve optimally the corresponding subproblem of length 2 (cost 9, as computed earlier), a total cost of 11, or (b) schedule next operation C (cost 3) and then solve optimally the corresponding subproblem of length 2 (cost 5, as computed earlier), a total cost of 8. The second possibility is optimal, and the corresponding cost of the tail subproblem is 8, as shown next to node A in Fig. 1.1.6.

State C: Here the possibilities are to (a) schedule next operation A (cost 4) and then solve optimally the corresponding subproblem of length 2 (cost 3, as computed earlier), a total cost of 7, or (b) schedule next operation D (cost 6) and then solve optimally the corresponding subproblem of length 2 (cost 5, as computed earlier), a total cost of 11. The first possibility is optimal, and the corresponding cost of the tail subproblem is 7, as shown next to node A in Fig. 1.1.6.

Original Problem of Length 4: The possibilities here are (a) start with operation A (cost 5) and then solve optimally the corresponding subproblem of length 3 (cost 8, as computed earlier), a total cost of 13, or (b) start with operation C (cost 3) and then solve optimally the corresponding subproblem of length 3 (cost 7, as computed earlier), a total cost of 10. The second possibility is optimal, and the corresponding optimal cost is 10, as shown next to the initial state node in Fig. 1.1.6.

Note that having computed the optimal cost of the original problem through the solution of all the tail subproblems, we can construct the optimal schedule: we begin at the initial node and proceed forward, each time choosing the optimal operation, i.e., the one that starts the optimal schedule for the corresponding tail subproblem. In this way, by inspection of the graph and the computational results of Fig. 1.1.6, we determine that CABD is the optimal schedule.

Finding an Optimal Control Sequence by DP

We now state the DP algorithm for deterministic finite horizon problem by translating into mathematical terms the heuristic argument underlying the principle of optimality. The algorithm constructs functions

$$J_N^*(x_N), J_{N-1}^*(x_{N-1}), \ldots, J_0^*(x_0),$$

sequentially, starting from J_N^*, and proceeding backwards to J_{N-1}^*, J_{N-2}^*, etc.

DP Algorithm for Deterministic Finite Horizon Problems

Start with

$$J_N^*(x_N) = g_N(x_N), \qquad \text{for all } x_N, \tag{1.3}$$

and for $k = 0, \ldots, N-1$, let

$$J_k^*(x_k) = \min_{u_k \in U_k(x_k)} \Big[g_k(x_k, u_k) + J_{k+1}^*\big(f_k(x_k, u_k)\big) \Big], \qquad \text{for all } x_k. \tag{1.4}$$

Note that at stage k, the calculation in (1.4) must be done for all states x_k before proceeding to stage $k-1$. The key fact about the DP algorithm is that for every initial state x_0, the number $J_0^*(x_0)$ obtained at the last step, is equal to the optimal cost $J^*(x_0)$. Indeed, a more general fact can be shown, namely that for all $k = 0, 1, \ldots, N-1$, and all states x_k at time k, we have

$$J_k^*(x_k) = \min_{\substack{u_m \in U_m(x_m) \\ m=k,\ldots,N-1}} J(x_k; u_k, \ldots, u_{N-1}), \tag{1.5}$$

where

$$J(x_k; u_k, \ldots, u_{N-1}) = g_N(x_N) + \sum_{m=k}^{N-1} g_m(x_m, u_m), \tag{1.6}$$

i.e., $J_k^*(x_k)$ is the optimal cost for an $(N-k)$-stage tail subproblem that starts at state x_k and time k, and ends at time N.† Based on this fact, we

† We can prove this by induction. The assertion holds for $k = N$ in view of the initial condition $J_N^*(x_N) = g_N(x_N)$. To show that it holds for all k, we use Eqs. (1.5) and (1.6) to write

$$\begin{aligned} J_k^*(x_k) &= \min_{\substack{u_m \in U_m(x_m) \\ m=k,\ldots,N-1}} \left[g_N(x_N) + \sum_{m=k}^{N-1} g_m(x_m, u_m) \right] \\ &= \min_{u_k \in U_k(x_k)} \left[g_k(x_k, u_k) \vphantom{\sum_{m=k+1}^{N-1}} \right. \\ &\qquad \left. + \min_{\substack{u_m \in U_m(x_m) \\ m=k+1,\ldots,N-1}} \left[g_N(x_N) + \sum_{m=k+1}^{N-1} g_m(x_m, u_m) \right] \right] \\ &= \min_{u_k \in U_k(x_k)} \Big[g_k(x_k, u_k) + J_{k+1}^*\big(f_k(x_k, u_k)\big) \Big], \end{aligned}$$

where for the last equality we use the induction hypothesis. A subtle mathematical point here is that, through the minimization operation, the cost-to-go call $J_k^*(x_k)$ the *optimal cost-to-go* at state x_k and time k, and refer to J_k^* as the *optimal cost-to-go function* or *optimal cost function* at time k. In maximization problems the DP algorithm (1.4) is written with maximization in place of minimization, and then J_k^* is referred to as the *optimal value function* at time k.

Note that the algorithm solves every tail subproblem, i.e., the minimization of the cost accumulated additively starting from an intermediate state up to the end of the horizon. Once the functions $J_0^*,\ldots,J_N^*$ have been obtained, we can use the following forward algorithm to construct an optimal control sequence $\{u_0^*,\ldots,u_{N-1}^*\}$ and corresponding state trajectory $\{x_1^*,\ldots,x_N^*\}$ for the given initial state x_0.

Construction of Optimal Control Sequence $\{u_0^*,\ldots,u_{N-1}^*\}$

Set

$$u_0^* \in \arg\min_{u_0\in U_0(x_0)} \Big[g_0(x_0,u_0) + J_1^*\big(f_0(x_0,u_0)\big)\Big],$$

and

$$x_1^* = f_0(x_0,u_0^*).$$

Sequentially, going forward, for $k=1,2,\ldots,N-1$, set

$$u_k^* \in \arg\min_{u_k\in U_k(x_k^*)} \Big[g_k(x_k^*,u_k) + J_{k+1}^*\big(f_k(x_k^*,u_k)\big)\Big], \tag{1.7}$$

and

$$x_{k+1}^* = f_k(x_k^*,u_k^*).$$

The same algorithm can be used to find an optimal control sequence for any tail subproblem. Figure 1.1.6 traces the calculations of the DP algorithm for the scheduling Example 1.1.1. The numbers next to the nodes, give the corresponding cost-to-go values, and the thick-line arcs give the construction of the optimal control sequence using the preceding algorithm.

1.1.3 Approximation in Value Space

The preceding forward optimal control sequence construction is possible only after we have computed $J_k^*(x_k)$ by DP for all x_k and k. Unfortunately, in practice this is often prohibitively time-consuming, because of the number of possible x_k and k can be very large. However, a similar

functions J_k^* may take the value $-\infty$ for some x_k. Still the preceding induction argument is valid even if this is so.

forward algorithmic process can be used if the optimal cost-to-go functions J_k^* are replaced by some approximations $\tilde J_k$. This is the basis for *approximation in value space*, which will be central in our future discussions. It constructs a suboptimal solution $\{\tilde u_0,\ldots,\tilde u_{N-1}\}$ in place of the optimal $\{u_0^*,\ldots,u_{N-1}^*\}$, based on using $\tilde J_k$ in place of J_k^* in the DP procedure (1.7).

Approximation in Value Space - Use of $\tilde J_k$ in Place of J_k^*

Start with

$$\tilde u_0 \in \arg\min_{u_0\in U_0(x_0)} \Big[g_0(x_0,u_0)+\tilde J_1\big(f_0(x_0,u_0)\big)\Big],$$

and set

$$\tilde x_1 = f_0(x_0,\tilde u_0).$$

Sequentially, going forward, for $k=1,2,\ldots,N-1$, set

$$\tilde u_k \in \arg\min_{u_k\in U_k(\tilde x_k)} \Big[g_k(\tilde x_k,u_k)+\tilde J_{k+1}\big(f_k(\tilde x_k,u_k)\big)\Big], \tag{1.8}$$

and

$$\tilde x_{k+1} = f_k(\tilde x_k,\tilde u_k).$$

The construction of suitable approximate cost-to-go functions $\tilde J_k$ is a major focal point of the RL methodology. There are several possible methods, depending on the context, and they will be taken up starting with the next chapter.

Q-Factors and Q-Learning

The expression

$$\tilde Q_k(x_k,u_k) = g_k(x_k,u_k)+\tilde J_{k+1}\big(f_k(x_k,u_k)\big),$$

which appears in the right-hand side of Eq. (1.8) is known as the (approximate) *Q-factor of* (x_k,u_k).† In particular, the computation of the

† The term "Q-learning" and some of the associated algorithmic ideas were introduced in the thesis by Watkins [Wat89] (after the symbol "Q" that he used to represent Q-factors). The term "Q-factor" was used in the book [BeT96], and is adopted here as well. Watkins [Wat89] used the term "action value" (at a given state). The terms "state-action value" and "Q-value" are also common in the literature.

approximately optimal control (1.8) can be done through the Q-factor minimization

$$\tilde{u}_k \in \arg \min_{u_k \in U_k(\tilde{x}_k)} \tilde{Q}_k(\tilde{x}_k, u_k).$$

This suggests the possibility of using Q-factors in place of cost functions in approximation in value space schemes. Methods of this type use as starting point an alternative (and equivalent) form of the DP algorithm, which instead of the optimal cost-to-go functions J_k^*, generates the *optimal Q-factors*, defined for all pairs (x_k, u_k) and k by

$$Q_k^*(x_k, u_k) = g_k(x_k, u_k) + J_{k+1}^*\big(f_k(x_k, u_k)\big). \tag{1.9}$$

Thus the optimal Q-factors are simply the expressions that are minimized in the right-hand side of the DP equation (1.4). Note that this equation implies that the optimal cost function J_k^* can be recovered from the optimal Q-factor Q_k^* by means of

$$J_k^*(x_k) = \min_{u_k \in U_k(x_k)} Q_k^*(x_k, u_k).$$

Moreover, using the above relation, the DP algorithm can be written in an essentially equivalent form that involves Q-factors only

$$Q_k^*(x_k, u_k) = g_k(x_k, u_k) + \min_{u_{k+1} \in U_{k+1}(f_k(x_k, u_k))} Q_{k+1}^*\big(f_k(x_k, u_k), u_{k+1}\big).$$

We will discuss later exact and approximate forms of related algorithms in the context of a class of RL methods known as *Q-learning*.

1.2 STOCHASTIC DYNAMIC PROGRAMMING

The stochastic finite horizon optimal control problem differs from the deterministic version primarily in the nature of the discrete-time dynamic system that governs the evolution of the state x_k. This system includes a random "disturbance" w_k, which is characterized by a probability distribution $P_k(\cdot \mid x_k, u_k)$ that may depend explicitly on x_k and u_k, but not on values of prior disturbances $w_{k-1}, \ldots, w_0$. The system has the form

$$x_{k+1} = f_k(x_k, u_k, w_k), \qquad k = 0, 1, \ldots, N-1,$$

where as before x_k is an element of some state space S_k, the control u_k is an element of some control space. The cost per stage is denoted $g_k(x_k, u_k, w_k)$ and also depends on the random disturbance w_k; see Fig. 1.2.1. The control

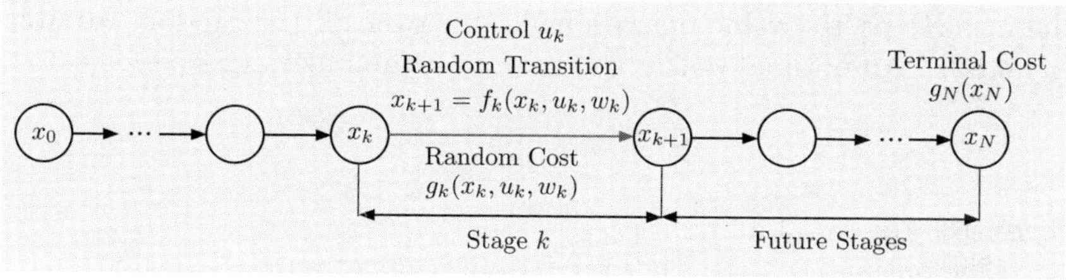

Figure 1.2.1 Illustration of an N-stage stochastic optimal control problem. Starting from state x_k, the next state under control u_k is generated randomly, according to

$$x_{k+1} = f_k(x_k, u_k, w_k),$$

where w_k is the random disturbance, and a random stage cost $g_k(x_k, u_k, w_k)$ is incurred.

u_k is constrained to take values in a given subset $U_k(x_k)$, which depends on the current state x_k.

An important difference is that we optimize not over control sequences $\{u_0, \ldots, u_{N-1}\}$, but rather over *policies* (also called *closed-loop control laws*, or *feedback policies*) that consist of a sequence of functions

$$\pi = \{\mu_0, \ldots, \mu_{N-1}\},$$

where μ_k maps states x_k into controls $u_k = \mu_k(x_k)$, and satisfies the control constraints, i.e., is such that $\mu_k(x_k) \in U_k(x_k)$ for all $x_k \in S_k$. Policies are more general objects than control sequences, and in the presence of stochastic uncertainty, they can result in improved cost, since they allow choices of controls u_k that incorporate knowledge of the state x_k. Without this knowledge, the controller cannot adapt appropriately to unexpected values of the state, and as a result the cost can be adversely affected. This is a fundamental distinction between deterministic and stochastic optimal control problems.

Another important distinction between deterministic and stochastic problems is that in the latter, the evaluation of various quantities such as cost function values involves forming expected values. Consequently, several of the methods that we will discuss for stochastic problems will involve the use of Monte Carlo simulation.

Given an initial state x_0 and a policy $\pi = \{\mu_0, \ldots, \mu_{N-1}\}$, the future states x_k and disturbances w_k are random variables with distributions defined through the system equation

$$x_{k+1} = f_k\big(x_k, \mu_k(x_k), w_k\big), \qquad k = 0, 1, \ldots, N-1.$$

Thus, for given functions g_k, $k = 0, 1, \ldots, N$, the expected cost of π starting at x_0 is

$$J_\pi(x_0) = E\left\{ g_N(x_N) + \sum_{k=0}^{N-1} g_k\big(x_k, \mu_k(x_k), w_k\big) \right\},$$

where the expected value operation $E\{\cdot\}$ is over all the random variables w_k and x_k. An optimal policy π^* is one that minimizes this cost; i.e.,

$$J_{\pi^*}(x_0) = \min_{\pi \in \Pi} J_\pi(x_0),$$

where Π is the set of all policies.

The optimal cost depends on x_0 and is denoted by $J^*(x_0)$; i.e.,

$$J^*(x_0) = \min_{\pi \in \Pi} J_\pi(x_0).$$

It is useful to view J^* as a function that assigns to each initial state x_0 the optimal cost $J^*(x_0)$, and call it the *optimal cost function* or *optimal value function*.

Finite Horizon Stochastic Dynamic Programming

The DP algorithm for the stochastic finite horizon optimal control problem has a similar form to its deterministic version, and shares several of its major characteristics:

(a) Using tail subproblems to break down the minimization over multiple stages to single stage minimizations.

(b) Generating backwards for all k and x_k the values $J_k^*(x_k)$, which give the optimal cost-to-go starting at stage k at state x_k.

(c) Obtaining an optimal policy by minimization in the DP equations.

(d) A structure that is suitable for approximation in value space, whereby we replace J_k^* by approximations $\tilde{J}_k$, and obtain a suboptimal policy by the corresponding minimization.

DP Algorithm for Stochastic Finite Horizon Problems

Start with

$$J_N^*(x_N) = g_N(x_N),$$

and for $k = 0, \ldots, N-1$, let

$$J_k^*(x_k) = \min_{u_k \in U_k(x_k)} E\Big\{ g_k(x_k, u_k, w_k) + J_{k+1}^*\big(f_k(x_k, u_k, w_k)\big)\Big\}. \quad (1.10)$$

If $u_k^* = \mu_k^*(x_k)$ minimizes the right side of this equation for each x_k and k, the policy $\pi^* = \{\mu_0^*, \ldots, \mu_{N-1}^*\}$ is optimal.

The key fact is that for every initial state x_0, the optimal cost $J^*(x_0)$ is equal to the function $J_0^*(x_0)$, obtained at the last step of the above DP algorithm. This can be proved by induction similar to the deterministic case; we will omit the proof (see the discussion of Section 1.3 in the textbook [Ber17]).†

Simultaneously with the off-line computation of the optimal cost-to-go functions $J_0^*, \ldots, J_N^*$, we can compute and store an optimal policy $\pi^* = \{\mu_0^*, \ldots, \mu_{N-1}^*\}$ by minimization in Eq. (1.10). We can then use this policy on-line to retrieve from memory and apply the control $\mu_k^*(x_k)$ once we reach state x_k.

The alternative is to forego the storage of the policy π^* and to calculate the control $\mu_k^*(x_k)$ by executing the minimization (1.10) on-line. This method of on-line control calculation is called *one-step lookahead minimization*. Its main use is not so much in the context of exact DP, but rather in the context of approximate DP methods that involve approximation in value space. There, approximations $\tilde J_k$ are used in place of J_k^*, similar to the deterministic case; cf. Eqs. (1.7) and (1.8).

Approximation in Value Space - Use of $\tilde J_k$ in Place of J_k^*

At any state x_k encountered at stage k, compute and apply the control

$$\tilde\mu_k(x_k) \in \arg \min_{u_k \in U_k(x_k)} E\Big\{g_k(x_k, u_k, w_k) + \tilde J_{k+1}\big(f_k(x_k, u_k, w_k)\big)\Big\}. \tag{1.11}$$

As in deterministic problems, the motivation for approximation in value space is that the DP algorithm can be very time-consuming. The one-step lookahead minimization (1.11) needs to be performed only for the N states $x_0, \ldots, x_{N-1}$ that are encountered during the on-line control of the system, and not for every state within the potentially enormous state space. Of course this simplification entails the loss of optimality, and requires the construction of suitable approximate cost-to-go functions $\tilde J_k$. This is a major focal point of the RL methodology, and will be discussed at length in the following chapters.

Q-Factors for Stochastic Problems

We can define optimal Q-factors for a stochastic problem, similar to the

† There are some technical/mathematical difficulties here, having to do with the expected value operation in Eq. (1.10) being well-defined and finite. These difficulties are of no concern in practice, and disappear completely when the disturbance spaces w_k can take only a finite number of values, in which case all expected values consist of sums of finitely many real number terms. For a mathematical treatment, see the relevant discussion in Chapter 1 of [Ber17] and the book [BeS78].

case of deterministic problems [cf. Eq. (1.9)], as the expressions that are minimized in the right-hand side of the stochastic DP equation (1.10). They are given by

$$Q_k^*(x_k, u_k) = E\Big\{g_k(x_k, u_k, w_k) + J_{k+1}^*\big(f_k(x_k, u_k, w_k)\big)\Big\}.$$

The optimal cost-to-go functions J_k^* can be recovered from the optimal Q-factors Q_k^* by means of

$$J_k^*(x_k) = \min_{u_k \in U_k(x_k)} Q_k^*(x_k, u_k),$$

and the DP algorithm can be written in terms of Q-factors as

$$\begin{aligned} Q_k^*(x_k, u_k) = E\Big\{ & g_k(x_k, u_k, w_k) \\ & + \min_{u_{k+1} \in U_{k+1}(f_k(x_k, u_k, w_k))} Q_{k+1}^*\big(f_k(x_k, u_k, w_k), u_{k+1}\big)\Big\}. \end{aligned}$$

1.3 EXAMPLES, VARIATIONS, AND SIMPLIFICATIONS

In this section we provide some examples to illustrate problem formulation techniques, solution methods, and adaptations of the basic DP algorithm to various contexts. As a guide for formulating optimal control problems in a manner that is suitable for DP solution, the following two-stage process is suggested:

(a) Identify the controls/decisions u_k and the times k at which these controls are applied. Usually this step is fairly straightforward. However, in some cases there may be some choices to make. For example in deterministic problems, where the objective is to select an optimal sequence of controls $\{u_0, \ldots, u_{N-1}\}$, one may lump multiple controls to be chosen together, e.g., view the pair (u_0, u_1) as a single choice. This is usually not possible in stochastic problems, where distinct decisions are differentiated by the information/feedback available when making them.

(b) Select the states x_k. The basic guideline here is that x_k should encompass all the information that is known to the controller at time k and can be used with advantage in choosing u_k. In effect, at time k the state x_k should separate the past from the future, in the sense that anything that has happened in the past (states, controls, and disturbances from stages prior to stage k) is irrelevant to the choices of future controls as long we know x_k. Sometimes this is described by saying that the state should have a "Markov property" to express an analogy with states of Markov chains, where (by definition) the conditional probability distribution of future states depends on the past history of the chain only through the present state.

Note that there may be multiple possibilities for selecting the states, because information may be packaged in several different ways that are equally useful from the point of view of control. It is thus worth considering alternative ways to choose the states; for example try to use states that minimize the dimensionality of the state space. For a trivial example that illustrates the point, if a quantity x_k qualifies as state, then (x_{k-1}, x_k) also qualifies as state, since (x_{k-1}, x_k) contains all the information contained within x_k that can be useful to the controller when selecting u_k. However, using (x_{k-1}, x_k) in place of x_k, gains nothing in terms of optimal cost while complicating the DP algorithm that would have to be executed over a larger space. The concept of a *sufficient statistic*, which refers to a quantity that summarizes all the essential content of the information available to the controller, may be useful in reducing the size of the state space (see the discussion in Section 3.1.1, and in [Ber17], Section 4.3). Section 1.3.6 provides an example, and Section 3.1.1 contains further discussion.

Generally minimizing the dimension of the state makes sense but there are exceptions. A case in point is problems involving *partial* or *imperfect* state information, where we collect measurements to use for control of some quantity of interest y_k that evolves over time (for example, y_k may be the position/velocity vector of a moving vehicle). If I_k is the collection of all measurements and controls up to time k, it is correct to use I_k as state. However, a better alternative may be to use as state the conditional probability distribution $P_k(y_k \mid I_k)$, called *belief state*, which may subsume all the information that is useful for the purposes of choosing a control. On the other hand, the belief state $P_k(y_k \mid I_k)$ is an infinite-dimensional object, whereas I_k may be finite dimensional, so the best choice may be problem-dependent; see the textbooks [Ber17] and [Kri16] for further discussion of partial state information problems.

We refer to DP textbooks for extensive additional discussions of modeling and problem formulation techniques. The subsequent chapters do not rely substantially on the material of the present section, so the reader may selectively skip forward to the next chapter and return to this material later as needed.

1.3.1 Deterministic Shortest Path Problems

Consider a directed graph with a special node, called the *destination*. Let $\{1, 2, \ldots, N, t\}$ be the set of nodes of the graph, where t is the destination, and let a_{ij} be the cost of moving from node i to node j [also referred to as the *length* of the directed arc (i, j) that joins i and j]. By a path we mean a sequence of arcs such that the end node of each arc in the sequence is the start node of the next arc. The length of a path from a given node to another node is the sum of the lengths of the arcs on the path. We want to find a shortest (i.e., minimum length) path from each node i to node t.

We make an assumption relating to cycles, i.e., paths of the form $(i, j_1), (j_1, j_2), \ldots, (j_k, i)$ that start and end at the same node. In particular, we exclude the possibility that a cycle has negative total length. Otherwise, it would be possible to decrease the length of some paths to arbitrarily small values simply by adding more and more negative-length cycles. We thus assume that *all cycles have nonnegative length.* With this assumption, it is clear that an optimal path need not take more than N moves, so we may limit the number of moves to N. To conform to the N-stage DP framework, we formulate the problem as one where *we require exactly N moves but allow degenerate moves from a node i to itself with cost $a_{ii} = 0$.* We also assume that *for every node i there exists at least one path from i to t*, so that the problem has at least one solution.

We can formulate this problem as a deterministic DP problem with N stages, where the states at any stage $0, \ldots, N-1$ are the nodes $\{1, \ldots, N\}$, the destination t is the unique state at stage N, and the controls correspond to the arcs (i, j), including the self arcs (i, i). Thus at each state i we select a control (i, j) and move to state j at cost a_{ij}.

We can write the DP algorithm for our problem, with the optimal cost-to-go functions J_k^* having the meaning

$$J_k^*(i) = \text{optimal cost of getting from } i \text{ to } t \text{ in } N - k \text{ moves},$$

for $i = 1, \ldots, N$ and $k = 0, \ldots, N-1$. The cost of the optimal path from i to t is $J_0^*(i)$. The DP algorithm takes the intuitively clear form

$$\begin{aligned} &\text{optimal cost from } i \text{ to } t \text{ in } N - k \text{ moves} \\ &\quad = \min_{\text{All arcs } (i,j)} \big[a_{ij} + (\text{optimal cost from } j \text{ to } t \text{ in } N - k - 1 \text{ moves})\big], \end{aligned}$$

or

$$J_k^*(i) = \min_{\text{All arcs } (i,j)} \big[a_{ij} + J_{k+1}^*(j)\big], \qquad k = 0, 1, \ldots, N-2,$$

with

$$J_{N-1}^*(i) = a_{it}, \qquad i = 1, 2, \ldots, N.$$

This algorithm is also known as the *Bellman-Ford algorithm* for shortest paths. It is one of the most popular shortest path algorithms.

The optimal policy when at node i after k moves is to move to a node j^* that minimizes $a_{ij} + J_{k+1}^*(j)$ over all j such that (i, j) is an arc. If the optimal path obtained from the algorithm contains degenerate moves from a node to itself, this simply means that the path involves in reality less than N moves.

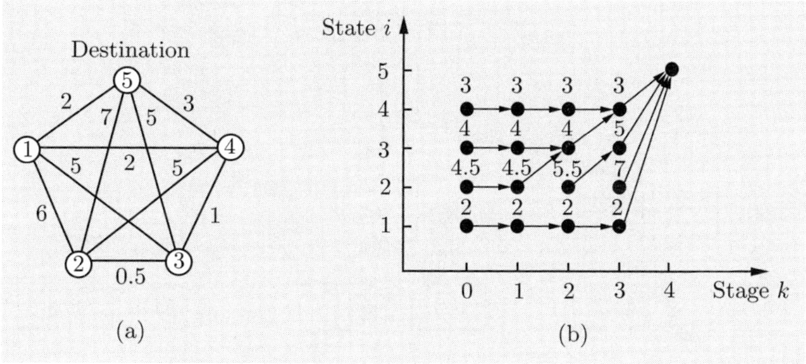

Figure 1.3.1 (a) Shortest path problem data. The destination is node 5. Arc lengths are equal in both directions and are shown along the line segments connecting nodes. (b) Costs-to-go generated by the DP algorithm. The number along stage k and state i is $J_k^*(i)$. Arrows indicate the optimal moves at each stage and node. The optimal paths that start from nodes 1,2,3,4 are $1 \to 5$, $2 \to 3 \to 4 \to 5$, $3 \to 4 \to 5$, $4 \to 5$, respectively.

Note that if for some $k > 0$, we have

$$J_k^*(i) = J_{k+1}^*(i), \qquad \text{for all } i,$$

then subsequent DP iterations will not change the values of the cost-to-go $[J_{k-m}^*(i) = J_k^*(i)$ for all $m > 0$ and $i]$, so the algorithm can be terminated with $J_k^*(i)$ being the shortest distance from i to t, for all i.

To demonstrate the algorithm, consider the problem shown in Fig. 1.3.1(a) where the costs a_{ij} with $i \neq j$ are shown along the connecting line segments (we assume that $a_{ij} = a_{ji}$). Figure 1.3.1(b) shows the optimal cost-to-go $J_k^*(i)$ at each i and k together with the optimal paths.

1.3.2 Discrete Deterministic Optimization

Discrete optimization problems can be typically formulated as DP problems by breaking down each feasible solution into a sequence of decisions/controls; see e.g., the scheduling Example 1.1.1. This formulation will often lead to an intractable DP computation because of an exponential explosion of the number of states. However, it brings to bear approximate DP methods, such as rollout and others that we will discuss in future chapters. We illustrate the reformulation by an example and then generalize.

Example 1.3.1 (The Traveling Salesman Problem)

An important model for scheduling a sequence of operations is the classical traveling salesman problem. Here we are given N cities and the travel time between each pair of cities. We wish to find a minimum time travel that visits

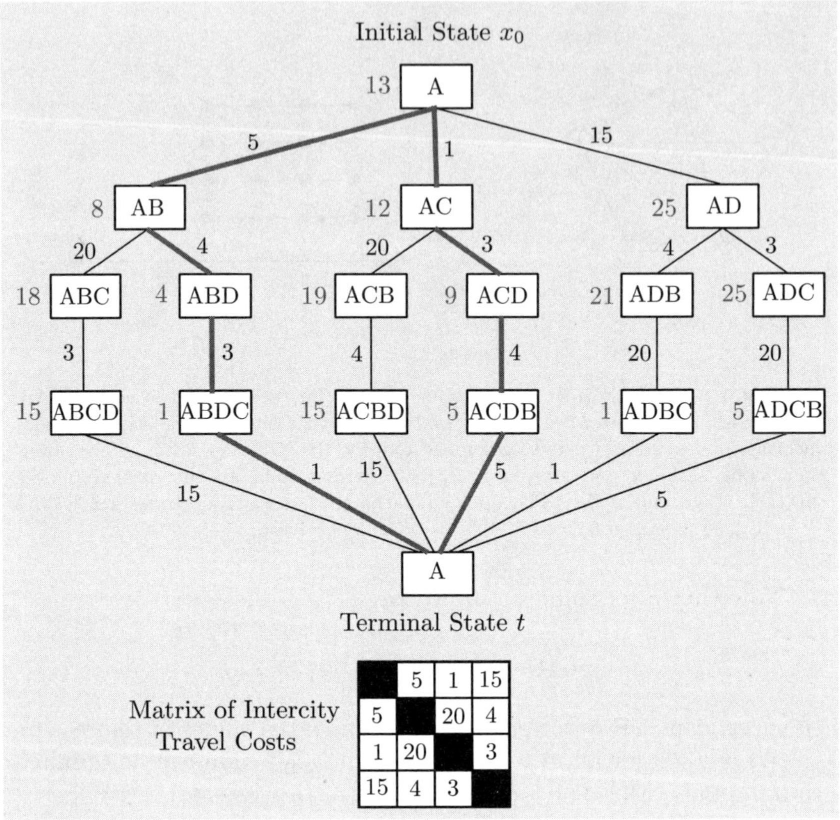

Figure 1.3.2 Example of a DP formulation of the traveling salesman problem. The travel times between the four cities A, B, C, and D are shown in the matrix at the bottom. We form a graph whose nodes are the k-city sequences and correspond to the states of the kth stage. The transition costs/travel times are shown next to the arcs. The optimal costs-to-go are generated by DP starting from the terminal state and going backwards towards the initial state, and are shown next to the nodes. There are two optimal sequences here (ABDCA and ACDBA), and they are marked with thick lines. Both optimal sequences can be obtained by forward minimization [cf. Eq. (1.7)], starting from the initial state x_0.

each of the cities exactly once and returns to the start city. To convert this problem to a DP problem, we form a graph whose nodes are the sequences of k distinct cities, where $k = 1, \ldots, N$. The k-city sequences correspond to the states of the kth stage. The initial state x_0 consists of some city, taken as the start (city A in the example of Fig. 1.3.2). A k-city node/state leads to a $(k+1)$-city node/state by adding a new city at a cost equal to the travel time between the last two of the $k+1$ cities; see Fig. 1.3.2. Each sequence of N cities is connected to an artificial terminal node t with an arc of cost equal to the travel time from the last city of the sequence to the starting city, thus

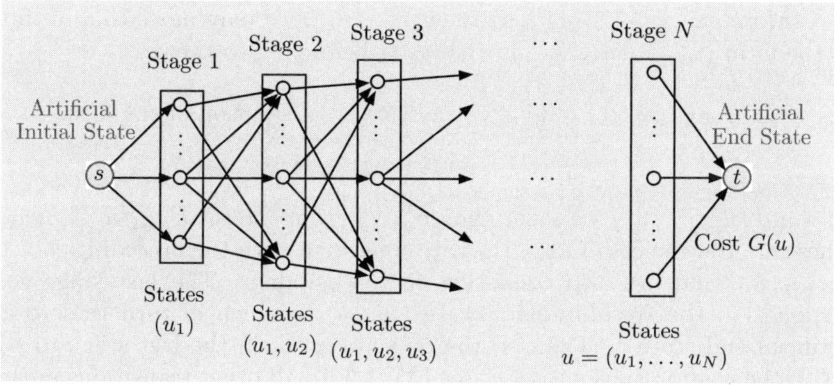

Figure 1.3.3. Formulation of a discrete optimization problem as a DP problem with $N + 1$ stages. There is a cost $G(u)$ only at the terminal stage on the arc connecting an N-solution $u = (u_1, \ldots, u_N)$ to the artificial terminal state. Alternative formulations may use fewer states by taking advantage of the problem's structure.

completing the transformation to a DP problem.

The optimal costs-to-go from each node to the terminal state can be obtained by the DP algorithm and are shown next to the nodes. Note, however, that the number of nodes grows exponentially with the number of cities N. This makes the DP solution intractable for large N. As a result, large traveling salesman and related scheduling problems are typically addressed with approximation methods, some of which are based on DP, and will be discussed as part of our subsequent development.

Let us now extend the ideas of the preceding example to the general discrete optimization problem:

$$\text{minimize } G(u)$$
$$\text{subject to } u \in U,$$

where U is a finite set of feasible solutions and $G(u)$ is a cost function. We assume that each solution u has N components; i.e., it has the form $u = (u_1, \ldots, u_N)$, where N is a positive integer. We can then view the problem as a sequential decision problem, where the components $u_1, \ldots, u_N$ are selected one-at-a-time. A k-tuple $(u_1, \ldots, u_k)$ consisting of the first k components of a solution is called a *k-solution*. We associate k-solutions with the kth stage of the finite horizon DP problem shown in Fig. 1.3.3. In particular, for $k = 1, \ldots, N$, we view as the states of the kth stage all the k-tuples $(u_1, \ldots, u_k)$. The initial state is an artificial state denoted s. From this state we may move to any state (u_1), with u_1 belonging to the set

$$U_1 = \big\{\tilde{u}_1 \mid \text{there exists a solution of the form } (\tilde{u}_1, \tilde{u}_2, \ldots, \tilde{u}_N) \in U\big\}.$$

Thus U_1 is the set of choices of u_1 that are consistent with feasibility.

More generally, from a state $(u_1, \ldots, u_k)$, we may move to any state of the form $(u_1, \ldots, u_k, u_{k+1})$, with u_{k+1} belonging to the set

$$U_{k+1}(u_1, \ldots, u_k) = \big\{\tilde{u}_{k+1} \mid \text{there exists a solution of the form } (u_1, \ldots, u_k, \tilde{u}_{k+1}, \ldots, \tilde{u}_N) \in U\big\}.$$

At state $(u_1, \ldots, u_k)$ we must choose u_{k+1} from the set $U_{k+1}(u_1, \ldots, u_k)$. These are the choices of u_{k+1} that are consistent with the preceding choices $u_1, \ldots, u_k$, and are also consistent with feasibility. The last stage corresponds to the N-solutions $u = (u_1, \ldots, u_N)$, which in turn lead to an artificial end state t. The cost-to-go from each u in the last stage to t is $G(u)$, the cost of the solution u; see Fig. 1.3.3. All other transitions in this DP problem formulation have cost 0.

Let $J_k^*(u_1, \ldots, u_k)$ denote the optimal cost starting from the k-solution $(u_1, \ldots, u_k)$, i.e., the optimal cost of the problem over solutions whose first k components are constrained to be equal to u_i, $i = 1, \ldots, k$, respectively. The DP algorithm is described by the equation

$$J_k^*(u_1, \ldots, u_k) = \min_{u_{k+1} \in U_{k+1}(u_1, \ldots, u_k)} J_{k+1}^*(u_1, \ldots, u_k, u_{k+1}), \tag{1.12}$$

with the terminal condition

$$J_N^*(u_1, \ldots, u_N) = G(u_1, \ldots, u_N).$$

The algorithm (1.12) executes backwards in time: starting with the known function $J_N^* = G$, we compute J_{N-1}^*, then J_{N-2}^*, and so on up to computing J_1^*. An optimal solution $(u_1^*, \ldots, u_N^*)$ is then constructed by going forward through the algorithm

$$u_{k+1}^* \in \arg \min_{u_{k+1} \in U_{k+1}(u_1^*, \ldots, u_k^*)} J_{k+1}^*(u_1^*, \ldots, u_k^*, u_{k+1}), \qquad k = 0, \ldots, N-1, \tag{1.13}$$

first compute u_1^*, then u_2^*, and so on up to u_N^*; cf. Eq. (1.7).

Of course here the number of states typically grows exponentially with N, but we can use the DP minimization (1.13) as a starting point for the use of approximation methods. For example we may try to use approximation in value space, whereby we replace J_{k+1}^* with some suboptimal $\tilde{J}_{k+1}$ in Eq. (1.13). One possibility is to use as

$$\tilde{J}_{k+1}(u_1^*, \ldots, u_k^*, u_{k+1}),$$

the cost generated by a heuristic method that solves the problem suboptimally with the values of the first $k + 1$ decision components fixed at $u_1^*, \ldots, u_k^*, u_{k+1}$. This is called a *rollout algorithm*, and it is a very simple and effective approach for approximate combinatorial optimization. It will be discussed later in this book, in Chapter 2 for finite horizon stochastic problems, and in Chapter 5 for infinite horizon problems, where it will be related to the method of policy iteration and self-learning ideas.

Finally, let us mention that shortest path and discrete optimization problems with a sequential character can be addressed by a variety of approximate shortest path methods. These include the so called *label correcting, A^*, and branch and bound methods* for which extensive discussions can be found in the literature. The author's DP textbook [Ber17] (Chapter 2) contains a substantial account, which connects with the material of this section, as well as with a much more detailed discussion of shortest path methods in the author's network optimization textbook [Ber98].

1.3.3 Problems with a Termination State

Many DP problems of interest involve a *termination state*, i.e., a state t that is cost-free and absorbing in the sense that for all k,

$$g_k(t, u_k, w_k) = 0, \quad f_k(t, u_k, w_k) = t, \qquad \text{for all } w_k \text{ and } u_k \in U_k(t).$$

Thus the control process essentially terminates upon reaching t, even if this happens before the end of the horizon. One may reach t by choice if a special stopping decision is available, or by means of a random transition from another state.

Generally, when it is known that an optimal policy will reach the termination state within at most some given number of stages N, the DP problem can be formulated as an N-stage horizon problem.† The reason is that even if the termination state t is reached at a time $k < N$, we can extend our stay at t for an additional $N - k$ stages at no additional cost. An illustration of this was given in the deterministic shortest path problem that we discussed in Section 1.3.1.

Discrete deterministic optimization problems generally have a close connection to shortest path problems as we have seen in Section 1.3.2. In the problem discussed in that section, the termination state is reached after exactly N stages (cf. Fig. 1.3.3), but in other problems it is possible that termination can happen earlier. The following well known puzzle is an example.

Example 1.3.2 (The Four Queens Problem)

Four queens must be placed on a 4×4 portion of a chessboard so that no queen can attack another. In other words, the placement must be such that every row, column, or diagonal of the 4×4 board contains at most one queen.

---

† When an upper bound on the number of stages to termination is not known, the problem must be formulated as an infinite horizon problem, as will be discussed in a subsequent chapter.

Equivalently, we can view the problem as a sequence of problems; first, placing a queen in one of the first two squares in the top row, then placing another queen in the second row so that it is not attacked by the first, and similarly placing the third and fourth queens. (It is sufficient to consider only the first two squares of the top row, since the other two squares lead to symmetric positions; this is an example of a situation where we have a choice between several possible state spaces, but we select the one that is smallest.)

We can associate positions with nodes of an acyclic graph where the root node s corresponds to the position with no queens and the terminal nodes correspond to the positions where no additional queens can be placed without some queen attacking another. Let us connect each terminal position with an artificial terminal node t by means of an arc. Let us also assign to all arcs cost zero except for the artificial arcs connecting terminal positions with less than four queens with the artificial node t. These latter arcs are assigned a cost of 1 (see Fig. 1.3.4) to express the fact that they correspond to dead-end positions that cannot lead to a solution. Then, the four queens problem reduces to finding a minimal cost path from node s to node t, with an optimal sequence of queen placements corresponding to cost 0.

Note that once the states/nodes of the graph are enumerated, the problem is essentially solved. In this 4×4 problem the states are few and can be easily enumerated. However, we can think of similar problems with much larger state spaces. For example consider the problem of placing N queens on an $N \times N$ board without any queen attacking another. Even for moderate values of N, the state space for this problem can be extremely large (for $N = 8$ the number of possible placements with exactly one queen in each row is $8^8 = 16,777,216$). It can be shown that there exist solutions to the N queens problem for all $N \geq 4$ (for $N = 2$ and $N = 3$, clearly there is no solution).

There are also several variants of the N queens problem. For example finding the minimal number of queens that can be placed on an $N \times N$ board so that they either occupy or attack every square; this is known as the *queen domination problem*. The minimal number can be found in principle by DP, and it is known for some N (for example the minimal number is 5 for $N = 8$), but not for all N (see e.g., the paper by Fernau [Fer10]).

1.3.4 Forecasts

Consider a situation where at time k the controller has access to a forecast y_k that results in a reassessment of the probability distribution of the subsequent disturbance w_k and, possibly, future disturbances. For example, y_k may be an exact prediction of w_k or an exact prediction that the probability distribution of w_k is a specific one out of a finite collection of distributions. Forecasts of interest in practice are, for example, probabilistic predictions on the state of the weather, the interest rate for money, and the demand for inventory.

Generally, forecasts can be handled by introducing additional state

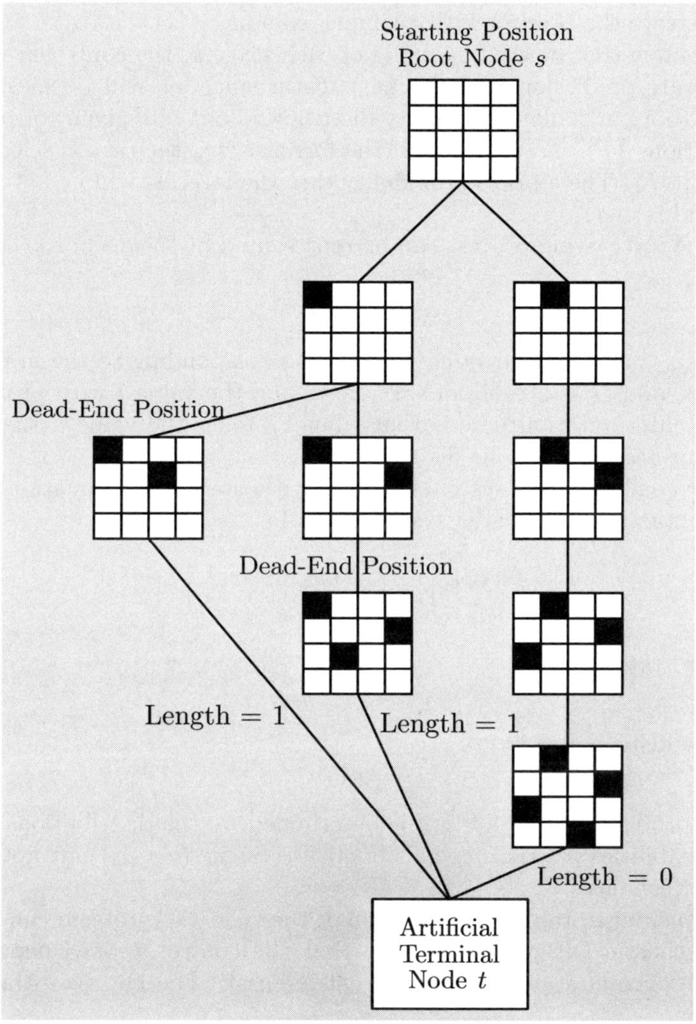

Figure 1.3.4 Discrete optimization formulation of the four queens problem. Symmetric positions resulting from placing a queen in one of the rightmost squares in the top row have been ignored. Squares containing a queen have been darkened. All arcs have length zero except for those connecting dead-end positions to the artificial terminal node.

variables corresponding to the information that the forecasts provide.† We

---

† The device of introducing additional states to modify a given problem so that it fits the DP formalism is known as *state augmentation*. It finds extensive use in the DP formulation of practical problems (see e.g. [Ber17]). Examples of problem reformulations that are based on state augmentation arise when

will illustrate the process with a simple example.

Assume that at the beginning of each stage k, the controller receives an accurate prediction that the next disturbance w_k will be selected according to a particular probability distribution out of a given collection of distributions $\{P_1, \ldots, P_m\}$; i.e., if the forecast is i, then w_k is selected according to P_i. The a priori probability that the forecast will be i is denoted by p_i and is given.

The forecasting process can be represented by means of the equation

$$y_{k+1} = \xi_k,$$

where y_{k+1} can take the values $1, \ldots, m$, corresponding to the m possible forecasts, and ξ_k is a random variable taking the value i with probability p_i. The interpretation here is that when ξ_k takes the value i, then w_{k+1} will occur according to the distribution P_i.

By combining the system equation with the forecast equation $y_{k+1} = \xi_k$, we obtain an augmented system given by

$$\begin{pmatrix} x_{k+1} \\ y_{k+1} \end{pmatrix} = \begin{pmatrix} f_k(x_k, u_k, w_k) \\ \xi_k \end{pmatrix}.$$

The new state is

$$\tilde{x}_k = (x_k, y_k).$$

The new disturbance is

$$\tilde{w}_k = (w_k, \xi_k),$$

and its probability distribution is determined by the distributions P_i and the probabilities p_i, and depends explicitly on $\tilde{x}_k$ (via y_k) but not on the prior disturbances.

Thus, by suitable reformulation of the cost, the problem can be cast as a stochastic DP problem. Note that the control applied depends on both the current state and the current forecast. The DP algorithm takes the form

$$J_N^*(x_N, y_N) = g_N(x_N),$$

$$\begin{aligned} J_k^*(x_k, y_k) = \min_{u_k \in U_k(x_k)} \mathop{E}_{w_k} \Big\{ & g_k(x_k, u_k, w_k) \\ & + \sum_{i=1}^{m} p_i J_{k+1}^*\big(f_k(x_k, u_k, w_k), i\big) \;\big|\; y_k \Big\}, \end{aligned} \tag{1.14}$$

the disturbances exhibit correlations over time. They also arise in the presence of *post-decision* states, where the system function $f_k(x_k, u_k, w_k)$ has the form $\tilde{f}_k\big(h_k(x_k, u_k), w_k\big)$, where $\tilde{f}_k$ is some function, and $h_k(x_k, u_k)$ represents an intermediate "state" that occurs after the control is applied. With post-decision states, the DP algorithm may be reformulated to a simpler form (see [Ber12], Section 6.1.5).

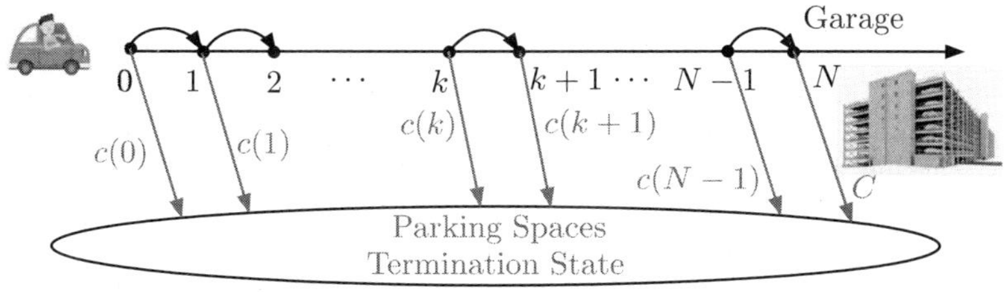

Figure 1.3.5 Cost structure of the parking problem. The driver may park at space $k = 0, 1, \ldots, N-1$ at cost $c(k)$, if the space is free, or continue to the next space $k+1$ at no cost. At space N (the garage) the driver must park at cost C.

where y_k may take the values $1, \ldots, m$, and the expectation over w_k is taken with respect to the distribution P_{y_k}.

It should be clear that the preceding formulation admits several extensions. One example is the case where forecasts can be influenced by the control action (e.g., pay extra for a more accurate forecast) and involve several future disturbances. However, the price for these extensions is increased complexity of the corresponding DP algorithm.

1.3.5 Problems with Uncontrollable State Components

In many problems of interest the natural state of the problem consists of several components, some of which cannot be affected by the choice of control. In such cases the DP algorithm can be simplified considerably, and be executed over the controllable components of the state. Before describing how this can be done in generality, let us consider an example.

Example 1.3.3 (Parking)

A driver is looking for inexpensive parking on the way to his destination. The parking area contains N spaces, numbered $0, \ldots, N-1$, and a garage following space $N-1$. The driver starts at space 0 and traverses the parking spaces sequentially, i.e., from space k he goes next to space $k+1$, etc. Each parking space k costs $c(k)$ and is free with probability $p(k)$ independently of whether other parking spaces are free or not. If the driver reaches the last parking space $N-1$ and does not park there, he must park at the garage, which costs C. The driver can observe whether a parking space is free only when he reaches it, and then, if it is free, he makes a decision to park in that space or not to park and check the next space. The problem is to find the minimum expected cost parking policy.

We formulate the problem as a DP problem with N stages, corresponding to the parking spaces, and an artificial termination state t that corresponds to having parked; see Fig. 1.3.5. At each stage $k = 1, \ldots, N-1$, we have three states: the artificial termination state t, and the two states (k, F) and $(k, \overline{F})$, corresponding to space k being free or taken, respectively.

At stage 0, we have only two states, $(0,F)$ and $(0,\overline{F})$, and at the final stage there is only one state, the termination state t. The decision/control is to park or continue at state (k,F) [there is no choice at states $(k,\overline{F})$ and states (k,t), $k=1,\ldots,N-1$]. The termination state t is reached at cost $c(k)$ when a parking decision is made at the states (k,F), $k=0,\ldots,N-1$, at cost C, when the driver continues at states $(N-1,F)$ or $(N-1,\overline{F})$, and at no cost at (k,t), $k=0,\ldots,N-1$.

Let us now derive the form of the DP algorithm, denoting:

$J_k^*(F)$: The optimal cost-to-go upon arrival at a space k that is free.

$J_k^*(\overline{F})$: The optimal cost-to-go upon arrival at a space k that is taken.

$J_k^*(t)$: The cost-to-go of the "parked"/termination state t.

The DP algorithm for $k=0,\ldots,N-1$ takes the form

$$J_k^*(F)=\begin{cases}\min\Big[c(k),\,p(k+1)J_{k+1}^*(F)+\big(1-p(k+1)\big)J_{k+1}^*(\overline{F})\Big] & \text{if } k<N-1,\\ \min\big[c(N-1),\,C\big] & \text{if } k=N-1,\end{cases}$$

$$J_k^*(\overline{F})=\begin{cases}p(k+1)J_{k+1}^*(F)+\big(1-p(k+1)\big)J_{k+1}^*(\overline{F}) & \text{if } k<N-1,\\ C & \text{if } k=N-1,\end{cases}$$

for the states other than the termination state t, while for t we have

$$J_k^*(t)=0,\qquad k=1,\ldots,N.$$

While this algorithm is easily executed, it can be written in a simpler and equivalent form, which takes advantage of the fact that the second component (F or $\overline{F}$) of the state is uncontrollable. This can be done by introducing the scalars

$$\hat{J}_k=p(k)J_k^*(F)+\big(1-p(k)\big)J_k^*(\overline{F}),\qquad k=0,\ldots,N-1,$$

which can be viewed as the optimal expected cost-to-go upon arriving at space k but *before verifying its free or taken status*.

Indeed, from the preceding DP algorithm, we have

$$\hat{J}_{N-1}=p(N-1)\min\big[c(N-1),\,C\big]+\big(1-p(N-1)\big)C,$$

$$\hat{J}_k=p(k)\min\big[c(k),\,\hat{J}_{k+1}\big]+\big(1-p(k)\big)\hat{J}_{k+1},\qquad k=0,\ldots,N-2.$$

From this algorithm we can also obtain the optimal parking policy, which is to park at space $k=0,\ldots,N-1$ if it is free and we have $c(k)\le\hat{J}_{k+1}$.

Figure 1.3.6 provides a plot for $\hat{J}_k$ for the case where

$$p(k)\equiv 0.05,\quad c(k)=N-k,\quad C=100,\quad N=200. \tag{1.15}$$

The optimal policy is to travel to space 165 and then to park at the first available space. The reader may verify that this type of policy, characterized by a single threshold distance, is optimal not just for the form of $c(k)$ given

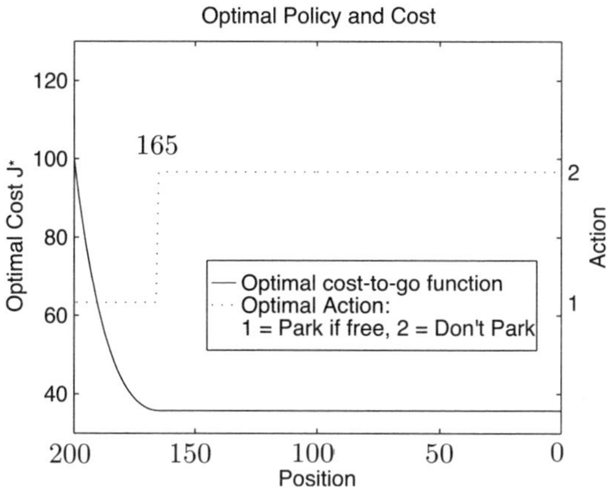

Figure 1.3.6 Optimal cost-to-go and optimal policy for the parking problem with the data in Eq. (1.15). The optimal policy is to travel from space 0 to space 165 and then to park at the first available space.

above, but also for any form of $c(k)$ that is monotonically decreasing as k increases.

We will now formalize the procedure illustrated in the preceding example. Let the state of the system be a composite (x_k, y_k) of two components x_k and y_k. The evolution of the main component, x_k, is affected by the control u_k according to the equation

$$x_{k+1} = f_k(x_k, y_k, u_k, w_k),$$

where the distribution $P_k(w_k \mid x_k, y_k, u_k)$ is given. The evolution of the other component, y_k, is governed by a given conditional distribution $P_k(y_k \mid x_k)$ and cannot be affected by the control, except indirectly through x_k. One is tempted to view y_k as a disturbance, but there is a difference: y_k is observed by the controller before applying u_k, while w_k occurs after u_k is applied, and indeed w_k may probabilistically depend on u_k.

It turns out that we can formulate a DP algorithm that is executed over the controllable component of the state, with the dependence on the uncontrollable component being "averaged out" as in the preceding example. In particular, let $J_k^*(x_k, y_k)$ denote the optimal cost-to-go at stage k and state (x_k, y_k), and define

$$\hat{J}_k(x_k) = \underset{y_k}{E}\big\{J_k^*(x_k, y_k) \mid x_k\big\}.$$

Then, similar to the preceding parking example, a DP algorithm that generates $\hat{J}_k(x_k)$ can be obtained, and has the following form:†

$$\hat{J}_k(x_k) = \mathop{E}_{y_k}\Big\{ \min_{u_k\in U_k(x_k,y_k)} \mathop{E}_{w_k}\Big\{ g_k(x_k,y_k,u_k,w_k) + \hat{J}_{k+1}\big(f_k(x_k,y_k,u_k,w_k)\big) \mid x_k,y_k,u_k\Big\} \Big| \, x_k\Big\}. \tag{1.16}$$

Note that the minimization in the right-hand side of the preceding equation must still be performed for all values of the full state (x_k, y_k) in order to yield an optimal control law as a function of (x_k, y_k). Nonetheless, the equivalent DP algorithm (1.16) has the advantage that it is executed over a significantly reduced state space. Later, when we consider approximation in value space, we will find that it is often more convenient to approximate $\hat{J}_k(x_k)$ than to approximate $J_k^*(x_k, y_k)$; see the following discussions of forecasts and of the game of tetris.

As an example, consider the augmented state resulting from the incorporation of forecasts, as described earlier in Section 1.3.4. Then, the forecast y_k represents an uncontrolled state component, so that the DP algorithm can be simplified as in Eq. (1.16). In particular, using the notation of Section 1.3.4, by defining

$$\hat{J}_k(x_k) = \sum_{i=1}^m p_i J_k^*(x_k, i), \qquad k = 0, 1, \ldots, N-1,$$

and

$$\hat{J}_N(x_N) = g_N(x_N),$$

we have, using Eq. (1.14),

$$\hat{J}_k(x_k) = \sum_{i=1}^m p_i \min_{u_k\in U_k(x_k)} \mathop{E}_{w_k}\Big\{ g_k(x_k,u_k,w_k) + \hat{J}_{k+1}\big(f_k(x_k,u_k,w_k)\big) \mid y_k = i\Big\},$$

† This is a consequence of the calculation

$$\begin{aligned}
\hat{J}_k(x_k) &= E_{y_k}\big\{J_k^*(x_k,y_k) \mid x_k\big\} \\
&= E_{y_k}\Big\{ \min_{u_k\in U_k(x_k,y_k)} E_{w_k,x_{k+1},y_{k+1}}\big\{ g_k(x_k,y_k,u_k,w_k) \\
&\qquad + J_{k+1}^*(x_{k+1},y_{k+1}) \mid x_k,y_k,u_k\big\} \mid x_k\Big\} \\
&= E_{y_k}\Big\{ \min_{u_k\in U_k(x_k,y_k)} E_{w_k,x_{k+1}}\Big\{ g_k(x_k,y_k,u_k,w_k) \\
&\qquad + E_{y_{k+1}}\big\{J_{k+1}^*(x_{k+1},y_{k+1}) \mid x_{k+1}\big\} \mid x_k,y_k,u_k\Big\} \mid x_k\Big\}.
\end{aligned}$$

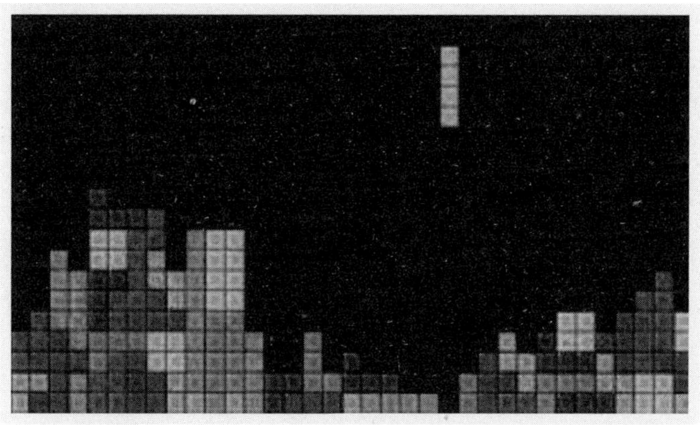

Figure 1.3.7 Illustration of a tetris board.

which is executed over the space of x_k rather than x_k *and* y_k. Note that this is a simpler algorithm to approximate than the one of Eq. (1.14).

Uncontrollable state components often occur in arrival systems, such as queueing, where action must be taken in response to a random event (such as a customer arrival) that cannot be influenced by the choice of control. Then the state of the arrival system must be augmented to include the random event, but the DP algorithm can be executed over a smaller space, as per Eq. (1.16). Here is an example of this type.

Example 1.3.4 (Tetris)

Tetris is a popular video game played on a two-dimensional grid. Each square in the grid can be full or empty, making up a "wall of bricks" with "holes" and a "jagged top" (see Fig. 1.3.7). The squares fill up as blocks of different shapes fall from the top of the grid and are added to the top of the wall. As a given block falls, the player can move horizontally and rotate the block in all possible ways, subject to the constraints imposed by the sides of the grid and the top of the wall. The falling blocks are generated independently according to some probability distribution, defined over a finite set of standard shapes. The game starts with an empty grid and ends when a square in the top row becomes full and the top of the wall reaches the top of the grid. When a row of full squares is created, this row is removed, the bricks lying above this row move one row downward, and the player scores a point. The player's objective is to maximize the score attained (total number of rows removed) within N steps or up to termination of the game, whichever occurs first.

We can model the problem of finding an optimal tetris playing strategy as a stochastic DP problem. The control, denoted by u, is the horizontal positioning and rotation applied to the falling block. The state consists of two components:

(1) The board position, i.e., a binary description of the full/empty status of each square, denoted by x.

(2) The shape of the current falling block, denoted by y.

There is also an additional termination state which is cost-free. Once the state reaches the termination state, it stays there with no change in cost.

The shape y is generated according to a probability distribution $p(y)$, independently of the control, so it can be viewed as an uncontrollable state component. The DP algorithm (1.16) is executed over the space of board positions x and has the intuitive form

$$\hat{J}_k(x) = \sum_y p(y) \max_u \Big[g(x, y, u) + \hat{J}_{k+1}\big(f(x, y, u)\big) \Big], \qquad \text{for all } x,$$

where

$g(x, y, u)$ is the number of points scored (rows removed),

$f(x, y, u)$ is the next board position (or termination state),

when the state is (x, y) and control u is applied, respectively. Note, however, that despite the simplification in the DP algorithm achieved by eliminating the uncontrollable portion of the state, the number of states x is still enormous, and the problem can only be addressed by suboptimal methods, which will be discussed later in this book.

1.3.6 Partial State Information and Belief States

We have assumed so far that the controller has access to the exact value of the current state x_k, so a policy consists of a sequence of functions $\mu_k(x_k)$, $k = 0, \ldots, N-1$. However, in many practical settings this assumption is unrealistic, because some components of the state may be inaccessible for measurement, the sensors used for measuring them may be inaccurate, or the cost of obtaining accurate measurements may be prohibitive.

Often in such situations the controller has access to only some of the components of the current state, and the corresponding measurements may also be corrupted by stochastic uncertainty. For example in three-dimensional motion problems, the state may consist of the six-tuple of position and velocity components, but the measurements may consist of noise-corrupted radar measurements of the three position components. This gives rise to problems of *partial* or *imperfect* state information, which have received a lot of attention in the optimization and artificial intelligence literature (see e.g., [Ber17], [RuN16]). Even though there are DP algorithms for partial information problems, these algorithms are far more computationally intensive than their perfect information counterparts. For this reason, in the absence of an analytical solution, partial information problems are typically solved suboptimally in practice.

On the other hand it turns out that conceptually, partial state information problems are no different than the perfect state information problems we have been addressing so far. In fact by various reformulations, we

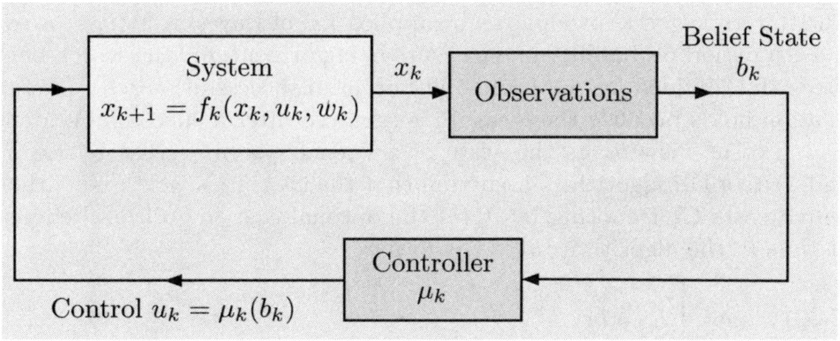

Figure 1.3.8 Schematic illustration of a control system with imperfect state observations. The belief state b_k is the conditional probability distribution of x_k given all the observations up to time k.

can reduce a partial state information problem to one with perfect state information (see [Ber17], Ch. 4).

The most common approach is to replace the state x_k with a *belief state*, which we will often denote by b_k. It is the probability distribution of x_k given all the observations that have been obtained by the controller up to time k (see Fig. 1.3.8), and it can serve as "state" in an appropriate DP algorithm. The belief state can in principle be computed and updated by a variety of methods that are based on Bayes' rule, such as *Kalman filtering* (see e.g., [AnM79], [KuV86], [Kri16], [ChC17]) and *particle filtering* (see e.g., [GSS93], [DoJ09], [Can16], [Kri16]). We illustrate the process of belief state updating and use in DP with a simple example.

Example 1.3.5 (Treasure Hunting)

In a classical problem of search, one has to decide at each of N periods whether to search a site that may contain a treasure. If a treasure is present, the search reveals it with probability ξ, in which case the treasure is removed from the site. Here the state x_k has two values: either a treasure is present in the site or it is not. The control u_k takes two values: search and not search. If the site is searched, we obtain an observation, which takes one of two values: treasure found or not found. If the site is not searched, no information is obtained.

Denote

b_k : probability a treasure is present at the beginning of period k given the search results so far.

This is the belief state at time k and it evolves according to the equation

$$b_{k+1} = \begin{cases} b_k & \text{if the site is not searched at time } k, \\ 0 & \text{if the site is searched and a treasure is found,} \\ \frac{b_k(1-\xi)}{b_k(1-\xi)+1-b_k} & \text{if the site is searched but no treasure is found.} \end{cases} \tag{1.17}$$

The third relation above follows by application of Bayes' rule (b_{k+1} is equal to the kth period probability of a treasure being present *and* the search being unsuccessful, divided by the probability of an unsuccessful search). The second relation holds because the treasure is removed after a successful search.

Let us view b_k as the state of a "belief system" given by Eq. (1.17), and write a DP algorithm, assuming that the treasure's worth is V, that each search costs C. Denoting by $J_k^*(b)$ the optimal cost-to-go from belief state b at time k, the algorithm takes the form

$$J_k^*(b_k) = \max \Bigg[J_{k+1}^*(b_k), \\ -C + b_k \xi V + (1 - b_k \xi) J_{k+1}^* \left(\frac{b_k(1-\xi)}{b_k(1-\xi) + 1 - b_k} \right) \Bigg], \tag{1.18}$$

with $J_N^*(b_N) = 0$. The two options in the maximization of Eq. (1.18) correspond to not searching (in which case b_k remains unchanged), and searching [in which case b_k evolves according to Eq. (1.17)].

Thanks to the simplicity of the problem, this DP algorithm can be used to obtain an analytical solution. In particular, it is straightforward to show by induction (starting with $k = N-1$) that the functions J_k^* satisfy $J_k^*(b_k) \geq 0$ for all $b_k \in [0,1]$ and

$$J_k^*(b_k) = 0 \qquad \text{if} \qquad b_k \leq \frac{C}{\xi V}.$$

From this it follows that it is optimal to search at period k if and only if

$$\frac{C}{\xi V} \leq b_k.$$

Thus, it is optimal to search if and only if the expected reward from the next search, $b_k \xi V$, is greater or equal to the cost C of the search - a myopic policy that focuses on just the next stage.

Of course the preceding example is extremely simple, involving a state x_k that takes just two values. As a result, the belief state b_k takes values within the interval $[0,1]$. Still there are infinitely many values in this interval, and if a computational solution were necessary, the belief state would have to be discretized and the DP algorithm (1.18) would have to be accordingly modified and executed over the discretized state space (discretization methods will be discussed in Chapter 6).

In problems where the state x_k can take a finite but large number of values, say n, the belief states comprise an n-dimensional simplex, so discretization becomes problematic. As a result, alternative suboptimal solution methods are often used in partial state information problems. Some of these methods will be described in future chapters; see Sections 2.3.2 and 5.7.3.

The following is a simple example of a partial state information problem whose belief state has large enough size to make an exact DP solution impossible.

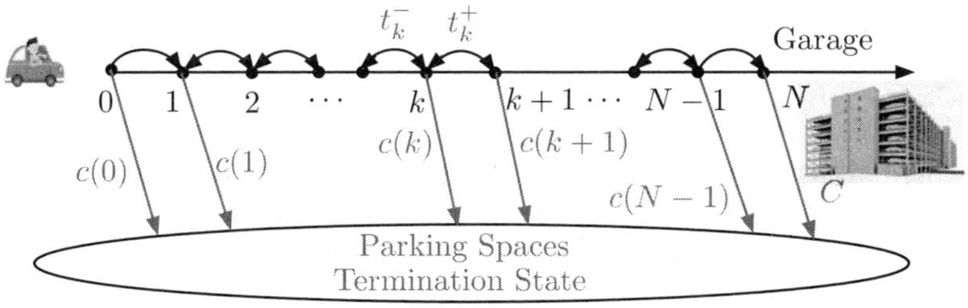

Figure 1.3.9 Cost structure and transitions of the bidirectional parking problem. The driver may park at space $k = 0, 1, \ldots, N-1$ at cost $c(k)$, if the space is free, can move to $k-1$ at cost t_k^- or can move to $k+1$ at cost t_k^+. At space N (the garage) the driver must park at cost C.

Example 1.3.6 (Bidirectional Parking)

Let us consider a more complex version of the parking problem of Example 1.3.3. As in that example, a driver is looking for inexpensive parking on the way to his destination, along a line of N parking spaces with a garage at the end. The difference is that the driver can move in either direction, rather than just forward towards the garage. In particular, at space i, the driver can park at cost $c(i)$ if i is free, can move to $i-1$ at a cost t_i^- or can move to $i+1$ at a cost t_i^+. Moreover, the driver records the free/taken status of the spaces previously visited and may return to any of these spaces; see Fig. 1.3.9.

Let us assume that the probability $p(i)$ of a space i being free changes over time, i.e., a space found free (or taken) at a given visit may get taken (or become free, respectively) by the time of the next visit. The initial probabilities $p(i)$, before visiting any spaces, are known, and the mechanism by which these probabilities change over time is also known to the driver. As an example, we may assume that at each time period, $p(i)$ increases by a certain known factor with some probability ξ and decreases by another known factor with the complementary probability $1-\xi$.

Here the belief state is the vector of current probabilities

$$\big(p(0), \ldots, p(N)\big),$$

and it is updated at each time based on the new observation: the free/taken status of the space visited at that time. This belief state can be computed exactly by the driver, given the parking status observations of the spaces visited thus far. While it is possible to state an exact DP algorithm that is defined over the set of belief states, and we will do so later, the algorithm is impossible to execute in practice.† Thus the problem can only be solved approximately, using methods that we will discuss in subsequent chapters.

† The problem as stated is an infinite horizon problem because there is nothing to prevent the driver from moving forever in the parking lot without ever parking. We can convert the problem to a similarly difficult finite horizon problem by restricting the number of moves to a given upper limit $\overline{N} > N$, and requiring that if the driver is at distance of k spaces from the garage at time $\overline{N} - k$, then driving in the direction away from the garage is not an option.

1.3.7 Linear Quadratic Optimal Control

In a few exceptional special cases the DP algorithm yields an analytical solution, which can be used among other purposes, as a starting point for approximate DP schemes to solve related problems. Prominent among such cases are various linear quadratic optimal control problems, which involve a linear (possibly multidimensional) system, a quadratic cost function, and no constraints on the control. Let us illustrate this with the deterministic scalar linear quadratic Example 1.1.2. We will apply the DP algorithm for the case of just two stages ($N = 2$), and illustrate the method for obtaining a nice analytical solution.

As defined in Example 1.1.2, the terminal cost is

$$g_2(x_2) = r(x_2 - T)^2.$$

Thus the DP algorithm starts with

$$J_2^*(x_2) = g_2(x_2) = r(x_2 - T)^2,$$

[cf. Eq. (1.3)].

For the next-to-last stage, we have [cf. Eq. (1.4)]

$$J_1^*(x_1) = \min_{u_1}\big[u_1^2 + J_2^*(x_2)\big] = \min_{u_1}\Big[u_1^2 + J_2^*\big((1-a)x_1 + au_1\big)\Big].$$

Substituting the previous form of J_2^*, we obtain

$$J_1^*(x_1) = \min_{u_1}\Big[u_1^2 + r\big((1-a)x_1 + au_1 - T\big)^2\Big]. \tag{1.19}$$

This minimization will be done by setting to zero the derivative with respect to u_1. This yields

$$0 = 2u_1 + 2ra\big((1-a)x_1 + au_1 - T\big),$$

and by collecting terms and solving for u_1, we obtain the optimal temperature for the last oven as a function of x_1:

$$\mu_1^*(x_1) = \frac{ra\big(T - (1-a)x_1\big)}{1 + ra^2}. \tag{1.20}$$

By substituting the optimal u_1 in the expression (1.19) for J_1^*, we obtain

$$\begin{aligned} J_1^*(x_1) &= \frac{r^2a^2\big((1-a)x_1 - T\big)^2}{(1+ra^2)^2} + r\left((1-a)x_1 + \frac{ra^2\big(T - (1-a)x_1\big)}{1+ra^2} - T\right)^2 \\ &= \frac{r^2a^2\big((1-a)x_1 - T\big)^2}{(1+ra^2)^2} + r\left(\frac{ra^2}{1+ra^2} - 1\right)^2\big((1-a)x_1 - T\big)^2 \\ &= \frac{r\big((1-a)x_1 - T\big)^2}{1+ra^2}. \end{aligned}$$

We now go back one stage. We have [cf. Eq. (1.4)]

$$J_0^*(x_0) = \min_{u_0} \left[u_0^2 + J_1^*(x_1)\right] = \min_{u_0} \Big[u_0^2 + J_1^*\big((1-a)x_0 + au_0\big)\Big],$$

and by substituting the expression already obtained for J_1^*, we have

$$J_0^*(x_0) = \min_{u_0} \left[u_0^2 + \frac{r\big((1-a)^2 x_0 + (1-a)au_0 - T\big)^2}{1+ra^2}\right].$$

We minimize with respect to u_0 by setting the corresponding derivative to zero. We obtain

$$0 = 2u_0 + \frac{2r(1-a)a\big((1-a)^2 x_0 + (1-a)au_0 - T\big)}{1+ra^2}.$$

This yields, after some calculation, the optimal temperature of the first oven:

$$\mu_0^*(x_0) = \frac{r(1-a)a\big(T - (1-a)^2 x_0\big)}{1 + ra^2\big(1 + (1-a)^2\big)}. \tag{1.21}$$

The optimal cost is obtained by substituting this expression in the formula for J_0^*. This leads to a straightforward but lengthy calculation, which in the end yields the rather simple formula

$$J_0^*(x_0) = \frac{r\big((1-a)^2 x_0 - T\big)^2}{1 + ra^2\big(1 + (1-a)^2\big)}.$$

This completes the solution of the problem.

Note that the algorithm has simultaneously yielded an optimal policy $\{\mu_0^*, \mu_1^*\}$ via Eqs. (1.21) and (1.20): a rule that tells us the optimal oven temperatures $u_0 = \mu_0^*(x_0)$ and $u_1 = \mu_1^*(x_1)$ for every possible value of the states x_0 and x_1, respectively. Thus the DP algorithm (as expected) solves all the tail subproblems and provides a feedback policy.

A noteworthy feature in this example is the facility with which we obtained an analytical solution. A little thought while tracing the steps of the algorithm will convince the reader that what simplifies the solution is the quadratic nature of the cost and the linearity of the system equation [see the derivation of Eq. (1.20)]. Indeed, it can be shown in generality that when the system is linear and the cost is quadratic, the optimal policy and cost-to-go function are given by closed-form expressions, regardless of the number of stages N (see [Ber17], Section 3.1).

Stochastic Linear Quadratic Problems - Certainty Equivalence

Let us now introduce a zero-mean stochastic additive disturbance in the linear system equation. Remarkably, it turns out that the optimal policy remains unaffected. To see this, assume that the material's temperature evolves according to

$$x_{k+1} = (1 - a)x_k + au_k + w_k, \qquad k = 0, 1,$$

where w_0 and w_1 are independent random variables with given distribution, zero mean

$$E\{w_0\} = E\{w_1\} = 0,$$

and finite variance. Then the equation for J_1^* [cf. Eq. (1.4)] becomes

$$\begin{aligned} J_1^*(x_1) &= \min_{u_1} \mathop{E}_{w_1} \left\{ u_1^2 + r\big((1 - a)x_1 + au_1 + w_1 - T\big)^2 \right\} \\ &= \min_{u_1} \Big[u_1^2 + r\big((1 - a)x_1 + au_1 - T\big)^2 \\ &\qquad + 2rE\{w_1\}\big((1 - a)x_1 + au_1 - T\big) + rE\{w_1^2\} \Big]. \end{aligned}$$

Since $E\{w_1\} = 0$, we obtain

$$J_1^*(x_1) = \min_{u_1} \Big[u_1^2 + r\big((1 - a)x_1 + au_1 - T\big)^2 \Big] + rE\{w_1^2\}.$$

Comparing this equation with Eq. (1.19), we see that the presence of w_1 has resulted in an additional inconsequential constant term, $rE\{w_1^2\}$. Therefore, the optimal policy for the last stage remains unaffected by the presence of w_1, while $J_1^*(x_1)$ is increased by $rE\{w_1^2\}$. It can be seen that a similar situation also holds for the first stage. In particular, the optimal cost is given by the same expression as before except for an additive constant that depends on $E\{w_0^2\}$ and $E\{w_1^2\}$.

Generally, if the optimal policy is unaffected when the disturbances are replaced by their means, we say that *certainty equivalence* holds. This occurs in several types of problems involving a linear system and a quadratic cost; see [Ber17], Sections 3.1 and 4.2. For other problems, certainty equivalence can be used as a basis for problem approximation, e.g., assume that certainty equivalence holds (i.e., replace stochastic quantities by some typical values, such as their expected values) and apply exact DP to the resulting deterministic optimal control problem (see Section 2.3.2).

1.3.8 Systems with Unknown Parameters - Adaptive Control

We have been dealing so far with systems having a known system equation. In practice, however, there are many cases where the system parameters are either not known exactly or change over time.

As an example consider a car cruise control system. The car's velocity at time k, denoted by x_k, evolves according to

$$x_{k+1} = x_k + bu_k,$$

where u_k is the force that propels the car forward (u_k can be related to the pressure applied on the car's gas pedal). However, the coefficient b changes frequently and cannot be modeled with any precision because it depends on unpredictable time-varying conditions, such as the slope and condition of the road, and the weight of the car (which is affected by the number of passengers). This points to the need of controllers that yield satisfactory performance over a potentially broad range of system parameters.

To construct a formal optimization framework for dealing with such a situation, we may embed the problem within an imperfect state information framework by modeling the unknown parameters as unobservable states. Indeed, let the system equation be of the form

$$x_{k+1} = f_k(x_k, \theta, u_k, w_k),$$

where θ is a vector of unknown parameters, which for simplicity we assume to be fixed over time. We introduce an additional state variable $y_k = \theta$ and obtain a system equation of the form

$$\begin{pmatrix} x_{k+1} \\ y_{k+1} \end{pmatrix} = \begin{pmatrix} f_k(x_k, y_k, u_k, w_k) \\ y_k \end{pmatrix}.$$

This equation can be written compactly as

$$\tilde{x}_{k+1} = \tilde{f}_k(\tilde{x}_k, u_k, w_k),$$

where $\tilde{x}_k = (x_k, y_k)$ is the new state, and $\tilde{f}_k$ is an appropriate function. The initial state is

$$\tilde{x}_0 = (x_0, \theta).$$

Unfortunately, however, since y_k (i.e., θ) is unobservable, the problem is one of partial state information even if the controller knows the state x_k exactly. This makes the exact solution by DP intractable. To address this situation, several suboptimal solution approaches have been suggested.

An apparently reasonable form of suboptimal control is to separate the control process into two phases, a *parameter estimation (or identification) phase* and a *control phase*. In the first phase the unknown parameters are identified, while the control takes no account of the interim results of estimation. The final parameter estimates from the first phase are then used to implement an optimal control law in the second phase. This alternation of estimation and control phases may be repeated several times during any system run in order to take into account subsequent changes of the parameters.

One drawback of this approach is that it is not always easy to determine when to terminate one phase and start the other. A second difficulty, of a more fundamental nature, is that the unknown parameters must often be estimated as the system is being controlled. Then, unfortunately, the control process may make some of the unknown parameters invisible to the estimation process. This is known as the problem of parameter *identifiability*, which is discussed in the context of optimal control in several sources, including [BoV79] and [Kum83]; see also [Ber17], Section 6.7. For a simple example, consider the scalar system

$$x_{k+1} = ax_k + bu_k, \qquad k = 0, \ldots, N-1,$$

and the quadratic cost

$$\sum_{k=1}^{N} (x_k)^2.$$

Assuming perfect state information, if the parameters a and b are known, it can be seen that the optimal control law is

$$\mu_k^*(x_k) = -\frac{a}{b} x_k,$$

which sets all future states to 0. Assume now that the parameters a and b are unknown, and consider the two-phase method. During the first phase the control law

$$\tilde{\mu}_k(x_k) = \gamma x_k \tag{1.22}$$

is used (γ is some scalar; for example, $\gamma = -\overline{a}/\overline{b}$, where $\overline{a}$ and $\overline{b}$ are some a priori estimates of a and b, respectively). At the end of the first phase, the control law is changed to

$$\overline{\mu}_k(x_k) = -\frac{\hat{a}}{\hat{b}} x_k,$$

where $\hat{a}$ and $\hat{b}$ are the estimates obtained from the estimation process. However, with the control law (1.22), the closed-loop system is

$$x_{k+1} = (a + b\gamma)x_k,$$

so the estimation process can at best yield the value of $(a + b\gamma)$ but not the values of both a and b. In other words, the estimation process cannot discriminate between pairs of values (a_1, b_1) and (a_2, b_2) such that $a_1 + b_1\gamma = a_2 + b_2\gamma$. Therefore, a and b are not identifiable when feedback control of the form (1.22) is applied.

The issues discussed above and the methods to address them are part of a broad field known as *adaptive control*, which deals with the design of controllers for systems with unknown parameters. This is a rich subject with many and diverse applications. We will not discuss adaptive control in this book, and we refer to textbook treatments, such as Aström and Wittenmark [AsW94], Goodwin and Sin [GoS84], Ioannou and Sun [IoS96], Krstic, Kanellakopoulos, and Kokotovic [KKK95], Kumar and Varaiya [KuV86], Narendra and Annaswamy [NaA12], Sastry and Bodson [SaB11], and Slotine and Li [SlL91].

We note, however, a simple and popular methodology, *PID (Proportional-Integral-Derivative) control*, which can be used for problems involving an unknown or changing mathematical model; see e.g., the books by Aström and Hagglund [AsH95], [AsH06]. In particular, PID control aims to maintain the output of a single-input single-output dynamic system around a set point or to follow a given trajectory, as the system parameters change within a relatively broad range. In its simplest form, the PID controller is parametrized by three scalar parameters, which may be determined by a variety of methods, some of them manual/heuristic. We will later discuss briefly PID control in Section 5.7, and point out that the automatic choice of its parameters can be considered within the context of the broader methodology of approximation in policy space.

1.4 REINFORCEMENT LEARNING AND OPTIMAL CONTROL - SOME TERMINOLOGY

There has been intense interest in DP-related approximations in view of their promise to deal with the *curse of dimensionality* (the explosion of the computation as the number of states increases is dealt with the use of approximate cost functions) and the *curse of modeling* (a simulator/computer model may be used in place of a mathematical model of the problem). The current state of the subject owes much to an enormously beneficial cross-fertilization of ideas from optimal control (with its traditional emphasis on sequential decision making and formal optimization methodologies), and from artificial intelligence (and its traditional emphasis on learning through observation and experience, heuristic evaluation functions in game-playing programs, and the use of feature-based and other representations).

The boundaries between these two fields are now diminished thanks to a deeper understanding of the foundational issues, and the associated methods and core applications. Unfortunately, however, there have been substantial differences in language and emphasis in RL-based discussions (where artificial intelligence-related terminology is used) and DP-based discussions (where optimal control-related terminology is used). This includes the typical use of maximization/value function/reward in the former field and the use of minimization/cost function/cost per stage in the latter field, and goes much further.

The terminology used in this book is standard in DP and optimal control, and in an effort to forestall confusion of readers that are accustomed to either the RL or the optimal control terminology, we provide a list of terms commonly used in RL, and their optimal control counterparts.

(a) **Environment** = System.

(b) **Agent** = Decision maker or controller.

(c) **Action** = Decision or control.

(d) **Reward of a stage** = (Opposite of) Cost of a stage.

(e) **State value** = (Opposite of) Cost starting from a state.

(f) **Value (or reward) function** = (Opposite of) Cost function.

(g) **Maximizing the value function** = Minimizing the cost function.

(h) **Action (or state-action) value** = Q-factor (or Q-value) of a state-control pair. (Q-value is also used often in RL.)

(i) **Planning** = Solving a DP problem with a known mathematical model.

(j) **Learning** = Solving a DP problem without using an explicit mathematical model. (This is the principal meaning of the term "learning" in RL. Other meanings are also common.)

(k) **Self-learning** (or self-play in the context of games) = Solving a DP problem using some form of policy iteration.

(l) **Deep reinforcement learning** = Approximate DP using value and/or policy approximation with deep neural networks.

(m) **Prediction** = Policy evaluation.

(n) **Generalized policy iteration** = Optimistic policy iteration.

(o) **State abstraction** = State aggregation.

(p) **Temporal abstraction** = Time aggregation.

(q) **Learning a model** = System identification.

(r) **Episodic task or episode** = Finite-step system trajectory.

(s) **Continuing task** = Infinite-step system trajectory.

(t) **Experience replay** = Reuse of samples in a simulation process.

(u) **Bellman operator** = DP mapping or operator.

(v) **Backup** = Applying the DP operator at some state.

(w) **Sweep** = Applying the DP operator at all states.

(x) **Greedy policy with respect to a cost function J** = Minimizing policy in the DP expression defined by J.

(y) **Afterstate** = Post-decision state.

(z) **Ground truth** = Empirical evidence or information provided by direct observation.

Some of the preceding terms will be introduced in future chapters. The reader may then wish to return to this section as an aid in connecting with the relevant RL literature.

Notation

Unfortunately the confusion arising from different terminology has been exacerbated by the use of different notations. The present textbook roughly follows the "standard" notation of the Bellman/Pontryagin optimal control era; see e.g., the classical books by Athans and Falb [AtF66], Bellman [Bel67], and Bryson and Ho [BrH75]. This notation is consistent with the author's other DP books.

A summary of our most prominently used symbols is as follows:

(a) x: state.

(b) u: control.

(c) J: cost function.

(d) g: cost per stage.

(e) f: system function.

(f) i: discrete state.

(g) $p_{ij}(u)$: transition probability from state i to state j under control u.

(h) α: discount factor in discounted problems.

The x-u-J notation is standard in optimal control textbooks (e.g., the books by Athans and Falb [AtF66], and Bryson and Ho [BrH75], as well as the more recent book by Liberzon [Lib11]). The notations f and g are also used most commonly in the literature of the early optimal control period as well as later (unfortunately the more natural symbol "c" has not been used much in place of "g" for the cost per stage). The discrete system notations i and $p_{ij}(u)$ are very common in the discrete-state Markov decision problem and operations research literature, where discrete-state problems have been treated extensively [sometimes the alternative notation $p(j \mid i, u)$ is used for the transition probabilities].

The RL literature addresses for the most part finite-state Markov decision problems, most frequently the discounted and stochastic shortest path infinite horizon problems that are discussed in Chapter 4. The most commonly used notation is s for state, a for action, $r(s, a, s')$ for reward per stage, $p(s' \mid s, a)$ or $P_{s,a}(s')$ for transition probability from s to s' under action a, and γ for discount factor (see e.g., Sutton and Barto [SuB18]).

1.5 NOTES AND SOURCES

Our discussion of exact DP in this chapter has been brief since our focus in this book will be on approximate DP and RL. The author's DP textbook [Ber17] provides an extensive discussion of finite horizon exact DP, and its applications to discrete and continuous spaces problems, using a notation and style that is consistent with the present book. The books by Puterman [Put94] and by the author [Ber12] provide detailed treatments of infinite horizon finite-state Markovian decision problems. Continuous spaces infinite horizon problems are covered in the author's book [Ber12], while some of the more complex mathematical aspects of exact DP are discussed in the monograph by Bertsekas and Shreve [BeS78] (particularly the probabilistic/measure-theoretic issues associated with stochastic optimal control).

The author's abstract DP monograph [Ber18a] aims at a unified development of the core theory and algorithms of total cost sequential decision problems, and addresses simultaneously stochastic, minimax, game, risk-sensitive, and other DP problems, through the use of the abstract DP operator (or Bellman operator as it is often called in RL). The idea here is to gain insight through abstraction. In particular, the structure of a DP model is encoded in its abstract Bellman operator, which serves as the "mathematical signature" of the model. Thus, characteristics of this operator (such as monotonicity and contraction) largely determine the analytical results and computational algorithms that can be applied to that model. It is likely that some of the approximation algorithms of the present book are transferable to a broad variety of DP models, well beyond the ones studied here. The design and analysis of these algorithms can then benefit from the broad algorithmic principles that have been developed through an abstract viewpoint.

The approximate DP and RL literature has expanded tremendously since the connections between DP and RL became apparent in the late 80s and early 90s. We restrict ourselves to mentioning textbooks, research monographs, and broad surveys, which supplement our discussions, express related viewpoints, and collectively provide a guide to the literature. Moreover, inevitably our referencing reflects a cultural bias, and an overemphasis on sources that are familiar to the author and are written in a similar style to the present book (including the author's own works). Thus we wish to apologize in advance for the many omissions of important research references that are somewhat outside our own understanding and view of the field.

Two books were written on our subject in the 1990s, setting the tone for subsequent developments in the field. One in 1996 by Bertsekas and Tsitsiklis [BeT96], which reflects a decision, control, and optimization viewpoint, and another in 1998 by Sutton and Barto, which reflects an artificial intelligence viewpoint (a 2nd edition, [SuB18], was published in 2018). We refer to the former book and also to the author's DP textbooks [Ber12], [Ber17] for a broader discussion of some of the topics of the present book, including algorithmic convergence issues and additional DP models, such as those based on average cost optimization. For historical accounts of the early development of the subject, see [BeT96], Section 6.7, and [SuB18], Section 1.7.

More recent books are by Gosavi [Gos15] (a much expanded 2nd edition of his 2003 monograph), which emphasizes simulation-based optimization and RL algorithms, Cao [Cao07], which focuses on a sensitivity approach to simulation-based methods, Chang, Fu, Hu, and Marcus [CFH13] (a 2nd edition of their 2007 monograph), which emphasizes finite-horizon/limited lookahead schemes and adaptive sampling, Busoniu et al. [BBD10], which focuses on function approximation methods for continuous space systems and includes a discussion of random search methods, Powell [Pow11], which emphasizes resource allocation and operations research applications, Vrabie, Vamvoudakis, and Lewis [VVL13], which discusses neural network-based methods, on-line adaptive control, and continuous-time optimal control applications, Kochenderfer et al. [KAC15], which selectively discusses applications and approximations in DP and the treatment of uncertainty, Jiang and Jiang [JiJ17], which develops adaptive control within an approximate DP framework, and Liu et al. [LWW17], which deals with forms of adaptive dynamic programming, and topics in both RL and optimal control. The book by Krishnamurthy [Kri16] focuses on partial state information problems, with discussion of both exact DP, and approximate DP/RL methods. The book by Haykin [Hay08] discusses approximate DP in the broader context of neural network-related subjects. The book by Borkar [Bor08] is an advanced monograph that addresses rigorously many of the convergence issues of iterative stochastic algorithms in approximate DP, mainly using the so called ODE approach. The book by Meyn [Mey07] is broader in its coverage, but touches upon some of the approximate DP algorithms that we discuss.

Influential early surveys were written, from an artificial intelligence viewpoint, by Barto, Bradtke, and Singh [BBS95] (which dealt with the methodologies of real-time DP and its antecedent, real-time heuristic search [Kor90], and the use of asynchronous DP ideas [Ber82], [Ber83], [BeT89] within their context), and by Kaelbling, Littman, and Moore [KLM96] (which focused on general principles of RL). The volume by White and Sofge [WhS92] also contains several surveys describing early work in the field.

Several overview papers in the volume by Si, Barto, Powell, and Wunsch [SBP04] describe some approximation methods that we will not be covering in much detail in this book: linear programming approaches (De Farias [DeF04]), large-scale resource allocation methods (Powell and Van Roy [PoV04]), and deterministic optimal control approaches (Ferrari and Stengel [FeS04], and Si, Yang, and Liu [SYL04]). Updated accounts of these and other related topics are given in the survey collections by Lewis, Liu, and Lendaris [LLL08], and Lewis and Liu [LeL13].

Recent extended surveys and short monographs are Borkar [Bor09] (a methodological point of view that explores connections with other Monte

Carlo schemes), Lewis and Vrabie [LeV09] (a control theory point of view), Szepesvari [Sze10] (which discusses approximation in value space from a RL point of view), Deisenroth, Neumann, and Peters [DNP11], and Grondman et al. [GBL12] (which focus on policy gradient methods), Browne et al. [BPW12] (which focuses on Monte Carlo Tree Search), Mausam and Kolobov [MaK12] (which deals with Markovian decision problems from an artificial intelligence viewpoint), Schmidhuber [Sch15], Arulkumaran et al. [ADB17], Li [Li17], Busoniu et al. [BDT18], and Caterini and Chang [CaC18] (which deal with reinforcement learning schemes that are based on the use of deep neural networks), the author's [Ber05a] (which focuses on rollout algorithms and model predictive control), [Ber11a] (which focuses on approximate policy iteration), and [Ber18b] (which focuses on aggregation methods), and Recht [Rec18a] (which focuses on continuous spaces optimal control).

2

Approximation in Value Space

Contents

As we noted in Chapter 1, the exact solution of optimal control problems by DP is often impossible. To a great extent, the reason lies in what Bellman has called the "curse of dimensionality." This refers to a rapid increase of the required computation and memory storage as the size of the problem increases. Moreover, there are many circumstances where the structure of the given problem is known well in advance, but some of the problem data, such as various system parameters, may be unknown until shortly before control is needed, thus seriously constraining the amount of time available for the DP computation. These difficulties motivate suboptimal control schemes that strike a reasonable balance between convenient implementation and adequate performance.

The philosophy that guides our subsequent presentation is based on the author's view of the present state of the art in the field: there are no RL methods that are guaranteed to work for all or even most DP problems, but there are enough methods to try on a given problem with a reasonable chance of success in the end. With this in mind, we will present a broad variety of methods, aiming to provide intuition into their inner workings, as well as an understanding of their analytical and computational properties. We present some analysis, usually based on DP principles, but we do not provide solid performance guarantees for most of our methods. We also discuss the potential merits of different methods within particular practical contexts, but our discussion is speculative and should be taken with a grain of salt and/or adapted to the problem at hand.

In this chapter, we first lay out in Section 2.1 a broad landscape of approximation approaches in the contexts of the finite horizon deterministic and stochastic DP problems of Chapter 1. We then focus on approximation in value space, the key ideas of one-step and multistep lookahead, and policy improvement by rollout, as well as various possibilities for their implementation. Many of the finite horizon methods will be adapted to infinite horizon DP later, together with some additional methods that are specific to the infinite horizon context.

2.1 APPROXIMATION APPROACHES IN REINFORCEMENT LEARNING

There are two general types of approximation in DP-based suboptimal control. The first is *approximation in value space*, where we aim to approximate the optimal cost function or the cost function of a given policy. The second is *approximation in policy space*, where we select the policy by using optimization over a suitable class of policies. In this section we provide a broad overview of these approximation approaches.

Approximation in Value Space

In approximation in value space, we approximate the optimal cost-to-go functions J_k^* with some other functions $\tilde{J}_k$. We then replace J_k^* in the DP equation with $\tilde{J}_k$. In particular, at state x_k, we use the control obtained from the minimization

$$\tilde{\mu}_k(x_k) \in \arg \min_{u_k \in U_k(x_k)} E\Big\{g_k(x_k, u_k, w_k) + \tilde{J}_{k+1}\big(f_k(x_k, u_k, w_k)\big)\Big\}. \tag{2.1}$$

This defines a suboptimal policy $\{\tilde{\mu}_0, \ldots, \tilde{\mu}_{N-1}\}$. There are several possibilities for selecting or computing the functions $\tilde{J}_k$, which are discussed in this chapter, and also in subsequent chapters.

Note that the expected value expression appearing in the right-hand side of Eq. (2.1) can be viewed as an approximate Q-factor,

$$\tilde{Q}_k(x_k, u_k) = E\Big\{g_k(x_k, u_k, w_k) + \tilde{J}_{k+1}\big(f_k(x_k, u_k, w_k)\big)\Big\},$$

and the minimization in Eq. (2.1) can be written as

$$\tilde{\mu}_k(x_k) \in \arg \min_{u_k \in U_k(x_k)} \tilde{Q}_k(x_k, u_k),$$

(cf. Section 1.2). This also suggests a variant of approximation in value space, which is based on using Q-factor approximations that may be obtained directly, i.e., without the intermediate step of obtaining the cost function approximations $\tilde{J}_k$. In what follows in this chapter, we will focus on cost function approximation, but we will occasionally digress to discuss direct Q-factor approximation.

Approximation in Value Space - Multistep Lookahead

Approximation in value space based on the minimization (2.1) is commonly referred to as *one-step lookahead*, because the future costs are approximated by $\tilde{J}_{k+1}$, after a single step. An important variation is *multistep lookahead*, whereby at state x_k we minimize the cost of the first $\ell > 1$ stages with the future costs approximated by a function $\tilde{J}_{k+\ell}$. For example, in two-step lookahead the function $\tilde{J}_{k+1}$ is given by

$$\tilde{J}_{k+1}(x_{k+1}) = \min_{u_{k+1} \in U_{k+1}(x_{k+1})} E\Big\{g_{k+1}(x_{k+1}, u_{k+1}, w_{k+1}) + \tilde{J}_{k+2}\big(f_{k+1}(x_{k+1}, u_{k+1}, w_{k+1})\big)\Big\},$$

where $\tilde{J}_{k+2}$ is some approximation of the optimal cost-to-go function J_{k+2}^*.

Actually, as the preceding two-step lookahead case illustrates, one may view ℓ-step lookahead as the special case of one-step lookahead where the lookahead function is the optimal cost function of an $(\ell - 1)$-stage DP problem with a terminal cost $(\tilde{J}_{k+\ell})$ at the end of the $\ell - 1$ stages.

However, it is often important to discuss ℓ-step lookahead separately, in order to address special implementation issues that do not arise in the context of one-step lookahead.

The motivation for ℓ-step lookahead is that for increasing values of ℓ, one may require a less accurate approximation $\tilde{J}_{k+\ell}$ to obtain good performance. Otherwise expressed, for the same quality of cost function approximation, better performance may be obtained as ℓ becomes larger. This makes intuitive sense, since in this case, the cost of more stages is treated exactly, i.e., with optimization. Indeed this expectation is typically realized in practice, although one can construct artificial examples when this is not so (see Section 2.2.1). Generally, one should at least try to use the largest value of ℓ for which the overhead for solving the ℓ-step lookahead minimization problem on-line is acceptable.

In our initial discussion of approximation in value space of the present section, we will focus on one-step lookahead. There are straightforward extensions of the main ideas to the multistep context, which we will discuss in the next section (Section 2.2).

Approximation in Policy Space

The major alternative to approximation in value space is *approximation in policy space*, whereby we select the policy from a suitably restricted class of policies, usually a parametric class of some form. In particular, we can introduce a parametric family of policies

$$\mu_k(x_k, r_k), \qquad k = 0, \ldots, N-1,$$

where r_k is a parameter, such as a family represented by a neural network, and then estimate the parameters r_k using some type of optimization.

An important advantage of approximation in policy space is that the computation of controls during on-line operation of the system is often much easier compared with the lookahead minimization (2.1). However, this advantage can also be gained by combining approximation in value space with policy approximation, as we will describe in Section 2.1.5.

In this chapter we discuss primarily approximation in value space, although some of the ideas are also relevant to approximation in policy space. We focus on finite horizon problems, postponing the discussion of infinite horizon problems for Chapter 4 and later. However, the finite horizon ideas are relevant to the infinite horizon setting, and many of the methods of the present chapter and Chapter 3 also apply with small modifications to infinite horizon problems.

Model-Based Versus Model-Free Implementation

Generally, a finite horizon DP problem is defined by the state, control, and disturbance spaces, the functions f_k and g_k, the control constraint sets

$U_k(x_k)$, and the probability distributions of the disturbances. We refer to these as the *mathematical model* of the problem. Of course there is only one mathematical model for a given problem, but there may be several different implementations of a given method for solving the problem exactly or approximately. Some of these implementations may rely exclusively on analytical calculations using the mathematical model, but others may additionally or exclusively rely on Monte Carlo simulation.

In this book, we will use an unambiguous technical definition of a model-free method (or more accurately a model-free implementation of a given method).† In particular, for us the key relevant attribute of an implementation is whether an analytical formula-based calculation or a Monte Carlo simulation is used to compute expected values such as those arising in one-step and multistep lookahead expressions. We thus distinguish between two types of implementations of the various methods given in this book:

(a) In the *model-based* case, a major assumption is that the functions g_k and f_k are available in closed form, and moreover the conditional probability distribution of w_k, given (x_k, u_k), is also available. By this we mean that the value of $p_k(w_k \mid x_k, u_k)$ is available for any triplet (x_k, u_k, w_k). In a model-based implementation, *some* expected values, such as the one in the lookahead expression of Eq. (2.1), are obtained with algebraic calculations as opposed to Monte Carlo simulation.

(b) In the *model-free* case, the calculation of *all* the expected values in the Eq. (2.1), and other related expressions, is done with Monte Carlo simulation. There may be two possible reasons for this.

  (1) An analytical expression of the probabilities $p_k(w_k \mid x_k, u_k)$ is not available, but instead there is a computer program/simulator that for any given state x_k and control u_k, simulates sample probabilistic transitions to a successor state x_{k+1}, and generates the corresponding transition costs. In this case the expected value in Eq. (2.1) and related expressions can be computed approximately by Monte Carlo simulation.‡

  (2) The probabilities $p_k(w_k \mid x_k, u_k)$ are available for any triplet (x_k, u_k, w_k), but for reasons of computational efficiency we pre-

---

† There is considerable variance in the use of the term "model-free" in the literature. For example, some authors define model-based RL methods as those that estimate costs-to-go of states, and model-free those that just estimate Q-factors of state-control pairs.

‡ The term Monte Carlo simulation mostly refers to the use of a software simulator. However, a hardware or a combined hardware/software simulator may also be used in some practical situations to generate samples that are used for Monte Carlo averaging.

fer to compute the expected value in the expression (2.1) by using sampling and Monte Carlo simulation. Thus the expected value is computed as in the case where an analytical expression of the probabilities $p_k(w_k \mid x_k, u_k)$ is not available, but instead there is a computer simulator.†

Note that for deterministic problems there is no expected value to compute, so methods for these problems typically come under the model-based category, even if values of the functions g_k and f_k become available through complicated computer calculations. Still however, Monte Carlo simulation may enter the solution process of a deterministic problem for a variety of reasons. For example the games of chess and Go are perfectly deterministic, but the AlphaGo and AlphaZero programs (Silver et al. [SHM16], [SHS17]) use randomized policies and rely heavily on Monte Carlo tree search techniques, which use sampling and will be discussed in Section 2.4.2. The same is true for some policy gradient methods, which will be discussed in Chapter 5.

To summarize, *the use of sampling and Monte Carlo simulation is the defining attribute for an implementation to be model-based or model-free in the terminology of this book.* This view of a model-free approach aims to avoid ambiguities in cases where the mathematical model is available to use for closed form calculations, but Monte Carlo simulation is used anyway for reasons of convenience or computational efficiency.

2.1.1 General Issues of Approximation in Value Space

There are two major issues in a value space approximation scheme, and each of the two can be considered separately from the other:

(1) *Obtaining* $\tilde{J}_k$, i.e., the method to compute the lookahead functions $\tilde{J}_k$ that are involved in the lookahead minimization (2.1). There are quite a few approaches here (see Fig. 2.1.1). Several of them are discussed in this chapter, and more will be discussed in subsequent chapters.

(2) *Control selection*, i.e., the method to perform the minimization (2.1) and implement the suboptimal policy $\tilde{\mu}_k$. Again there are several exact and approximate methods for control selection, some of which will be discussed in this chapter (see Fig. 2.1.1).

In this section we will provide a high level discussion of these issues, focusing for simplicity on just the case of one-step lookahead.

---

† The idea of using Monte Carlo simulation to compute complicated integrals or even sums of many numbers is used widely in various types of numerical computations. It encompasses efficient Monte Carlo techniques known as *Monte Carlo integration* and *importance sampling*; see textbooks such as [Liu01], [AsG10], [RoC10], [Gla13] for detailed developments.

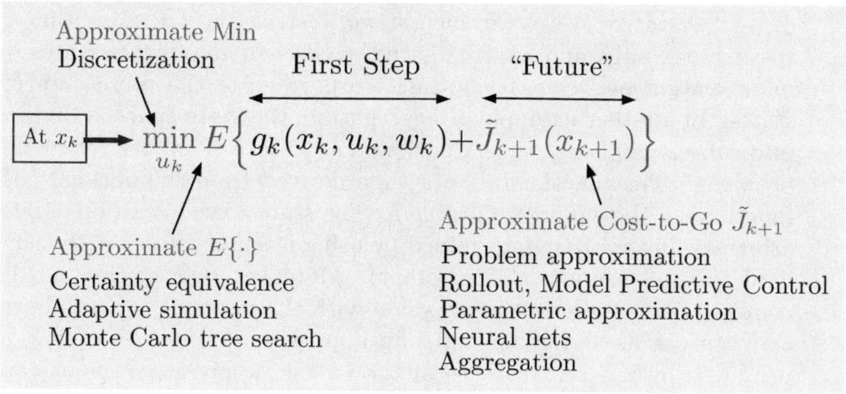

Figure 2.1.1 Schematic illustration of various options for approximation in value space with one-step lookahead. The lookahead function values $\tilde J_{k+1}(x_{k+1})$ approximate the optimal cost-to-go values $J^*_{k+1}(x_{k+1})$, and can be computed by a variety of methods. There may be additional approximations in the minimization over u_k and in the computation of the expected value over w_k; see Section 2.1.1.

Regarding the computation of $\tilde J_k$, we will consider four types of methods:

(a) *Problem approximation (Section 2.3)*: Here the functions $\tilde J_k$ in Eq. (2.1) are obtained as the optimal or nearly optimal cost functions of a simplified optimization problem, which is more convenient for computation. Simplifications may include exploiting decomposable structure, ignoring various types of uncertainties, and reducing the size of the state space. Another form of simplification is *aggregation*, which is discussed separately in Chapter 6.

(b) *On-line approximate optimization (Sections 2.4, 2.5)*: These methods often involve the use of a suboptimal policy or heuristic, which is applied on-line when needed to approximate the true optimal cost-to-go values. The suboptimal policy may be obtained by any other method, e.g., problem approximation. *Rollout algorithms* and *model predictive control* are prime examples of these methods.

(c) *Parametric cost approximation (Chapter 3)*: Here the functions $\tilde J_k$ in Eq. (2.1) are obtained from a given parametric class of functions $\tilde J_k(x_k, r_k)$, where r_k is a parameter vector, selected by a suitable algorithm. The parametric class is typically obtained by using prominent characteristics of x_k called *features*, which can be obtained either through insight into the problem at hand, or by using training data and some form of neural network.

(d) *Aggregation (Chapter 6)*: This is a special but rather sophisticated form of problem approximation. A simple example is to select a set of representative states for each stage, restrict the DP algorithm to these states only, and approximate the costs-to-go of other states by interpolation between the optimal costs-to-go of the representative states. In another example of aggregation, the state space is divided into subsets, and each subset is viewed as a state of an "aggregate DP problem." The functions $\tilde{J}_k$ are then derived from the optimal cost functions of the aggregate problem. The state space partition can be arbitrary, but is often determined by using features (states with "similar" features are grouped together). Moreover, aggregation can be combined in complementary fashion with the methods (a)-(c) above, and can use as a starting point an approximate cost-to-go function produced by any one of these methods; e.g., apply a parametric approximation method and enhance the resulting cost function through local corrections obtained by aggregation.

Additional variations of the above methods are obtained when used in combination with approximate minimization over u_k in Eq. (2.1), and also when the expected value over w_k is computed approximately via a certainty equivalence approximation (cf. Section 2.3.2) or adaptive simulation and Monte Carlo tree search (cf. Section 2.4.2).

2.1.2 Off-Line and On-Line Methods

In value space approximation, an important consideration is whether the functions $\tilde{J}_{k+1}$ and the corresponding suboptimal policy $\{\tilde{\mu}_0, \ldots, \tilde{\mu}_{N-1}\}$ of Eq. (2.1) are computed *off-line* (i.e., before the control process begins, and for all x_k and k), or *on-line* (i.e., after the control process begins, when needed, and for just the states x_k to be encountered).

Usually, for challenging problems, the controls $\tilde{\mu}_k(x_k)$ are computed on-line, since their storage may be difficult when the state space is large. However, the on-line or off-line computation of $\tilde{J}_{k+1}$ is an important design choice. We thus distinguish between:

(i) *Off-line methods*, where the entire function $\tilde{J}_{k+1}$ in Eq. (2.1) is computed for every k, before the control process begins. The values $\tilde{J}_{k+1}(x_{k+1})$ are either stored in memory or can be obtained with a simple and fast computation, as needed in order to compute controls by one-step lookahead. The advantage of this is that most of the computation is done off-line, before the first control is applied at time 0. Once the control process starts, no extra computation is needed to obtain $\tilde{J}_{k+1}(x_{k+1})$ for implementing the corresponding suboptimal policy.

(ii) *On-line methods*, where most of the computation is performed just after the current state x_k becomes known, the values $\tilde{J}_{k+1}(x_{k+1})$ are computed only at the relevant next states x_{k+1}, and are used to compute the control to be applied via Eq. (2.1). In contrast with the off-line approximation methods, these methods are well-suited for *on-line replanning*, whereby the problem data may change over time. A similar situation where on-line methods may have an advantage is when the initial state and other problem data may become known just before the control process begins.

Examples of typically off-line schemes are neural network and other parametric approximations, as well as aggregation. Examples of typically on-line schemes are rollout and model predictive control. Schemes based on problem approximation may be either on-line or off-line depending on other problem-related factors. Of course there are also hybrid methods, where significant computation is done off-line to expedite the on-line computation of needed values of $\tilde{J}_{k+1}$. An example is the truncated rollout schemes to be discussed in Sections 2.4 and 5.1.

2.1.3 Model-Based Simplification of the Lookahead Minimization

We will now consider ways to facilitate the calculation of the suboptimal control $\tilde{\mu}_k(x_k)$ at state x_k via the minimization of the one-step lookahead expression

$$E\Big\{g_k(x_k,u_k,w_k)+\tilde{J}_{k+1}\big(f_k(x_k,u_k,w_k)\big)\Big\}, \tag{2.2}$$

once the cost-to-go approximating functions $\tilde{J}_{k+1}$ have been selected. In this section, we will assume that we have a mathematical model, i.e., that the functions g_k and f_k are available in essentially closed form, and that the conditional probability distribution of w_k, given (x_k, u_k), is also available. Moreover, Monte Carlo simulation is not used to compute the expected value in Eq. (2.2). We will address the model-free case in the next section.

Important issues here are the computation of the expected value (if the problem is stochastic) and the minimization over $u_k \in U_k(x_k)$ in Eq. (2.2). Both of these operations may involve substantial work, which is of particular concern when the minimization is to be performed on-line.

One possibility to eliminate the expected value from the expression (2.2) is (assumed) *certainty equivalence*. Here we choose a typical value $\tilde{w}_k$ of w_k, and use the control $\tilde{\mu}_k(x_k)$ that solves the deterministic problem

$$\min_{u_k\in U_k(x_k)}\Big[g_k(x_k,u_k,\tilde{w}_k)+\tilde{J}_{k+1}\big(f_k(x_k,u_k,\tilde{w}_k)\big)\Big]. \tag{2.3}$$

The approach of turning a stochastic problem into a deterministic one by replacing uncertain quantities with single typical values highlights the possibility that $\tilde{J}_{k+1}$ may itself be obtained by using deterministic methods. We will discuss this approach and its variations in greater detail later in this chapter (see Section 2.3).

Let us now consider the issue of algorithmic minimization over $U_k(x_k)$ in Eqs. (2.2) and (2.3). If $U_k(x_k)$ is a finite set, the minimization can be done by brute force, through exhaustive computation and comparison of the relevant cost expressions. This of course can be very time consuming, particularly for multistep lookahead, but parallel computation can be used with great effect for this purpose [as well as for the calculation of the expected value in the expression (2.2)]. For some discrete control problems, *integer programming* techniques may also be used. Moreover, for deterministic problems with multistep lookahead, sophisticated exact or approximate *shortest path methods* may be considered; several methods of this type are available, such as label correcting methods, A^* methods, and their variants (see the author's textbooks [Ber98] and [Ber17] for detailed accounts, which are consistent with the context of this chapter).

When the control constraint set is infinite, it may be replaced by a finite set through discretization. However, a more efficient alternative may be to use continuous space *nonlinear programming* techniques. This possibility can be attractive for deterministic problems, which lend themselves better to continuous space optimization; an example is the model predictive control context (see Section 2.5).

For stochastic problems with continuous control spaces and either one-step or multistep lookahead, the methodology of *stochastic programming* may be useful. This methodology bears a close connection with linear and nonlinear programming methods. We refer to the textbook [Ber17] for a discussion of its application to the approximate DP context, and references to the literature. Still another possibility to simplify the one-step lookahead minimization (2.2) is based on Q-factor approximation, which is also suitable for model-free policy implementation, as we discuss next.

2.1.4 Model-Free Off-Line Q-Factor Approximation

One of the major aims of this book is to discuss methods where a mathematical model [the system functions f_k, the probability distribution of w_k, and the one-stage cost functions g_k] is not used because it is either hard to construct, or simply inconvenient. We assume instead that the system and cost structure can be simulated in software far more easily (think, for example, of control of a queueing network with complicated but well-defined service disciplines at the queues).†

In this section, we will review some of the high-level ideas of passing from model-based to model-free policy implementations for stochastic problems. In particular, we assume that:

(a) There is a computer program/simulator that for any given state x_k and control $u_k \in U_k(x_k)$, simulates sample probabilistic transitions to a successor state x_{k+1}, and generates the corresponding transition costs.

---

† Another possibility is to use the real system to provide the next state and transition cost, but we will not deal explicitly with this case in this book.

(b) A cost function approximation $\tilde{J}_{k+1}$ is available. Approaches to obtain $\tilde{J}_{k+1}$ in model-free fashion will be discussed in the context of specific methods later. For example $\tilde{J}_{k+1}$ may be obtained by solving a simpler problem for which a model is available, or it may be separately obtained without a mathematical model, by using a simulator.

We want to use the functions $\tilde{J}_{k+1}$ and the simulator to compute or approximate the Q-factors

$$E\Big\{g_k(x_k, u_k, w_k) + \tilde{J}_{k+1}\big(f_k(x_k, u_k, w_k)\big)\Big\},$$

for all $u_k \in U_k(x_k)$, and then find the minimal Q-factor and corresponding one-step lookahead control.

Given a state x_k, we may use the simulator to compute these Q-factors for all the pairs (x_k, u_k), $u_k \in U_k(x_k)$, and then select the minimizing control. However, in many cases this can be very time-consuming. To deal with this difficulty, we may introduce a parametric family (or approximation architecture, as we will call it in Chapter 3) of Q-factor functions,

$$\tilde{Q}_k(x_k, u_k, r_k),$$

where r_k is the parameter vector and use a least squares fit/regression to approximate the expected value that is minimized in Eq. (2.2). One possibility is to use a neural network parametric architecture; see Chapter 3, where we discuss methods for selecting and training parametric architectures. The steps are as follows:

Summary of Q-Factor Approximation Based on Approximation in Value Space

Assume that the value of $\tilde{J}_{k+1}(x_{k+1})$ is available for any given x_{k+1}:

(a) Use the simulator to collect a large number of "representative" quadruplets $(x_k^s, u_k^s, x_{k+1}^s, g_k^s)$, and corresponding Q-factors

$$\beta_k^s = g_k^s + \tilde{J}_{k+1}(x_{k+1}^s), \qquad s = 1, \ldots, q. \tag{2.4}$$

Here x_{k+1}^s is the simulator's output of the next state

$$x_{k+1}^s = f_k(x_k^s, u_k^s, w_k^s)$$

that corresponds to some disturbance w_k^s. This disturbance also determines the one-stage-cost sample

$$g_k^s = g_k(x_k^s, u_k^s, w_k^s).$$

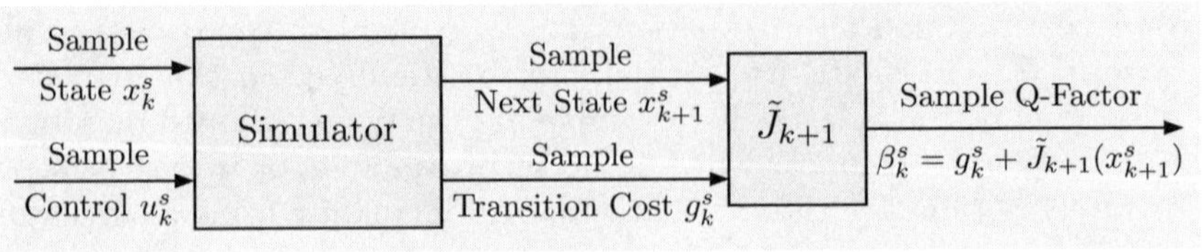

Figure 2.1.2 Schematic illustration of the simulator used for a model-free Q-factor approximation, assuming approximate cost functions $\tilde{J}_{k+1}$ are known. The input to the simulator are sample state-control pairs (x_k^s, u_k^s), and the outputs are a next state sample x_{k+1}^s and cost sample g_k^s. These correspond to a disturbance w_k^s according to

$$x_{k+1}^s = f_k(x_k^s, u_k^s, w_k^s), \qquad g_k^s = g_k(x_k^s, u_k^s, w_k^s).$$

The actual value of w_k^s need not be output by the simulator. The sample Q-factors β_k^s are generated according to Eq. (2.4), and are used in the least squares regression (2.5) to yield a parametric Q-factor approximation $\tilde{Q}_k$ and the policy implementation (2.6).

The simulator need not output w_k^s; only the sample next state x_{k+1}^s and sample cost g_k^s are needed (see Fig. 2.1.2). Moreover, the simulator may output β_k^s directly, assuming it has the means to compute it.

(b) Compute the parameter $\bar{r}_k$ by the least-squares regression

$$\bar{r}_k \in \arg\min_{r_k} \sum_{s=1}^{q} \big(\tilde{Q}_k(x_k^s, u_k^s, r_k) - \beta_k^s\big)^2. \tag{2.5}$$

(c) Use the policy

$$\tilde{\mu}_k(x_k) \in \arg \min_{u_k \in U_k(x_k)} \tilde{Q}_k(x_k, u_k, \bar{r}_k). \tag{2.6}$$

Note some important points about the preceding procedure:

(1) It is model-free in the sense that it is based on Monte Carlo simulation. Moreover, it does not need the functions f_k and g_k, and the probability distribution of w_k to generate the policy $\tilde{\mu}_k$ through the least squares regression (2.5) and the Q-factor minimization (2.6). Using the simulator to collect the samples (2.4) and the cost function approximation $\tilde{J}_{k+1}$ suffices.

(2) Two approximations are potentially required: One to compute $\tilde{J}_{k+1}$,

which is needed for the samples β_k^s [cf. Eq. (2.4)], and another to compute $\tilde{Q}_k$ through the regression (2.5). The approximation methods to obtain $\tilde{J}_{k+1}$ and $\tilde{Q}_k$ may be unrelated.

(3) The policy $\tilde{\mu}_k$ obtained through the minimization (2.6) is not the same as the one obtained through the minimization (2.2). There are two reasons for this. One is the approximation error introduced by the Q-factor architecture $\tilde{Q}_k$, and the other is the simulation error introduced by the finite-sample regression (2.5). We have to accept these sources of error as the price to pay for the convenience of not requiring a mathematical model for policy implementation.

Let us also mention a variant of the least squares minimization in Eq. (2.5), which is to use a *regularized minimization* where a quadratic regularization term is added to the least squares objective. This term is a multiple of the squared deviation $\|r - \hat{r}\|^2$ of r from some initial guess $\hat{r}$. Moreover, in some cases, a nonquadratic minimization may be used in place of Eq. (2.5) to determine $\bar{r}_k$, but in this book we will focus on least squares exclusively. We refer to Chapter 3 for further discussion.

2.1.5 Approximation in Policy Space on Top of Approximation in Value Space

A common approach for approximation in policy space, is to introduce a parametric family of policies $\tilde{\mu}_k(x_k, r_k)$, where r_k is a parameter vector. The parametrization may involve a neural network as we will discuss in Chapter 3. Alternatively, the parametrization may involve problem-specific features, exploiting the special structure of the problem at hand.

A general scheme for parametric approximation in policy space is to obtain a large number of sample state-control pairs (x_k^s, u_k^s), $s = 1, \ldots, q$, such that for each s, u_k^s is a "good" control at state x_k^s. We can then choose the parameter r_k by solving the least squares/regression problem

$$\min_{r_k} \sum_{s=1}^{q} \left\| u_k^s - \tilde{\mu}_k(x_k^s, r_k) \right\|^2 \tag{2.7}$$

(possibly with added regularization).† In particular, we may determine u_k^s using a human or a software "expert" that can choose "near-optimal"

† It is implicitly assumed here (and in similar situations later) that the controls are members of a Euclidean space so that the distance between two controls can be measured by their normed difference. Otherwise the norm should be replaced by some distance metric within the control space. Regression problems of this type arise in the training of *parametric classifiers* based on data, including the use of neural networks (see Section 3.5). Assuming a finite control space, the classifier is trained using the data $\big(x_k^s, u_k^s\big)$, $s = 1, \ldots, q$, which are viewed as state-category pairs, and then a state x_k is classified as being of "category"

controls at given states, so $\tilde{\mu}_k$ is trained to match the behavior of the expert. Methods of this type are commonly referred to as *supervised learning* in artificial intelligence (see also the discussion in Section 5.7.2).

A special case of the above procedure, which connects with approximation in value space, is to generate the sample state-control pairs (x_k^s, u_k^s) through a one-step lookahead minimization of the form

$$u_k^s \in \arg \min_{u \in U_k(x_k)} E\Big\{g_k(x_k^s, u, w_k) + \tilde{J}_{k+1}\big(f_k(x_k^s, u, w_k)\big)\Big\}, \tag{2.8}$$

where $\tilde{J}_{k+1}$ is a suitable (separately obtained) approximation in value space; cf. Eq. (2.2), or an approximate Q-factor based minimization

$$u_k^s \in \arg \min_{u_k \in U_k(x_k^s)} \tilde{Q}_k(x_k^s, u_k, \bar{r}_k), \tag{2.9}$$

[cf. Eq. (2.6)]. In this case, we collect the sample state-control pairs (x_k^s, u_k^s), $s = 1, \ldots, q$, by using approximation in value space through Eq. (2.8) or Q-factor approximation through Eq. (2.9), and then apply approximation in policy space through Eq. (2.7) (i.e., approximation in policy space is built on top of approximation in value space).

A major advantage of schemes based on the minimization (2.7) is that once the parametrized policy is obtained, the on-line implementation of the policy is fast and does not involve extensive calculations such as minimizations of the form (2.8) or (2.9). This advantage is generally shared by schemes that are based on approximation in policy space.

2.1.6 When is Approximation in Value Space Effective?

An important question is what constitutes good approximating functions $\tilde{J}_k$ in a one-step or multistep lookahead scheme. An answer that suggests itself is that $\tilde{J}_k$ should be "close" to the optimal cost-to-go function J_k^* for all k. This guarantees a certain degree of quality of the approximation scheme, but is neither necessary nor is it satisfied by all or even most good practical schemes.

For example if the approximating values $\tilde{J}_k(x_k)$ differ from the optimal values $J_k^*(x_k)$ uniformly by the same constant, the policy obtained by the approximation in value space scheme (2.1) is optimal. This suggests that a better condition might be that relative values of $\tilde{J}_k$ and J_k^* should be "close" to each other, i.e.,

$$\tilde{J}_k(x_k) - \tilde{J}_k(x_k') \approx J_k^*(x_k) - J_k^*(x_k'),$$

$\tilde{\mu}_k(x_k, r_k)$. Parametric approximation architectures, and their training through the use of data and regression techniques are described in Chapter 3.

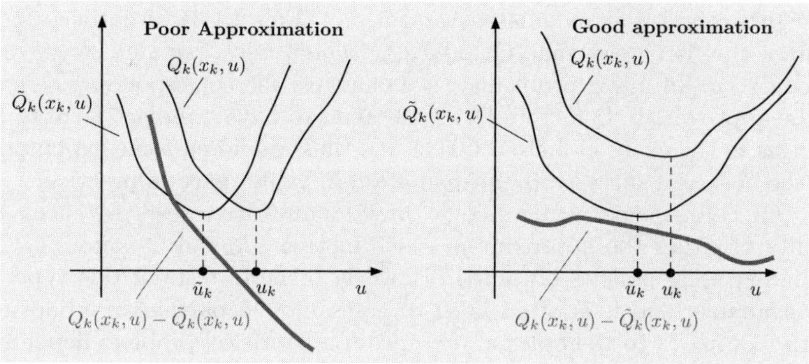

Figure 2.1.3 Schematic illustration of the "slope" of the Q-factor approximation error as a predictor of quality of an approximation in value space scheme. At a given state x_k, let u_k be optimal, so that it minimizes $Q_k(x_k,u)$ over $u \in U_k(x_k)$, and let $\tilde{u}_k$ be generated by approximation in value space, so that it minimizes $\tilde{Q}_k(x_k,u)$ over $u \in U_k(x_k)$. In the figure on the right the approximation error $Q_k(x_k,u) - \tilde{Q}_k(x_k,u)$ changes gradually (i.e., has small "slope"), and $\tilde{u}_k$ is a good choice, because $Q_k(x_k,\tilde{u}_k)$ is close the optimal $Q_k(x_k,u_k)$. In the figure on the left the approximation error $Q_k(x_k,u) - \tilde{Q}_k(x_k,u)$ changes fast (i.e., has large "slope"), and $\tilde{u}_k$ is a poor choice. In the extreme case where the Q-factors $Q_k(x_k,u)$ and $\tilde{Q}_k(x_k,u)$ differ by a constant, minimization of either one of them yields the same result.

for all pairs of states x_k and x_k'. Still, however, this guideline neglects the role of the first stage cost (or the cost of the first ℓ stages in the case of ℓ-step lookahead).

A more accurate predictor of good quality of the suboptimal policy obtained is that *the Q-factor approximation error* $Q_k(x_k,u) - \tilde{Q}_k(x_k,u)$ *changes gradually as* u *changes*, where Q_k and $\tilde{Q}_k$ denote the exactly optimal Q-factor and its approximation, respectively. For a heuristic explanation, suppose that approximation in value space generates a control $\tilde{u}_k$ at a state x_k where another control u_k is optimal. Then we have

$$\tilde{Q}_k(x_k,u_k) - \tilde{Q}_k(x_k,\tilde{u}_k) \geq 0, \tag{2.10}$$

since $\tilde{u}_k$ minimizes $\tilde{Q}_k(x_k,\cdot)$, and

$$Q_k(x_k,\tilde{u}_k) - Q_k(x_k,u_k) \geq 0, \tag{2.11}$$

since u_k minimizes $Q_k(x_k,\cdot)$. If $\tilde{u}_k$ is far from optimal, the Q-factor difference in Eq. (2.11) will be large, and by adding Eq. (2.10), it follows that the expression

$$\big(Q_k(x_k,\tilde{u}_k) - \tilde{Q}_k(x_k,\tilde{u}_k)\big) - \big(Q_k(x_k,u_k) - \tilde{Q}_k(x_k,u_k)\big)$$

will be even larger. This is not likely to happen if the approximation error $Q_k(x_k,u) - \tilde{Q}_k(x_k,u)$ changes gradually (i.e., has small "slope") for u in a neighborhood that includes u_k and $\tilde{u}_k$ (cf. Fig. 2.1.3). Another way to express this is to say that *Q_k and $\tilde{Q}_k$ should have "similar slope" with respect to u.* In many problems, as u changes, the corresponding changes in the approximate Q-factors $\tilde{Q}_k(x_k, u)$ tend to have "similar" form to the changes in the exact Q-factors $Q_k(x_k, u)$, thus providing some explanation for the observed success of approximation in value space in practice.

Of course, one would like to have quantitative tests to check the quality of either the approximate cost functions $\tilde{J}_k$ and Q-factors $\tilde{Q}_k$, or the suboptimal policies obtained. However, general tests of this type are not available, and it is often hard to assess how a particular suboptimal policy compares to the optimal, except on a heuristic, problem-dependent basis. Unfortunately, this is a recurring difficulty in approximate DP/RL.

2.2 MULTISTEP LOOKAHEAD

In the preceding section, we discussed approximation in value space with one-step lookahead. In this section, we focus on the more ambitious, but also computationally more intensive multistep lookahead scheme.

To illustrate, in *two-step lookahead*, at time k and state x_k, we use the control obtained from the minimization

$$\tilde{\mu}_k(x_k) \in \arg \min_{u_k \in U_k(x_k)} E\Big\{g_k(x_k, u_k, w_k) + \tilde{J}_{k+1}\big(f_k(x_k, u_k, w_k)\big)\Big\},$$

where for all possible states x_{k+1} that can be generated via the system equation starting from x_k, we have

$$\begin{aligned} \tilde{J}_{k+1}(x_{k+1}) = \min_{u_{k+1} \in U_{k+1}(x_{k+1})} & E\Big\{g_{k+1}(x_{k+1}, u_{k+1}, w_{k+1}) \\ & + \tilde{J}_{k+2}\big(f_{k+1}(x_{k+1}, u_{k+1}, w_{k+1})\big)\Big\}, \end{aligned}$$

where $\tilde{J}_{k+2}$ is some approximation of the optimal cost-to-go function J^*_{k+2}.

Thus two-step lookahead amounts to *solving a two-stage version of the DP problem with x_k as the initial state and $\tilde{J}_{k+2}$ as the terminal cost function.* Given x_k, the solution of this DP problem yields a two-stage policy that consists of the single control u_k for the first lookahead stage, plus a control $\mu_{k+1}(x_{k+1})$ for each value of $x_{k+1} = f_k(x_k, u_k, w_k)$ that can occur at the second lookahead stage $k + 1$. However, once this two-stage policy is computed, the controls $\mu_{k+1}(x_{k+1})$ are discarded, and only u_k is used as the control applied by the two-step lookahead policy at x_k. At the next stage, this process is repeated, i.e., we solve a two-stage DP problem with x_{k+1} as the initial state and $\tilde{J}_{k+3}$ as the terminal cost function.

Policies with lookahead of $\ell > 2$ stages are similarly defined: at state x_k, we solve an ℓ-stage version of the DP problem with x_k as the initial state and $\tilde{J}_{k+\ell}$ as the terminal cost function, and use the first control of the

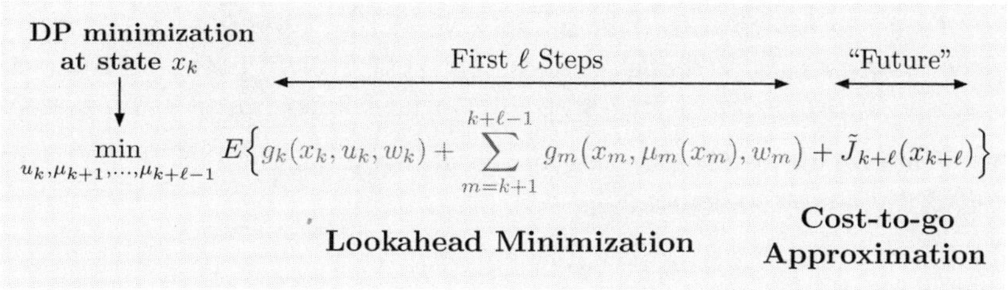

Figure 2.2.1 Illustration of approximation in value space with ℓ-step lookahead. At state x_k, we solve an ℓ-stage version of the problem, with $\tilde J_{k+\ell}$ as terminal cost function approximation. We use at x_k the first control of the ℓ-stage policy thus obtained, while discarding the others.

ℓ-stage policy thus obtained, while discarding the others; see Fig. 2.2.1. Of course, in the final stages where $k > N - \ell$, we should shorten the size of lookahead to $N-k$. Note that the simplifications in the one-step lookahead minimization discussed in the preceding section (assumed certainty equivalence, adaptive sampling, model-free policy implementation, etc) extend to multistep lookahead.

2.2.1 Multistep Lookahead and Rolling Horizon

There are several ways to compute the lookahead functions $\tilde J_{k+\ell}$ in ℓ-step lookahead, similar to the one-step lookahead case. However, there may also be additional possibilities: with sufficiently long lookahead, we may capture enough of the character of the DP problem at hand so that a sophisticated choice of $\tilde J_{k+\ell}$ may not be needed.

In particular, we may set $\tilde J_{k+\ell}(x_{k+\ell}) \equiv 0$, or set

$$\tilde J_{k+\ell}(x_{k+\ell}) = g_N(x_{k+\ell}).$$

The idea is to use a sufficiently large number of lookahead steps ℓ to ensure a reasonably faithful approximation of the optimal Q-factors Q_k or cost-to-go functions $J^*_{k+\ell}$ within a constant.† This is also referred to as the *rolling horizon approach* (or sometimes *receding horizon approach*), but essentially it is the same as multistep lookahead with a simplified cost-to-go approximation. Note that the idea of a rolling horizon is well-suited and applies with few modifications to infinite horizon problems as well.‡

† See the discussion in Section 2.1.6. Generally, rolling horizon schemes tend to work well if the probability distribution of the state $k+\ell$ steps ahead is roughly independent of the current state and control, or is concentrated around "low cost" states.

‡ For infinite horizon problems the cost-to-go approximations $\tilde J_k$ will typically be the same at all stages k, i.e., $\tilde J_k \equiv \tilde J$ for some $\tilde J$. As a result, the

Typically, *as the size ℓ of the lookahead is chosen larger, the need for a good choice of $\tilde{J}_{k+\ell}$ tends to diminish.* The reason is that the effective cost-to-go approximation in ℓ-step lookahead consists of two components:

(a) The cost of an $(\ell-1)$ step problem involving the last $(\ell-1)$ stages of the ℓ-step lookahead.

(b) The terminal cost approximation $\tilde{J}_{k+\ell}$.

Since the $(\ell-1)$-step problem is treated by exact optimization, the overall approximation will be accurate if the contribution of the terminal cost approximation is relatively insignificant. This is likely to be true with large enough ℓ.

In view of the advantages offered by multistep lookahead, one is tempted to conjecture that if ℓ is increased, then the performance of the lookahead policy is improved. This, however, need not be true always, essentially because beyond the next ℓ stages, the policy may be "blind" to the presence of particularly "favorable" or "unfavorable" states. The following example is an illustration.

Example 2.2.1

This is an oversimplified example that demonstrates a pitfall of the multistep lookahead and rolling horizon approaches with approximate cost-to-go functions

$$\tilde{J}_{k+\ell}(x_{k+\ell}) \equiv 0.$$

It may happen that with longer lookahead the quality of the suboptimal control obtained is degraded.

Consider the 4-stage deterministic shortest problem of Fig. 2.2.2. At the initial state there are two possible controls, denoted u and u'. At all other states there is only one control available, so a policy is specified by just the initial choice between u and u'. The costs of the four transitions on the upper and the lower path are shown next to the corresponding arcs $(0,1,2,1$ for the upper path, and $0,2,0,10$ on the lower path). From the initial state, 2-step lookahead with terminal cost approximation $\tilde{J}_2 = 0$, compares $0+1$ with $0+2$ and prefers the optimal control u, while 3-step lookahead with terminal cost approximation $\tilde{J}_3 = 0$, compares $0+1+2$ with $0+2+0$ and prefers the suboptimal control u'.

Thus in this example, using a longer lookahead yields worse performance. The problem here has to do with "edge effects", i.e., large cost changes at the "edge" of the lookahead (a cost of 0 just after the 2-step lookahead, followed by a cost of 10 just after the 3-step lookahead).

---

limited lookahead scheme produces a stationary policy. In the case of discounted problems with an infinite horizon (see Chapter 4), a simple approach is to use a rolling horizon that is long enough so that the tail cost is negligible and can be replaced by zero, but it is also possible to use a small number of lookahead stages ℓ, as long as we compensate with a terminal cost function approximation $\tilde{J}$.

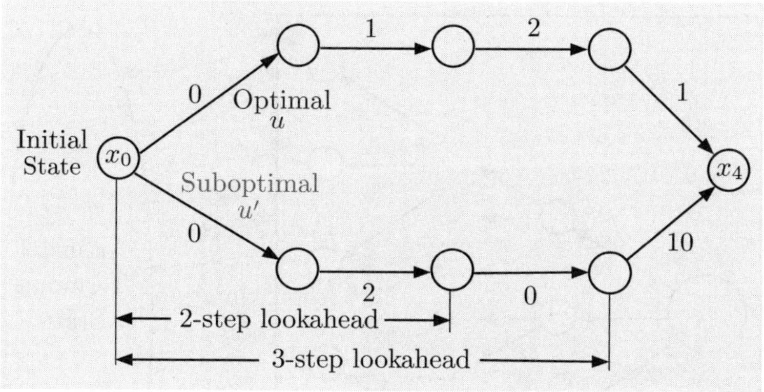

Figure 2.2.2 The 4-stage deterministic shortest problem of Example 2.2.1, illustrating how using a longer lookahead with a cost function approximation $\tilde{J}_{k+\ell}(x_{k+\ell}) \equiv 0$ may degrade the performance of the policy obtained.

2.2.2 Multistep Lookahead and Deterministic Problems

Generally, the implementation of multistep lookahead can be prohibitively time-consuming for stochastic problems, because it requires at each step the solution of a stochastic DP problem with a horizon that is equal to the size of the lookahead. However, in a deterministic setting, the lookahead problems are also deterministic, and can be solved by shortest path methods for a finite spaces problem, or even for an infinite spaces problem after some form of discretization; cf. Fig. 2.2.3. This makes deterministic problems particularly good candidates for the use of long-step lookahead.

Similarly, for a continuous-spaces deterministic optimal control problem, the lookahead minimization may be conveniently solvable by nonlinear programming methods. This idea finds wide application in the context of model predictive control (see the discussion in Section 2.5).

Partially Deterministic Form of Multistep Lookahead

When the problem is stochastic, one may consider an approximation to ℓ-step lookahead, which is based on deterministic computations. This is a hybrid, partially deterministic approach, whereby at state x_k, we *allow for a stochastic disturbance* w_k *at the current stage, but fix the future disturbances* $w_{k+1}, \ldots, w_{k+\ell-1}$, *up to the end of the lookahead horizon, to some typical values*. This allows us to bring to bear deterministic shortest path methods in the computation of approximate costs-to-go beyond the first stage.

In particular, with this approach, when at state x_k, the needed values $\tilde{J}_{k+1}(x_{k+1})$ are the ones corresponding to $x_{k+1} = f_k(x_k, u_k, w_k)$, for all $u_k \in U_k(x_k)$ and w_k. Each of these values will be computed by solving

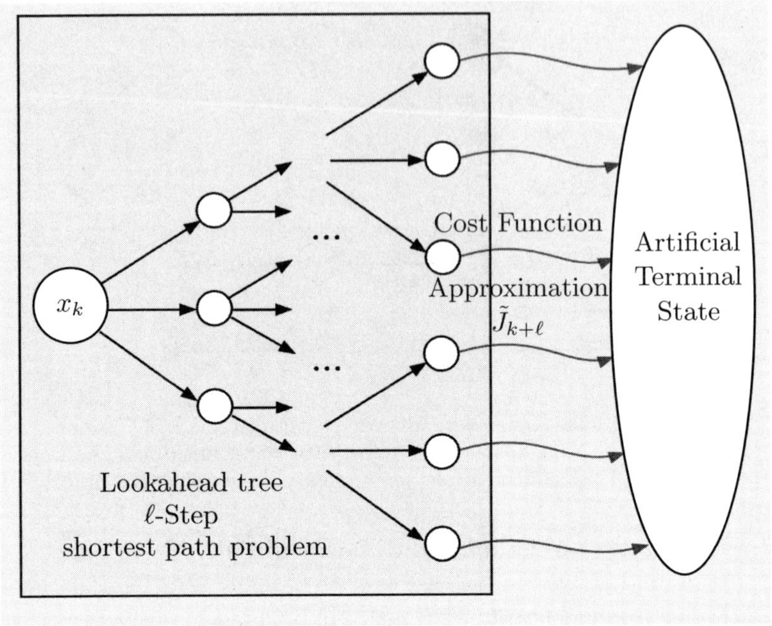

Figure 2.2.3 Multistep lookahead for a deterministic finite-state problem. The lookahead minimization is equivalent to a shortest path problem with an artificial terminal state after ℓ steps and terminal cost function $\tilde J_{k+\ell}$.

an $(\ell - 1)$-step deterministic shortest path problem that starts at x_{k+1} and involves the typical values of the disturbances $w_{k+1}, \ldots, w_{k+\ell-1}$. Once the values $\tilde J_{k+1}(x_{k+1})$ are obtained, they will be used to compute the approximate Q-factors of pairs (x_k, u_k) using the formula

$$\tilde Q_k(x_k, u_k) = E\Big\{g_k(x_k, u_k, w_k) + \tilde J_{k+1}\big(f_k(x_k, u_k, w_k)\big)\Big\},$$

which incorporates the first stage uncertainty. Finally, the control chosen by such a scheme at time k will be

$$\tilde\mu_k(x_k) \in \arg \min_{u_k \in U_k(x_k)} \tilde Q_k(x_k, u_k);$$

cf. Eq. (2.6).

The expected values involved in the multistep lookahead minimization can also be simplified in a variety of other ways, which may be suited to the special structure of the problem at hand. For an example application of this idea to POMDP, see the paper [YSH17].

The approach of fixing uncertain quantities to typical values for approximation purposes is generally referred to as (assumed) *certainty equivalence*, and will be discussed at length in Section 2.3.2. The idea of using multistep lookahead for deterministic problems will also be reexamined in the context of the rollout algorithm in Section 2.4.1.

2.3 PROBLEM APPROXIMATION

A key issue in implementing a limited lookahead policy is the choice of the cost-to-go approximation at the end of the lookahead. In this section, we discuss the problem approximation approach whereby we approximate the optimal cost function J_k^* with some function $\tilde{J}_k$ derived from a related but simpler problem (for example the optimal cost-to-go function of that problem). In the following subsections we consider two approaches:

(1) *Simplifying the structure of the problem through enforced decomposition*, e.g., replacing coupling constraints with simpler decoupled constraints or with Lagrange multiplier-related penalties.

(2) *Simplifying the probabilistic structure of the problem*, e.g., replacing stochastic disturbances with deterministic ones.

Another approach that can be viewed as problem approximation is *aggregation*, whereby the original problem is approximated with a related "aggregate" problem that has smaller dimension or fewer states. This problem is solved exactly to yield a cost-to-go approximation for the original problem. Aggregation is also related to the feature-based parametric approximation ideas of Chapter 3, and will be discussed in Chapter 6.

2.3.1 Enforced Decomposition

The simplification/approximation approach is often well-suited for problems involving a number of subsystems that are coupled through the system equation, or the cost function, or the control constraints, but the degree of coupling is "relatively weak." It is difficult to define precisely what constitutes "weak coupling," but in specific problem contexts, usually this type of structure is easily recognized. For such problems it is often sensible to introduce approximations by artificially decoupling the subsystems in some way, thereby creating either a simpler problem or a simpler cost calculation, where subsystems can be dealt with in isolation.

There are a number of different ways to effect enforced decomposition, and the best approach is often problem-dependent. Generally, for a deterministic problem, enforced decomposition can be applied both off-line and on-line to produce a suboptimal control sequence. For a stochastic problem, it can be applied with off-line computation of the approximate cost-to-go functions $\tilde{J}_k$ and on-line computation of the corresponding suboptimal policy. We will illustrate these two possibilities with various application contexts in what follows.

Optimization of One Subsystem at a Time

When a problem involves multiple subsystems, a potentially interesting approximation approach is to optimize one subsystem at a time. In this way the control computation at time k may become simpler.

As an example consider an N-stage deterministic problem, where the control u_k at state x_k consists of n components, $u_k = \{u_k^1, \ldots, u_k^n\}$, with u_k^i corresponding to the ith subsystem. Then to compute a cost-to-go approximation at a given state x_k, one may optimize over the control sequence of a single subsystem, while keeping the controls of the remaining subsystems at some nominal values. Thus, upon arriving at x_k, we first optimize over the control sequence $\{u_k^1, u_{k+1}^1, \ldots, u_{N-1}^1\}$ of the first subsystem, then optimize over the controls of the second subsystem, and so on, while keeping the controls of the other subsystem at the latest "optimal" values computed.

There are several possible variations; for example to make the ordering in which the subsystems are considered subject to optimization as well, or to repeat cycling through the subsystems multiple times, each time using the results of the latest computation as nominal values of subsystem controls. This is similar to "coordinate descent," used in other optimization contexts. A related one-at-a-time optimization approach arises in multiagent problems, and leads to very efficient computation; see [Ber19c].

Additional variations are obtained when we use approximate minimization over u_k in Eq. (2.1), and also when the expected value over w_k is computed approximately via adaptive simulation or a certainty equivalence approximation (cf. Section 2.1).

Example 2.3.1 (Vehicle Routing)

Consider n vehicles that move along the arcs of a given graph. Each node of the graph has a known "value" and the first vehicle that will pass through the node will collect its value, while vehicles that pass subsequently through the node do not collect any value. This may serve as a model of a situation where there are various valuable tasks to be performed at the nodes of a transportation network, and each task can be performed at most once and by a single vehicle. We assume that each vehicle starts at a given node and after at most a given number of arc moves, it must return to some other given node. The problem is to find a route for each vehicle satisfying these constraints, so that the total value collected by the vehicles is maximized.

This is a difficult combinatorial problem that in principle can be approached by DP. In particular, we can view as state the n-tuple of current positions of the vehicles together with the list of nodes that have been visited by some vehicle in the past, and have thus "lost" their value. Unfortunately, the number of these states is enormous (it increases exponentially with the number of nodes and the number of vehicles). The version of the problem that involves a single vehicle, while still difficult in principle, can often be solved in reasonable time either exactly by DP or fairly accurately using a suitable heuristic. Thus a one-step lookahead policy suggests itself, with the value-to-go approximation obtained by solving single vehicle problems.

In particular, in a one-step lookahead scheme, at a given time k and from a given state x_k we consider all possible n-tuples of moves by the n

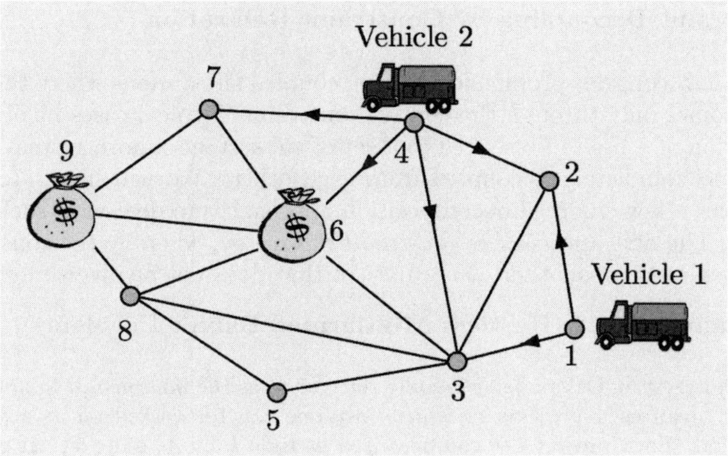

Figure 2.3.1 Schematic illustration of the vehicle routing problem and the one-vehicle-at-a-time approach. As an example, given the position pair $x_k = (1, 4)$ of the two vehicles and the current valuable tasks at positions 6 and 9, we consider moves to all possible position pairs x_{k+1}:

$$(2, 2), \ (2, 3), \ (2, 6), \ (2, 7), \ (3, 2), \ (3, 3), \ (3, 6), \ (3, 7).$$

From each of these pairs, we first compute the best route of vehicle 1 assuming vehicle 2 does not move, and then the best route vehicle 2, taking into account the previously computed route of vehicle 1. We then select the pair x_{k+1} that results in optimal value, and move the vehicles to the corresponding positions.

vehicles. At the resulting state x_{k+1} corresponding to each n-tuple of vehicle moves, we approximate the optimal value-to-go with the value corresponding to a suboptimal set of paths. These paths are obtained as follows: we fix an order of the vehicles and we calculate a path for the first vehicle, starting at x_{k+1}, assuming the other vehicles do not move. (This is done either optimally by DP, or near optimally using some heuristic.) Then we calculate a path for the second vehicle, taking into account the value collected by the first vehicle, and we similarly continue: for each vehicle, we calculate in the given order a path, taking into account the value collected by the preceding vehicles. We end up with a set of paths that have a certain total value associated with them. This is the value $\tilde{J}_{k+1}(x_{k+1})$ associated with the successor state x_{k+1}. We repeat with all successor states x_{k+1} corresponding to all the n-tuples of vehicle moves that are possible at x_k. We then use as suboptimal control at x_k the n-tuple of moves that yields the best value; see Fig. 2.3.1.

There are several enhancements and variations of the scheme just described. For example, we can consider multiple alternative orders for optimizing paths one-at-a-time, and choose the n-tuple of moves that corresponds to the best value obtained. Other variations may include travel costs between nodes of the graph, and constraints on how many tasks can be performed by each vehicle.

Constraint Decoupling by Constraint Relaxation

Let us now consider problems involving coupled subsystems where the coupling comes only through the control constraint. Typical cases involve the allocation of a limited resource to a set of subsystems whose system equations are completely decoupled from each other. We will illustrate with examples a few enforced decomposition approaches to deal with such situations. The first approach is *constraint relaxation*, whereby the constraint set is replaced by another constraint set that does not involve coupling.

Example 2.3.2 (Restless Multiarmed Bandit Problems)

An interesting DP model, generally referred to as the *multiarmed bandit problem*, involves n projects of which only one can be worked on at any time period. Each project i is characterized at time k by its state x_k^i. If project i is worked on at time k, one receives an expected reward $R^i(x_k^i)$, and the state x_k^i then evolves according to the equation

$$x_{k+1}^i = f^i(x_k^i, w_k^i), \qquad \text{if } i \text{ is worked on at time } k,$$

where w_k^i is a random disturbance with probability distribution depending on x_k^i but not on prior disturbances. If project i is not worked on, its state changes according to

$$x_{k+1}^i = \overline{f}^i(x_k^i, \overline{w}_k^i),$$

where $\overline{f}^i$ is a given function and $\overline{w}_k^i$ is a random disturbance with distribution depending on x_k^i but not on prior disturbances. Furthermore, a reward $\overline{R}^i(x_k^i)$ is earned, where $\overline{R}^i$ is a given function. The projects are coupled through the control constraint (only one of the projects may be worked on at any one period).† A suboptimal enforced decomposition approach is to consider the n

---

† In the classical and simplest version of the problem, the state of a project that is not worked on remains unchanged and produces no reward, i.e.,

$$x_{k+1}^i = x_k^i, \qquad \overline{R}^i(x_k^i) = 0, \qquad \text{if } i \text{ is not worked on at time } k.$$

This problem admits optimal policies with a nice structure that can be computationally exploited. The problem has a long history and is discussed in many sources; we refer to [Ber12] and the references quoted there. In particular, in favorable instances of the problem, optimal policies have the character of an *index rule*, which is structurally similar to the decoupled suboptimal decision rules discussed in this section, and has been analyzed extensively, together with its variations and special cases. The term "restless" in the title of the present example, introduced by Whittle [Whi88], refers to the fact that the states of the projects that are not worked on may change.

Another classical type of multiarmed bandit problem is related to adaptive control (Section 1.3.8) and involves a system with unknown parameters. Here, when selecting a project to work on, the controller must attend to a dual single project problems where a single project is worked on through the entire remaining horizon, and add the contributions of the n problems to form an optimal reward approximation.

In particular, suppose that the optimal reward function $J_k^*(x^1,\ldots,x^n)$ is approximated by a separable function of the form $\sum_{i=1}^n \tilde{J}_k^i(x^i)$, where each $\tilde{J}_k^i$ is a function that quantifies the contribution of the ith project to the total reward. The corresponding one-step lookahead policy selects at time k the project i that maximizes

$$R^i(x^i) + \sum_{j\neq i} \overline{R}^j(x^j) + E\left\{\tilde{J}_{k+1}^i\big(f^i(x^i,w^i)\big)\right\} + \sum_{j\neq i} E\left\{\tilde{J}_{k+1}^j\big(\overline{f}^j(x^j,\overline{w}^j)\big)\right\},$$

which can also be written as

$$R^i(x^i) - \overline{R}^i(x^i) + E\left\{\tilde{J}_{k+1}^i\big(f^i(x^i,w^i)\big) - \tilde{J}_{k+1}^i\big(\overline{f}^i(x^i,\overline{w}^i)\big)\right\}$$
$$+ \sum_{j=1}^n \left\{\overline{R}^j(x^j) + E\big\{\tilde{J}_{k+1}^j\big(\overline{f}^j(x^j,\overline{w}^j)\big)\big\}\right\}.$$

Noting that the last term in the above expression does not depend on i, it follows that the one-step lookahead policy takes the form

$$\text{work on project } i \qquad \text{if} \qquad \tilde{m}_k^i(x^i) = \max_j\big\{\tilde{m}_k^j(x^j)\big\},$$

where for all i,

$$\tilde{m}_k^i(x^i) = R^i(x^i) - \overline{R}^i(x^i) + E\big\{\tilde{J}_{k+1}^i\big(f^i(x^i,w^i)\big) - \tilde{J}_{k+1}^i\big(\overline{f}^i(x^i,\overline{w}^i)\big)\big\}.$$

An important question for the implementation of the preceding suboptimal control scheme is the determination of the separable reward function terms $\tilde{J}_{k+1}^i$. There are several possibilities here, and the best choice may strongly depend on the problem's structure. One possibility is to compute $\tilde{J}_{k+1}^i$ as the optimal cost-to-go function for a problem involving just project i, i.e., assuming that none of the other projects $j \neq i$ will be worked on for the remaining periods $k+1,\ldots,N-1$. This corresponds to restricting the control constraint set of the problem, and involves a single-project optimization that may be tractable.

An alternative possibility is to use a separable parametric approximation of the form

$$\sum_{i=1}^n \tilde{J}_{k+1}^i(x_{k+1}^i, r_{k+1}^i),$$

where r_{k+1}^i are vectors of "tunable" parameters. The values of r_{k+1}^i can be obtained by some training algorithm such as the ones to be discussed in Chapter 3.

---

objective: a high short term reward (exploitation) and a better parameter identification (exploration). While this type of bandit problem has received a lot of attention, and has motivated some interesting RL ideas (such as the UCB rule in Monte Carlo tree search, which will be discussed in Section 2.4.2), it will not receive much attention in this book.

Constraint Decoupling by Lagrangian Relaxation

Another approach to deal with coupled constraints is to replace them with linear Lagrange multiplier-related penalty functions that are added to the cost function. We illustrate this approach with an extension of the preceding multiarmed bandit Example 2.3.2.

Example 2.3.3 (Separable Lower Bound Approximation in Multiarmed Bandit Problems)

Let us consider a version of the multiple projects Example 2.3.2 involving a more general form of control. Here there are n subsystems, with a control u_k^i applied to subsystem i at time k. Instead of the requirement that only one subsystem is worked on at any one time, we assume a control constraint of the form

$$u_k = (u_k^1, \dots, u_k^n) \in U, \qquad k = 0, 1, \dots,$$

where the set U is given (Example 2.3.2 is obtained as the special case where U consists of the union of the coordinate vectors, i.e., those whose components are equal to 0, except for one component that is equal to 1). The ith subsystem is described by

$$x_{k+1}^i = f^i(x_k^i, u_k^i, w_k^i), \qquad i = 1, \dots, n,\ k = 0, 1, \dots,$$

where x_k^i is the state taking values in some space, u_k^i is the control, w_k^i is a random disturbance, and f^i is a given function. We assume that w_k^i is selected according to a probability distribution that may depend on x_k^i and u_k^i, but not on prior disturbances or the disturbances w_k^j of the other subsystems $j \neq i$. The cost incurred at the kth stage by the ith subsystem is

$$g^i(x_k^i, u_k^i, w_k^i), \tag{2.12}$$

where g^i is a given one-stage cost function. For notational convenience, we assume stationarity of the system equation, the constraints, and the cost per stage, but the approach to be discussed applies to the nonstationary case as well.

One possibility for a separable approximation of the problem is to replace the constraint $u_k \in U$ by a smaller or larger decoupled constraint, i.e., requiring that

$$u_k^i \in U^i, \qquad i = 1, \dots, n,\ k = 0, 1, \dots,$$

where the subsets $U^1, \dots, U^n$ satisfy $U^1 \times \cdots \times U^n \subset U$ or $U \subset U^1 \times \cdots \times U^n$, respectively.

We discuss another possibility for the case where the constraint set U includes linear inequality constraints. As a simple example, let us focus on a constraint set U of the form

$$U = \left\{ (u^1, \dots, u^n) \;\middle|\; u^i \in U^i \subset \Re,\ i = 1, \dots, n,\ \sum_{i=1}^n c^i u^i \le b \right\}, \tag{2.13}$$

where $c^1, \ldots, c^n$, and b are some scalars.† Here we replace the coupling constraint

$$\sum_{i=1}^{n} c^i u_k^i \leq b, \qquad k = 0, 1, \ldots, \tag{2.14}$$

by a "relaxed" (larger) constraint whereby we require that

$$\sum_{k=0}^{N-1} \sum_{i=1}^{n} c^i u_k^i \leq Nb. \tag{2.15}$$

Roughly speaking, the constraint (2.15) requires that the coupling constraint (2.14) is satisfied "on the average," over the N stages.

We may now obtain a lower bound approximation of the optimal cost of our problem by assigning a scalar Lagrange multiplier $\lambda \geq 0$ to the constraint (2.15), and add a Lagrangian term

$$\lambda \left(\sum_{k=0}^{N-1} \sum_{i=1}^{n} c^i u_k^i - Nb \right) \tag{2.16}$$

to the cost function. This amounts to replacing the kth stage cost (2.12) by

$$g^i(x_k^i, u_k^i, w_k^i) + \lambda c^i u_k^i,$$

while replacing the coupling constraint (2.13) with the decoupled constraint

$$u_k^i \in U^i, \qquad i = 1, \ldots, n,$$

cf. Eq. (2.13). This is a lower bound approximation, as is typical in Lagrange multiplier-based decomposition approaches in linear and nonlinear programming (see e.g., [BeT97], [Ber16b]). To see this, note that for every feasible solution of the original problem, the Lagrangian term (2.16) makes a nonpositive contribution when added to the cost function, while with the constraint relaxed, the resulting optimal cost can only be reduced further.

With the subsystems now decoupled, we may solve each single subsystem problem separately, thereby obtaining a separable lower bound approximation

$$\sum_{i=1}^{n} \tilde{J}_k^i(x_k^i, \lambda)$$

for every $k = 1, \ldots, N-1$. This approximation can in turn be used to obtain a suboptimal one-step lookahead policy. Note that we may also try to optimize the approximation over λ, either by ad hoc experimentation or by a

† More general cases, where u^i and b are multi-dimensional, and c^i are replaced by matrices of appropriate dimension, can be handled similarly, albeit with greater computational complications.

more systematic optimization method.† Another possibility is to use a more general Lagrangian term of the form

$$\left(\sum_{k=0}^{N-1} \lambda_k \sum_{i=1}^{n} c^i u_k^i - Nb \right),$$

in place of the term (2.16), where $\lambda_0, \ldots, \lambda_{N-1} \geq 0$ are time-varying scalar multipliers.

2.3.2 Probabilistic Approximation - Certainty Equivalent Control

We will now consider problem approximation based on modifying the underlying probabilistic structure. The most common example of this type is the *certainty equivalent controller* (CEC). It replaces the stochastic disturbances with deterministic variables that are fixed at some "typical" values. Thus it acts as if a form of the certainty equivalence principle were holding, cf. the discussion of linear quadratic problems in Section 1.3.7.‡

The advantage of the CEC is that it involves a much less demanding computation than the stochastic DP algorithm: it requires the solution of a *deterministic* optimal control problem at each stage. This problem yields an optimal control sequence, the first component of which is used at the current stage, while the remaining components are discarded. Thus the CEC is able to deal with stochastic and even partial information problems by using the more flexible and potent methodology of deterministic optimal control.

We will first describe the most common version of the CEC for the stochastic DP problem of Section 1.2. Suppose that for every state-control pair (x_k, u_k) we have selected a "typical" value of the disturbance, which we denote by $\tilde{w}_k(x_k, u_k)$. For example the expected values

$$\tilde{w}_k(x_k, u_k) = E\{w_k \mid x_k, u_k\},$$

† Maximization of the lower bound approximation over λ is an interesting possibility. This is common in duality-based optimization. Generally the approach of this example falls within the framework of *Lagrangian relaxation*, a decomposition method that is based on the use of Lagrange multipliers and duality theory; see e.g., [BeT97], [Ber15a], [Ber16a].

‡ Our use of the term "certainty equivalence" in this section is not very precise. It generally refers to the simplification of some approximation in value space computations by replacing some stochastic quantities by deterministic ones. In particular, we allow that only some (not all) of the stochastic quantities are replaced by deterministic ones. Moreover the stochastic to deterministic simplification process at time k may depend on k, and may depend on the results of the computation and/or the information collected up to the time k.

can serve as typical values, if the disturbance spaces are convex subsets of Euclidean spaces [so that they include $\tilde{w}_k(x_k, u_k)$]. Alternatively, $\tilde{w}_k(x_k, u_k)$ could be the disturbance with maximal conditional probability given (x_k, u_k).

To implement the CEC at state x_k and stage k we solve a deterministic optimal control problem obtained from the original problem by replacing all uncertain quantities by their typical values. In particular, we solve the $(N-k)$-stages problem

$$\min_{\substack{x_{i+1}=f_i(x_i,u_i,\tilde{w}_i(x_i,u_i))\\ u_i\in U_i(x_i),\ i=k,\ldots,N-1}} \left[g_N(x_N) + \sum_{i=k}^{N-1} g_i\big(x_i, u_i, \tilde{w}_i(x_i, u_i)\big)\right]. \tag{2.17}$$

If $\{\tilde{u}_k, \ldots, \tilde{u}_{N-1}\}$ is the optimal control sequence for this problem, we use the first control in this sequence and discard the remaining controls:

$$\tilde{\mu}_k(x_k) = \tilde{u}_k.$$

An alternative implementation of the CEC is to compute off-line an optimal policy

$$\big\{\mu_0^d(x_0), \ldots, \mu_{N-1}^d(x_{N-1})\big\}$$

for the following deterministic version of the original problem

$$\begin{aligned}
&\text{minimize} \quad g_N(x_N) + \sum_{k=0}^{N-1} g_k\big(x_k, \mu_k(x_k), \tilde{w}_k(x_k, u_k)\big)\\
&\text{subject to} \quad x_{k+1} = f_k\big(x_k, \mu_k(x_k), \tilde{w}_k(x_k, u_k)\big), \quad \mu_k(x_k) \in U_k(x_k), \quad k \geq 0,
\end{aligned} \tag{2.18}$$

by using the DP algorithm. Then the control input $\tilde{\mu}_k(x_k)$ applied by the CEC at time k is given by

$$\tilde{\mu}_k(x_k) = \mu_k^d(x_k).$$

The two variants of the CEC just given are equivalent in terms of performance. The main difference is that the first variant is well-suited for on-line replanning, while the second variant is more suitable for an off-line implementation.

Finally, let us note that the CEC can be extended to partial state information problems, where the state x_k is not known at time k, but instead an estimate of x_k is available, which is based on measurements that have been obtained up to time k. In this case, we find a suboptimal control similarly [cf. Eqs. (2.17) and (2.18)], but with x_k replaced by an estimate, as if this estimate were exact.

Certainty Equivalent Control with Heuristics

Even though the CEC approach simplifies a great deal the computations, it still requires the optimal solution of a deterministic tail subproblem at each stage [cf. Eq. (2.17)]. This problem may still be difficult, and a more convenient approach may be to solve it suboptimally using a heuristic algorithm. In particular, given the state x_k at time k, we may use some (easily implementable) heuristic to find a suboptimal control sequence $\{\tilde{u}_k, \tilde{u}_{k+1}, \ldots, \tilde{u}_{N-1}\}$ for the problem of Eq. (2.17), and then use $\tilde{u}_k$ as the control for stage k.

An important enhancement of this idea is to use minimization over the first control u_k and to use the heuristic only for the remaining stages $k+1, \ldots, N-1$. To implement this variant of the CEC, we must apply at time k a control $\tilde{u}_k$ that minimizes over $u_k \in U_k(x_k)$ the expression

$$g_k\big(x_k, u_k, \tilde{w}_k(x_k, u_k)\big) + H_{k+1}(x_{k+1}), \tag{2.19}$$

where

$$x_{k+1} = f_k\big(x_k, u_k, \tilde{w}_k(x_k, u_k)\big),$$

and H_{k+1} is the cost-to-go function corresponding to the heuristic, i.e., $H_{k+1}(x_{k+1})$ is the cost incurred over the remaining stages $k+1, \ldots, N-1$ starting from a state x_{k+1}, using the heuristic. This is a hybrid approach: it resembles one-step lookahead with lookahead function H_{k+1}, and it resembles certainty equivalence in that the uncertain quantities have been replaced by their typical values.

Note that for any next state x_{k+1}, it is not necessary to have a closed-form expression for the heuristic cost-to-go $H_{k+1}(x_{k+1})$. Instead we can generate this cost by running the heuristic from x_{k+1} and computing the corresponding cost. Thus all the possible next states x_{k+1} must be computed for all possible values of the control u_k, and then the heuristic must be run from each of these x_{k+1} to calculate $H_{k+1}(x_{k+1})$, which is needed in the minimization of the expression (2.19).

Example 2.3.4 (Parking with Probability Estimates)

Consider the one-directional parking problem of Example 1.3.3, where a driver is looking for a parking space along a line of N spaces, with a garage at the end of the line (position N). The driver starts at space 0 and traverses the parking spaces sequentially, i.e., continues to subsequent spaces, up to a decision to park in space k at cost $c(k)$, if space k is free, or upon reaching the garage where parking is mandatory at cost C. In Example 1.3.3, we assumed that space k is free with a given and fixed probability $p(k)$, independently of whether other parking spaces are free or not.

Assume instead that $p(k)$ is a belief estimate that is based on the driver's observations of the status of preceding spaces. For example, this estimation process may involve exploiting probabilistic relations that may exist between the free/taken status of different spaces. In particular, let us assume that the driver, upon arrival at space k, can estimate the belief state of the spaces that lie ahead, i.e., the vector of conditional probabilities $\big(p(k+1),\ldots,p(N)\big)$ given the observations of the spaces $0,\ldots,k$.

The problem now becomes very hard to solve by exact DP, because the state space is infinite: at time k the state consists of the free/taken status of the current position k, plus the belief state of the spaces that lie ahead. However, a simple suboptimal approach to the problem can be based on certainty equivalence: at time k, we fix the free/taken probabilities of the spaces that lie ahead to their current belief values, and act as if these values will not change in the future. Then upon reaching space k, the fast DP algorithm of Example 1.3.3 can be used to solve on-line the corresponding fixed probabilities problem, and to find the corresponding suboptimal decision.

As an illustration, let $p(k)$ be estimated by using the ratio $R(k)$ of the number of free spaces encountered up to reaching space k divided by the total number $k+1$. Knowing $R(k)$, the driver adjusts the current probabilities $p(m)$ for spaces $m > k$ to the level

$$\hat{p}\big(m, R(k)\big) = \gamma p(m) + (1-\gamma)R(k),$$

where γ is a known constant between 0 and 1. The problem can then be solved by exact DP, by using as state at time k the free/taken status of space k together with the ratio $R(k)$ [in the terminology of Section 1.3, $R(k)$ is a sufficient statistic, which contains all the relevant information for the purposes of control]. The number of possible values of $R(k)$ grows exponentially with k, so the solution by exact DP may become intractable for large N. However, by applying the probabilistic approximation approach of the present example, the corresponding suboptimal policy may be easily obtained and implemented on-line.

Partial Certainty Equivalent Control - Partially Observed States

In the preceding descriptions of the CEC all future and present uncertain quantities are fixed at their typical values. A useful variation is to fix at typical values only *some* of these quantities. For example, a partial state information problem may be treated as one of perfect state information, using an estimate $\tilde{x}_k$ of x_k as if it were exact, while fully taking into account the stochastic nature of the disturbances. Thus, if $\big\{\mu_0^p(x_0),\ldots,\mu_{N-1}^p(x_{N-1})\big\}$ is an optimal policy obtained from the DP algorithm for the stochastic perfect state information problem

$$\text{minimize } E\left\{g_N(x_N) + \sum_{k=0}^{N-1} g_k\big(x_k, \mu_k(x_k), w_k\big)\right\}$$

$$\text{subject to } x_{k+1} = f_k\big(x_k, \mu_k(x_k), w_k\big), \quad \mu_k(x_k) \in U_k, \quad k = 0,\ldots,N-1,$$

then the control input applied by this variant of CEC at time k is $\mu_k^p(\tilde{x}_k)$, where $\tilde{x}_k$ is the estimate of x_k given the information available up to time k. Let us provide an example.

Example 2.3.5 (The Unscrupulous Innkeeper)

Consider an innkeeper who charges one of m different rates $r_1, \ldots, r_m$ for a room as the day progresses, depending on whether he has many or few vacancies, so as to maximize his expected total income during the day. A quote of a rate r_i is accepted with probability p_i and is rejected with probability $1 - p_i$, in which case the customer departs, never to return during that day. If (somewhat unrealistically) the innkeeper knows the number y of customers that will ask for a room during the rest of the day (including the customer currently asking for a room) and the number of vacancies is x, the optimal expected income $\tilde{J}(x, y)$ of the innkeeper is given by the DP algorithm

$$\tilde{J}(x, y) = \max_{i=1,\ldots,m} \Big[p_i\big(r_i + \tilde{J}(x-1, y-1)\big) + (1 - p_i)\tilde{J}(x, y-1)\Big], \tag{2.20}$$

for all $x \geq 1$ and $y \geq 1$, with initial conditions

$$\tilde{J}(x, 0) = \tilde{J}(0, y) = 0, \qquad \text{for all } x \text{ and } y.$$

This algorithm can be used to obtain the values of $\tilde{J}(x, y)$ for all pairs (x, y).

Consider now the case where the innkeeper does not know y at the times of decision, but instead only maintains a probability distribution for y. Then, it can be seen that the problem becomes a difficult partial state information problem. The exact DP algorithm should then be executed over the set of the pairs of x and the belief state of y. Yet a reasonable partially stochastic CEC is based on approximating the optimal cost-to-go of subsequent decisions with $\tilde{J}(x-1, \tilde{y}-1)$ or $\tilde{J}(x, \tilde{y}-1)$, where the function $\tilde{J}$ is calculated by the preceding recursion (2.20) and $\tilde{y}$ is an estimate of y, such as the closest integer to the expected value of y. In particular, according to this one-step lookahead policy, when the innkeeper has a number of vacancies $x \geq 1$, he quotes to the current customer the rate that maximizes

$$p_i\big(r_i + \tilde{J}(x-1, \tilde{y}-1) - \tilde{J}(x, \tilde{y}-1)\big).$$

Thus in this suboptimal scheme, the innkeeper acts as if the state estimate $(x, \tilde{y})$ were exact.

In a variant of the preceding approach for partial state information problems, we may use the belief state b_k (the conditional distribution of x_k given the information obtained up to time k), instead of using an estimate $\tilde{x}_k$ of x_k as if it were exact. In particular, we may obtain u_k by minimizing

$$E_{x_k, w_k}\Big\{g_k(x_k, u_k, w_k) + J^p_{k+1}\big(f_k(x_k, u_k, w_k)\big)\Big\}, \tag{2.21}$$

where J^p_{k+1} is the optimal cost-to-go function of the corresponding perfect state information problem, and it is assumed that $x_{k+1}, \ldots, x_{N-1}$ will become exactly known. The expected value in Eq. (2.21) is taken with respect to b_k, the belief distribution of x_k, and the distribution of w_k given (x_k, u_k).

In this approach it is necessary that the control constraint set does not depend on x_k in order for the minimization of the expression (2.21) over u_k to be possible.†

In an alternative and simpler scheme, we may also fix the future disturbances $w_{k+1}, \ldots, w_{N-1}$ at nominal values $\tilde{w}_m(x_m, u_m)$, $m = k + 1, \ldots, N-1$, and obtain u_k by minimizing

$$E_{x_k, w_k}\Big\{g_k(x_k, u_k, w_k) + J^d_{k+1}\big(f_k(x_k, u_k, w_k)\big)\Big\}, \qquad (2.22)$$

where J^d_{k+1} is the optimal cost-to-go function of the corresponding *deterministic* problem where $w_{k+1}, \ldots, w_{N-1}$ are fixed at their nominal values and $x_{k+1}, \ldots, x_{N-1}$ are perfectly observed. Another possibility is to additionally replace w_k in Eq. (2.22) with a nominal value $\tilde{w}_k(x_k, u_k)$.

Other Variations of Certainty Equivalent Control

It is also possible to use more general approaches to implement one-step lookahead control based on simplification of the probabilistic structure of the problem. The possibilities are highly problem-dependent by nature, but we may distinguish a few general techniques, which we illustrate through examples.

Example 2.3.6 (Decoupling Disturbance Distributions)

Let us consider a CEC approach in the context of enforced decomposition (cf. Section 2.3.1) when the subsystems are coupled only through the disturbance. In particular, consider n subsystems of the form

$$x^i_{k+1} = f^i(x^i_k, u^i_k, w^i_k), \qquad i = 1, \ldots, n.$$

Here the ith subsystem has its own state x^i_k, control u^i_k, and cost per stage $g^i(x^i_k, u^i_k, w^i_k)$, but the probability distribution of w^i_k depends on the full state $x_k = (x^1_k, \ldots, x^n_k)$.

---

† Generally, in the partial state information problems considered in this section, regardless of the approximations used to obtain a suboptimal policy, one needs to maintain and update the belief state b_k. In particular, given b_k, one needs to compute b_{k+1} once a new measurement relating to x_{k+1} is obtained, but before computing and applying the control u_{k+1}. There are several types of state estimation algorithms that may be used for this purpose, depending on the nature of the problem. For example, methods based on *Kalman filtering* are used widely for continuous-state problems (see e.g., [AnM79], [KuV86], [Kri16], [ChC17]). Another important class of state estimation methods is *particle filtering* (see e.g., [GSS93], [DoJ09], [Can16], [Kri16]). These methods are well-suited for Monte-Carlo simulation and can be used for discrete-state problems as well.

A natural form of suboptimal control is to solve at each stage k and for each i, the ith subsystem optimization problem where *the probability distribution of the future disturbances* $w^i_{k+1},\ldots,w^i_{N-1}$ *is "decoupled,"* in the sense that it depends only on the corresponding "local" states $x^i_{k+1},\ldots,x^i_{N-1}$. This distribution may be derived by using some nominal values

$$\tilde{x}^j_{k+1},\ldots,\tilde{x}^j_{N-1}, \qquad j\neq i,$$

of the future states of the other subsystems, and these nominal values may in turn depend on the full current state x_k. The first control $\overline{u}^i_k$ in the optimal policy thus obtained is applied at the ith subsystem in stage k, and the remaining portion of this policy is discarded.

Example 2.3.7 (Approximation Using Weighted Scenarios)

We noted earlier the possibility to approximate the optimal cost-to-go with a CEC approach, whereby for a given state x_{k+1} at time $k+1$, we fix the remaining disturbances at some nominal values $\tilde{w}_{k+1},\ldots,\tilde{w}_{N-1}$, and we compute an optimal control or heuristic-based trajectory starting from x_{k+1} at time $k+1$.

This CEC approximation involves a single nominal trajectory of the remaining uncertainty. To strengthen this approach, it is natural to consider multiple trajectories of the uncertainty, called *scenarios*. We may then construct a "weighted scenarios" approximation to the optimal cost-to-go, which involves the cost of either an optimal or a heuristic policy for each scenario.

Mathematically, we assume that we have a method, which at a given state x_{k+1}, generates q uncertainty sequences

$$w^s(x_{k+1})=(w^s_{k+1},\ldots,w^s_{N-1}), \qquad s=1,\ldots,q.$$

These are the scenarios considered at state x_{k+1}. The optimal cost $J^*_{k+1}(x_{k+1})$ is then approximated by

$$\tilde{J}_{k+1}(x_{k+1},r)=\sum_{s=1}^{q} r_s C_s(x_{k+1}),$$

where $r=(r_1,\ldots,r_q)$ is a probability distribution, i.e., a vector of nonnegative parameters that add to 1, and $C_s(x_{k+1})$ is the cost corresponding to an occurrence of the scenario $w^s(x_{k+1})$, when starting from state x_{k+1} and using either an optimal or a given heuristic policy. For a related idea, see the paper [YSH17], which deals with the use of scenarios in POMDP.

There may be several problem-dependent ways to generate the scenarios, possibly including randomization and/or simulation. The parameters $r_1,\ldots,r_q$ may depend on the time index, and may be interpreted as "aggregate probabilities" that encode the aggregate effect on the cost-to-go function of uncertainty sequences that are similar to the scenario $w^s(x_{k+1})$. They may be computed using some ad hoc scheme, or some more systematic approach. See also the papers [CGC00], [WCG02], [YFG08], which deal with a probabilistic approximation technique called *hindsight optimization*. We finally note that the possibility of simplifying the probabilistic model of the system within the multistep lookahead process, also arises in the context of the rollout approach, which is the subject of the next section.

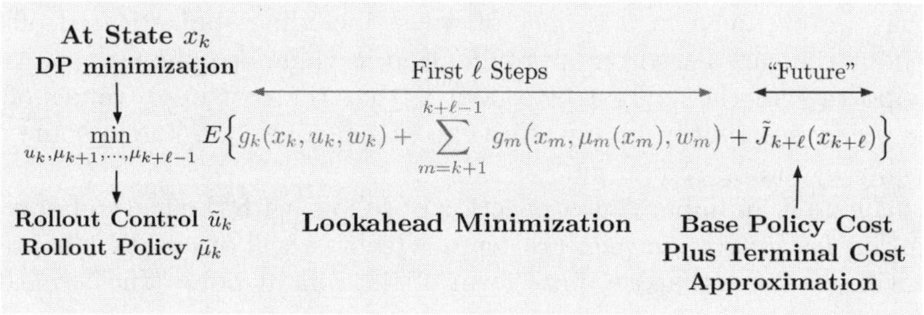

Figure 2.4.1 The structure of rollout with ℓ-step lookahead at state x_k. It involves:

(a) Multistep lookahead over ℓ stages (possibly $\ell = 1$).

(b) Running a heuristic algorithm/base policy from state $x_{k+\ell}$ for a number of steps, say m.

(c) Adding a cost function approximation at the end of the $\ell + m$ steps.

The approximate cost $\tilde J_{k+\ell}(x_{k+\ell})$ is the sum of the base policy costs of the m stages plus the terminal cost function approximation from state $x_{k+\ell+m}$. The minimization produces the rollout control $\tilde u_k$, and the rollout policy $\tilde\mu_k(x_k) = \tilde u_k$.

2.4 ROLLOUT AND THE POLICY IMPROVEMENT PRINCIPLE

The principal aim of rollout is *policy improvement*, i.e., start with a suboptimal/heuristic policy, called the *base policy* (or sometimes, the *default policy*, or the *base heuristic*), and produce an improved policy by limited lookahead minimization with use of the heuristic at the end. This latter policy is called the *rollout policy*, and the fact that it is indeed "improved" will be established, under various conditions; see what follows in this section, and also in Section 5.1.2, in the context of infinite horizon problems.

In its pure ℓ-step lookahead form, rollout can be defined very simply: it is approximation in value space with the approximate cost-to-go values $\tilde J_{k+1}(x_{k+\ell})$ calculated by running the base policy, starting from each possible next state $x_{k+\ell}$. There is also a "truncated" variant of rollout for problems involving a long horizon, where the base policy is used for a limited number of steps, and some cost function approximation is added at the end to take into account the cost of the remaining steps (see Fig. 2.4.1). This cost function approximation is often simple (for example equal to zero), particularly when the base policy is run for a fairly large number of steps (or to the end of the horizon). It may also be quite complex, and may be obtained with a sophisticated off-line training process, involving for example a neural network (see Chapter 3).

The choice of the base policy is of course important for the performance of the rollout approach. However, experimental evidence has shown that surprisingly good rollout performance may be attained even with a relatively poor base heuristic, particularly when long multistep lookahead is used. Generally speaking, there is no restriction on the base policy.

It may be obtained in a variety of ways, including sophisticated off-line methods that are based on approximation in value or policy space. What is important in the context of rollout is that the cost-to-go values of the base policy starting from any state can be calculated on-line in some way, possibly including simulation.

There is an important connection of rollout with the method of *policy iteration for infinite horizon problems*, which we will discuss in Chapter 4. We will see there that the pure form of the rollout policy (no truncation and no cost function approximation) can be viewed as the result of a *single policy iteration*, starting from the base policy. Reversely, policy iteration can be viewed as a *perpetual rollout algorithm*, whereby a sequence of policies is generated, each one being a rollout policy obtained by using the preceding one as a base policy (see the discussion of Section 5.7.3).

Finally, let us note that rollout with terminal cost function approximation from state $\tilde{J}(x_{k+\ell+m})$ after $\ell + m$ steps (cf. Fig. 2.4.1), can be viewed as an approximate $(\ell + m)$-lookahead scheme. The approximation is due to the use of the base policy for the last m stages of the $(\ell+m)$-step minimization, in place of an optimal policy.

We will first describe rollout for finite-state deterministic problems and one-step lookahead, and then for stochastic problems, in Sections 2.4.1 and 2.4.2, respectively. We will also develop some variants of rollout that involve cost function evaluation by a human or software expert in Section 2.4.3. In Section 2.5, we will discuss rollout for infinite-spaces problems, including the important case of model predictive control. Furthermore, we will revisit rollout schemes in Section 5.1.2, in the context of infinite horizon problems.

2.4.1 On-Line Rollout for Deterministic Discrete Optimization

Let us consider a deterministic DP problem with a finite number of controls and a given initial state (so the number of states is also finite); cf. Section 1.3.2. We first focus on the pure form of rollout that uses one-step lookahead and no terminal cost approximation. Given a state x_k at time k, this algorithm considers all the tail subproblems that start at every possible next state x_{k+1}, and solves them suboptimally by using some algorithm, referred to as *base heuristic*.†

Thus when at x_k, rollout generates on-line the next states x_{k+1} that correspond to all $u_k \in U_k(x_k)$, and uses the base heuristic to compute the sequence of states $\{x_{k+1}, \ldots, x_N\}$ and controls $\{u_{k+1}, \ldots, u_{N-1}\}$ such that

$$x_{i+1} = f_i(x_i, u_i), \qquad i = k, \ldots, N-1.$$

† For deterministic problems we prefer to use the term "base heuristic" rather than "base policy" for reasons to be explained later in this section, in the context of the notion of sequential consistency.

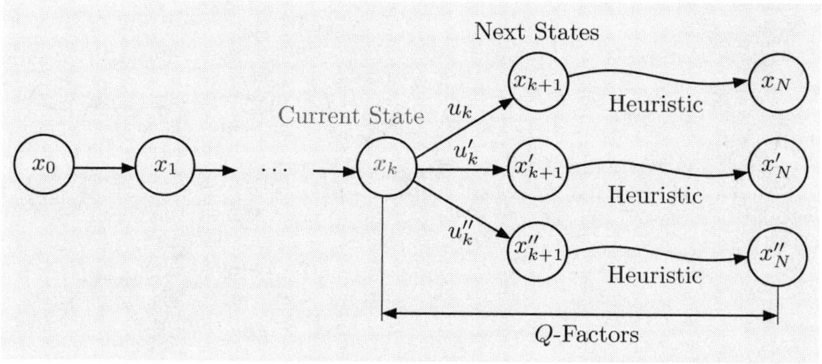

Figure 2.4.2 Schematic illustration of rollout with one-step lookahead for a deterministic problem. At state x_k, for every pair (x_k, u_k), $u_k \in U_k(x_k)$, the base heuristic generates a Q-factor $\tilde{Q}_k(x_k, u_k)$ [cf. Eq. (2.24)], and selects the control $\tilde{\mu}_k(x_k)$ with minimal Q-factor.

The rollout algorithm then applies the control that minimizes over $u_k \in U_k(x_k)$ the tail cost expression for stages k to N:

$$g_k(x_k, u_k) + g_{k+1}(x_{k+1}, u_{k+1}) + \cdots + g_{N-1}(x_{N-1}, u_{N-1}) + g_N(x_N). \quad (2.23)$$

Equivalently, and more succinctly, the rollout algorithm applies at state x_k the control $\tilde{\mu}_k(x_k)$ given by the minimization

$$\tilde{\mu}_k(x_k) \in \arg \min_{u_k \in U_k(x_k)} \tilde{Q}_k(x_k, u_k),$$

where $\tilde{Q}_k(x_k, u_k)$ is the approximate Q-factor defined by

$$\tilde{Q}_k(x_k, u_k) = g_k(x_k, u_k) + H_{k+1}\big(f_k(x_k, u_k)\big), \quad (2.24)$$

with $H_{k+1}(x_{k+1})$ denoting the cost of the base heuristic starting from state x_{k+1} [i.e., $H_{k+1}(x_{k+1})$ is the sum of all the terms in Eq. (2.23), except the first]; see Fig. 2.4.2. The rollout process defines a suboptimal policy $\tilde{\pi} = \{\tilde{\mu}_0, \ldots, \tilde{\mu}_{N-1}\}$, referred to as the *rollout policy*.

Example 2.4.1 (Traveling Salesman Problem)

Let us consider the traveling salesman problem, whereby a salesman wants to find a minimum mileage/cost tour that visits each of N given cities $c = 0, \ldots, N-1$ exactly once and returns to the city he started from (cf. Example 1.3.1). With each pair of distinct cities c, c', we associate a traversal cost $g(c, c')$. Note that we assume that we can go directly from every city to every other city. There is no loss of generality in doing so because we can assign a very high cost $g(c, c')$ to any pair of cities (c, c') that is precluded from participation in the solution. The problem is to find a visit order that goes through each city exactly once and whose sum of costs is minimum.

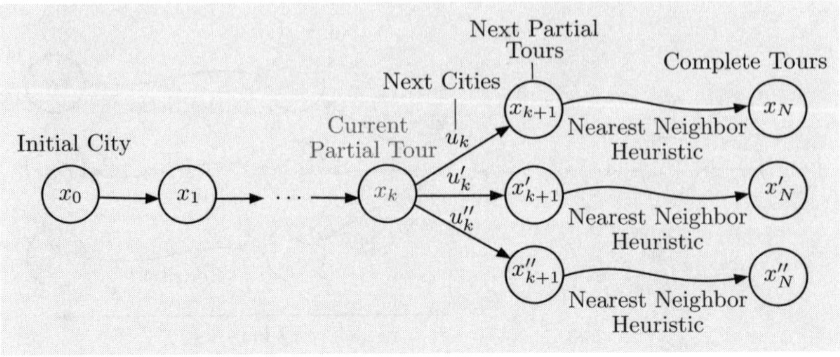

Figure 2.4.3 Schematic illustration of rollout with the nearest neighbor heuristic for the traveling salesman problem. The initial state x_0 consists of a single city. The final state x_N is a complete tour of N cities, containing each city exactly once.

There are many heuristic approaches for solving the traveling salesman problem. For illustration purposes, let us focus on the simple *nearest neighbor* heuristic, which starts with a partial tour, i.e., an ordered collection of distinct cities, and constructs a sequence of partial tours, adding to the each partial tour a new city that does not close a cycle and minimizes the cost of the enlargement. In particular, given a sequence $\{c_0, c_1, \ldots, c_k\}$ consisting of distinct cities, the nearest neighbor heuristic adds a city c_{k+1} that minimizes $g(c_k, c_{k+1})$ over all cities $c_{k+1} \neq c_0, \ldots, c_k$, thereby forming the sequence $\{c_0, c_1, \ldots, c_k, c_{k+1}\}$. Continuing in this manner, the heuristic eventually forms a sequence of N cities, $\{c_0, c_1, \ldots, c_{N-1}\}$, thus yielding a complete tour with total cost

$$g(c_0, c_1) + \cdots + g(c_{N-2}, c_{N-1}) + g(c_{N-1}, c_0). \tag{2.25}$$

We can formulate the traveling salesman problem as a DP problem as we discussed in Example 1.3.1. We choose a starting city, say c_0, as the initial state x_0. Each state x_k corresponds to a partial tour $(c_0, c_1, \ldots, c_k)$ consisting of distinct cities. The states x_{k+1}, next to x_k, are sequences of the form $(c_0, c_1, \ldots, c_k, c_{k+1})$ that correspond to adding one more unvisited city $c_{k+1} \neq c_0, c_1, \ldots, c_k$ (thus the unvisited cities are the feasible controls at a given partial tour/state). The terminal states x_N are the complete tours of the form $(c_0, c_1, \ldots, c_{N-1}, c_0)$, and the cost of the corresponding sequence of city choices is the cost of the corresponding complete tour given by Eq. (2.25).

Let us now use as a base heuristic the nearest neighbor method. The corresponding rollout algorithm operates as follows: After $k < N - 1$ iterations, we have a state x_k, i.e., a sequence $\{c_0, \ldots, c_k\}$ consisting of distinct cities. At the next iteration, we add one more city by running the nearest neighbor heuristic starting from each of the sequences of the form $\{c_0, \ldots, c_k, c\}$ where $c \neq c_0, \ldots, c_k$. We then select as next city c_{k+1} the city c that yielded the minimum cost tour under the nearest neighbor heuristic; see Fig. 2.4.3.

Cost Improvement with a Rollout Algorithm - Sequential Consistency

The definition of the rollout algorithm leaves open the choice of the base heuristic. There are several types of suboptimal solution methods that can be used as base heuristics, such as greedy algorithms, local search, genetic algorithms, tabu search, and others. Clearly we want to choose a base heuristic that strikes a good balance between quality of solutions produced and computational tractability.

Intuitively, we expect that the rollout policy's performance is no worse than the one of the base heuristic: since rollout optimizes over the first control before applying the heuristic, it makes sense to conjecture that it performs better than applying the heuristic without the first control optimization. However, some special conditions must hold in order to guarantee this cost improvement property. We provide two such conditions, *sequential consistency* and *sequential improvement*, and then show how to modify the algorithm to deal with the case where these conditions are not met.

We say that the base heuristic is *sequentially consistent* if it has the property that when it generates the sequence

$$\{x_k, x_{k+1}, \ldots, x_N\}$$

starting from state x_k, it also generates the sequence

$$\{x_{k+1}, \ldots, x_N\}$$

starting from state x_{k+1}. In other words, the base heuristic is sequentially consistent if it "stays the course": when the starting state x_k is moved forward to the next state x_{k+1} of its state trajectory, the heuristic will not deviate from the remainder of the trajectory.

As an example, the reader may verify that the nearest neighbor heuristic described in the traveling salesman Example 2.4.1 is sequentially consistent. Similar examples include the use of many types of greedy heuristics (see [Ber17], Section 6.4). Generally most heuristics used in practice satisfy the sequential consistency condition at "most" states x_k. However, some heuristics of interest may violate this condition at some states.

Note that *under sequential consistency the heuristic defines a legitimate DP policy*. This is the policy that moves from x_k to the state x_{k+1} that lies on the path $\{x_k, x_{k+1}, \ldots, x_N\}$ generated by the base heuristic.

We will now show that *the rollout algorithm obtained with a sequentially consistent base heuristic yields no worse cost than the base heuristic*. In particular, consider the rollout policy $\tilde{\pi} = \{\tilde{\mu}_0, \ldots, \tilde{\mu}_{N-1}\}$, and let $J_{k,\tilde{\pi}}(x_k)$ denote the cost obtained with $\tilde{\pi}$ starting from x_k. We claim that

$$J_{k,\tilde{\pi}}(x_k) \leq H_k(x_k), \qquad \text{for all } x_k \text{ and } k, \tag{2.26}$$

where $H_k(x_k)$ denotes the cost of the base heuristic starting from x_k.

We prove this inequality by induction. Clearly it holds for $k = N$, since $J_{N,\tilde{\pi}} = H_N = g_N$. Assume that it holds for index $k+1$. For any state x_k, let $\overline{u}_k$ be the control applied by the base heuristic at x_k. Then we have

$$\begin{aligned} J_{k,\tilde{\pi}}(x_k) &= g_k\big(x_k, \tilde{\mu}_k(x_k)\big) + J_{k+1,\tilde{\pi}}\Big(f_k\big(x_k, \tilde{\mu}_k(x_k)\big)\Big) \\ &\le g_k\big(x_k, \tilde{\mu}_k(x_k)\big) + H_{k+1}\Big(f_k\big(x_k, \tilde{\mu}_k(x_k)\big)\Big) \\ &= \min_{u_k \in U_k(x_k)} \Big[g_k(x_k, u_k) + H_{k+1}\big(f_k(x_k, u_k)\big)\Big] \\ &\le g_k(x_k, \overline{u}_k) + H_{k+1}\big(f_k(x_k, \overline{u}_k)\big) \\ &= H_k(x_k), \end{aligned} \tag{2.27}$$

where:

(a) The first equality is the DP equation for the rollout policy $\tilde{\pi}$.

(b) The first inequality holds by the induction hypothesis.

(c) The second equality holds by the definition of the rollout algorithm.

(d) The third equality is the DP equation for the policy that corresponds to the base heuristic (this is the step where we need sequential consistency).

This completes the induction proof of the cost improvement property (2.26).

Sequential Improvement

We will now show that the rollout policy has no worse performance than its base heuristic under a condition that is weaker than sequential consistency. Let us recall that the rollout algorithm $\tilde{\pi} = \{\tilde{\mu}_0, \ldots, \tilde{\mu}_{N-1}\}$ is defined by the minimization

$$\tilde{\mu}_k(x_k) \in \arg \min_{u_k \in U_k(x_k)} \tilde{Q}_k(x_k, u_k),$$

where $\tilde{Q}_k(x_k, u_k)$ is the approximate Q-factor defined by

$$\tilde{Q}_k(x_k, u_k) = g_k(x_k, u_k) + H_{k+1}\big(f_k(x_k, u_k)\big),$$

[cf. Eqs. (2.24)], and $H_{k+1}\big(f_k(x_k, u_k)\big)$ denotes the cost of the base heuristic starting from state $f_k(x_k, u_k)$.

We say that the base heuristic is *sequentially improving*, if for all x_k and k, we have

$$\min_{u_k \in U_k(x_k)} \tilde{Q}_k(x_k, u_k) \le H_k(x_k). \tag{2.28}$$

In words, the sequential improvement property (2.28) states that

$$\text{Minimal heuristic Q-factor at } x_k \leq \text{Heuristic cost at } x_k. \tag{2.29}$$

To show that a sequentially improving heuristic yields policy improvement, simply note that from the calculation of Eq. (2.27), replacing the last two steps (that rely on sequential consistency) with Eq. (2.28), we have

$$J_{k,\tilde{\pi}}(x_k) \leq H_k(x_k), \qquad \text{for all } x_k \text{ and } k.$$

Thus the rollout algorithm obtained with a sequentially improving base heuristic, will improve or at least will perform no worse than the base heuristic, from every starting state x_k. Note that *when the heuristic is sequentially consistent it is also sequentially improving*. This follows from Eq. (2.29), since for a sequentially consistent heuristic, the heuristic cost at x_k is equal to the Q-factor of the control $\overline{u}_k$ that the heuristic applies at x_k,

$$\tilde{Q}_k(x_k, \overline{u}_k) = g_k(x_k, \overline{u}_k) + H_{k+1}\big(f_k(x_k, \overline{u}_k)\big),$$

which is greater or equal to the minimal Q-factor at x_k. This implies Eq. (2.28).

Empirically, it has been observed that the cost improvement obtained by rollout with a sequentially improving heuristic is typically considerable and often dramatic. Generally, however, it is hard to provide solid theoretical support for this observation. Several case studies support the consistently good performance of rollout (at least in the pure form described above); see the end of chapter references. The textbook [Ber17], Section 6.4, provides some detailed worked-out examples. The price for the performance improvement is extra computation that is typically equal to the computation time of the base heuristic times a factor that is a low order polynomial of the problem size.

On the other hand the sequential improvement condition may not hold for a given base heuristic. Here is an example:

Example 2.4.2: (Sequential Improvement Violation)

Consider a rollout algorithm with a base heuristic that accidentally produces an optimal control sequence at the initial state x_0, and uses the (optimal) first control u_0 of this sequence to move to the (optimal) next state x_1. Suppose that the base heuristic produces a strictly suboptimal sequence from every successor state $x_2 = f_1(x_1, u_1)$, $u_1 \in U_1(x_1)$, so that the rollout yields a control u_1 that is strictly suboptimal (this can happen if the heuristic is not sequentially consistent). Then the trajectory produced by the rollout algorithm starting from the initial state x_0 is strictly inferior to the one produced by the base heuristic starting from x_0, while it can be seen that the sequential improvement condition does not hold.

The preceding example suggests a simple enhancement to the rollout algorithm, which we describe next. The main idea is to detect when the sequential improvement condition is violated, and to discard the suggestion of the rollout algorithm when this happens.

The Fortified Rollout Algorithm

We will describe a variant of the rollout algorithm that implicitly enforces the sequential improvement property. This variant, called the *fortified rollout algorithm*, starts at x_0, and generates step-by-step a sequence of states $\{x_0, x_1, \ldots, x_N\}$ and corresponding sequence of controls. Upon reaching state x_k it stores the trajectory

$$\overline{P}_k = \{x_0, u_0, \ldots, u_{k-1}, x_k\}$$

that has been constructed up to stage k, called *permanent* trajectory, and it also stores a *tentative* trajectory

$$\overline{T}_k = \{x_k, \overline{u}_k, \overline{x}_{k+1}, \overline{u}_{k+1}, \ldots, \overline{u}_{N-1}, \overline{x}_N\}$$

with corresponding cost

$$C(\overline{T}_k) = g_k(x_k, \overline{u}_k) + g_{k+1}(\overline{x}_{k+1}, \overline{u}_{k+1}) + \cdots + g_{N-1}(\overline{x}_{N-1}, \overline{u}_{N-1}) + g_N(\overline{x}_N).$$

The tentative trajectory is such that $\overline{P}_k \cup \overline{T}_k$ is the best end-to-end trajectory computed up to stage k of the algorithm. Initially, $\overline{T}_0$ is the trajectory generated by the base heuristic starting at the initial state x_0. The idea now is to discard the suggestion of the rollout algorithm at every state x_k where the base heuristic produces a trajectory that is inferior to $\overline{T}_k$, and use $\overline{T}_k$ instead.

In particular, upon reaching state x_k, we run the rollout algorithm as earlier, i.e., for every $u_k \in U_k(x_k)$ and next state $x_{k+1} = f_k(x_k, u_k)$, we run the base heuristic from x_{k+1}, and find the control $\tilde{u}_k$ that gives the best trajectory, denoted

$$\tilde{T}_k = \{x_k, \tilde{u}_k, \tilde{x}_{k+1}, \tilde{u}_{k+1}, \ldots, \tilde{u}_{N-1}, \tilde{x}_N\}$$

with corresponding cost

$$C(\tilde{T}_k) = g_k(x_k, \tilde{u}_k) + g_{k+1}(\tilde{x}_{k+1}, \tilde{u}_{k+1}) + \cdots + g_{N-1}(\tilde{x}_{N-1}, \tilde{u}_{N-1}) + g_N(\tilde{x}_N).$$

Whereas the ordinary rollout algorithm would choose control $\tilde{u}_k$ and move to $\tilde{x}_{k+1}$, the fortified algorithm compares $C(\overline{T}_k)$ and $C(\tilde{T}_k)$, and depending on which of the two is smaller, chooses $\overline{u}_k$ or $\tilde{u}_k$ and moves to $\overline{x}_{k+1}$ or to $\tilde{x}_{k+1}$, respectively. In particular, if $C(\overline{T}_k) \leq C(\tilde{T}_k)$ the algorithm sets the next state and corresponding tentative trajectory to

$$x_{k+1} = \overline{x}_{k+1}, \qquad \overline{T}_{k+1} = \{\overline{x}_{k+1}, \overline{u}_{k+1}, \ldots, \overline{u}_{N-1}, \overline{x}_N\},$$

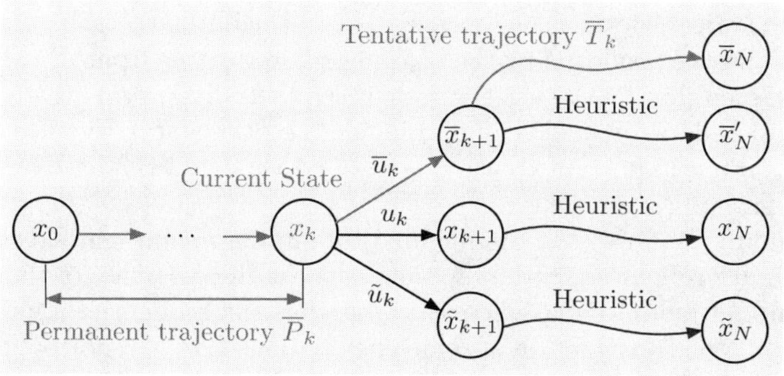

Figure 2.4.4 Schematic illustration of fortified rollout. After k steps, we have constructed the permanent trajectory

$$\overline{P}_k = \{x_0, u_0, \ldots, u_{k-1}, x_k\},$$

and the tentative trajectory

$$\overline{T}_k = \{x_k, \overline{u}_k, \overline{x}_{k+1}, \overline{u}_{k+1}, \ldots, \overline{u}_{N-1}, \overline{x}_N\}$$

such that $\overline{P}_k \cup \overline{T}_k$ is the best end-to-end trajectory computed so far. We now run the rollout algorithm at x_k, i.e., we find the control $\tilde{u}_k$ that minimizes over u_k the sum of $g_k(x_k, u_k)$ plus the heuristic cost from the state $x_{k+1} = f_k(x_k, u_k)$, and the corresponding trajectory

$$\tilde{T}_k = \{x_k, \tilde{u}_k, \tilde{x}_{k+1}, \tilde{u}_{k+1}, \ldots, \tilde{u}_{N-1}, \tilde{x}_N\}.$$

If the cost of the end-to-end trajectory $\overline{P}_k \cup \tilde{T}_k$ is lower than the cost of $\overline{P}_k \cup \overline{T}_k$, we add $(\tilde{u}_k, \tilde{x}_{k+1})$ to the permanent trajectory and set the tentative trajectory to

$$\overline{T}_{k+1} = \{\tilde{x}_{k+1}, \tilde{u}_{k+1}, \ldots, \tilde{u}_{N-1}, \tilde{x}_N\}.$$

Otherwise we add $(\overline{u}_k, \overline{x}_{k+1})$ to the permanent trajectory and set the tentative trajectory to

$$\overline{T}_{k+1} = \{\overline{x}_{k+1}, \overline{u}_{k+1}, \ldots, \overline{u}_{N-1}, \overline{x}_N\}.$$

Note that the fortified rollout will produce a different result than the ordinary rollout if the heuristic when started from $\overline{x}_{k+1}$ constructs a trajectory that is different than the tail portion of the tentative trajectory that starts at $\overline{x}_{k+1}$.

and if $C(\overline{T}_k) > C(\tilde{T}_k)$ it sets the next state and corresponding tentative trajectory to

$$x_{k+1} = \tilde{x}_{k+1}, \qquad \overline{T}_{k+1} = \{\tilde{x}_{k+1}, \tilde{u}_{k+1}, \ldots, \tilde{u}_{N-1}, \tilde{x}_N\}.$$

In other words the fortified rollout at x_k follows the current tentative trajectory $\overline{T}_k$ unless a lower cost trajectory $\tilde{T}_k$ is discovered by running the base heuristic from all possible next states x_{k+1}. It follows that at every state the trajectory that consists of the union of the permanent and the tentative trajectories, has no larger cost than the initial tentative trajectory, which is the one produced by the base heuristic starting from x_0. Moreover, it can be seen that if the base heuristic is sequentially improving, the rollout algorithm and its fortified version coincide. Experimental evidence suggests that it is often important to use the fortified version if the base heuristic is not sequentially improving.

Finally we note that the fortified rollout algorithm may be viewed as the ordinary rollout algorithm applied to a modified version of the original problem and modified base heuristic that has the sequential improvement property. This construction is somewhat tedious and will not be given; we refer to Bertsekas, Tsitsiklis, and Wu [BTW97], and [Ber17], Section 6.4.2.

Using Multiple Heuristics

In many problems, several promising heuristics may be available. It is then possible to use all of these heuristics in the rollout framework. The idea is to construct a *superheuristic*, which selects the best out of the trajectories produced by the entire collection of base heuristics. The superheuristic can then be used as the base heuristic for a rollout algorithm.

In particular, let us assume that we have M base heuristics, and that the mth of these, given a state x_{k+1}, produces a trajectory

$$\tilde{T}_{k+1}^m = \{x_{k+1}, \tilde{u}_{k+1}^m, \ldots, \tilde{u}_{N-1}^m, \tilde{x}_N^m\},$$

and corresponding cost $C(\tilde{T}_{k+1}^m)$. The superheuristic then produces at x_{k+1} the trajectory $\tilde{T}_{k+1}^m$ for which $C(\tilde{T}_{k+1}^m)$ is minimum.

An interesting property, which can be readily verified by using the definitions, is that *if all the base heuristics are sequentially improving, the same is true for the superheuristic.* Moreover, there is a fortified version of the rollout algorithm, which has the property that it produces a trajectory with no worse cost than all the trajectories produced by the M base heuristics when started at the initial state x_0.

Truncated Rollout Algorithms with Multistep Lookahead and Terminal Cost Function Approximation

We may incorporate multistep lookahead into the rollout framework. This may be essential in problems where at a given state, the effect of the first control may be small relative to the effect of the remaining controls.

To describe two-step lookahead for deterministic problems, in its most straightforward implementation, suppose that after k steps we have reached state x_k. We then consider the set of all two-step-ahead states x_{k+2}, we run the base heuristic starting from each of them, and compute the two-stage cost to get from x_k to x_{k+2}, plus the cost of the base heuristic from x_{k+2}. We select the state, say $\tilde{x}_{k+2}$, that is associated with minimum

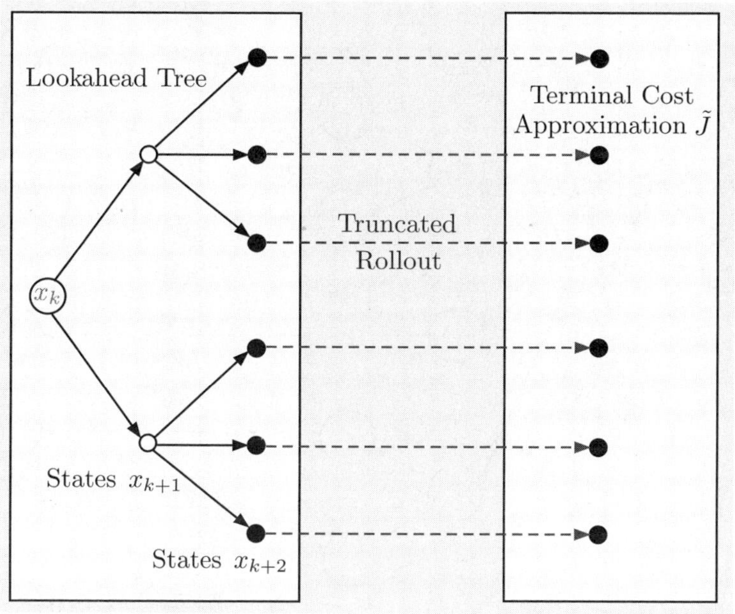

Figure 2.4.5 Illustration of truncated rollout with two-step lookahead and a terminal cost approximation $\tilde{J}$. The base heuristic is used for a limited number of steps and the terminal cost is added to compensate for the remaining steps.

cost, compute the controls $\tilde{u}_k$ and $\tilde{u}_{k+1}$ that lead from x_k to $\tilde{x}_{k+2}$, and choose $\tilde{u}_k$ as the next rollout control and $x_{k+1} = f_k(x_k, \tilde{u}_k)$ as the next state; see Fig. 2.4.5. The extension of the algorithm to lookahead of more than two steps is straightforward: instead of the two-step-ahead states x_{k+2} we run the base heuristic starting from all the possible ℓ-step ahead states $x_{k+\ell}$, etc. Moreover, the ℓ-step lookahead minimization may involve approximations aimed at simplifying the associated computations, similar to problem approximation methods (cf. Section 2.3).

An important variation for problems with a large number of stages, is *truncated rollout with terminal cost approximation.* Here the rollout trajectories are obtained by running the base heuristic from the leaf nodes of the lookahead tree, and they are truncated after a given number of steps, while a terminal cost approximation is added to the heuristic cost to compensate for the resulting error; see Fig. 2.4.5. One possibility that works well for many problems is to simply set the terminal cost approximation to zero. Alternatively, the terminal cost function approximation may be obtained by problem approximation or by using some sophisticated off-line training process that may involve an approximation architecture such as a neural network (see Chapter 3).

Among other variations of deterministic multistep rollout, let us mention a fortified version, which maintains a tentative trajectory, along the

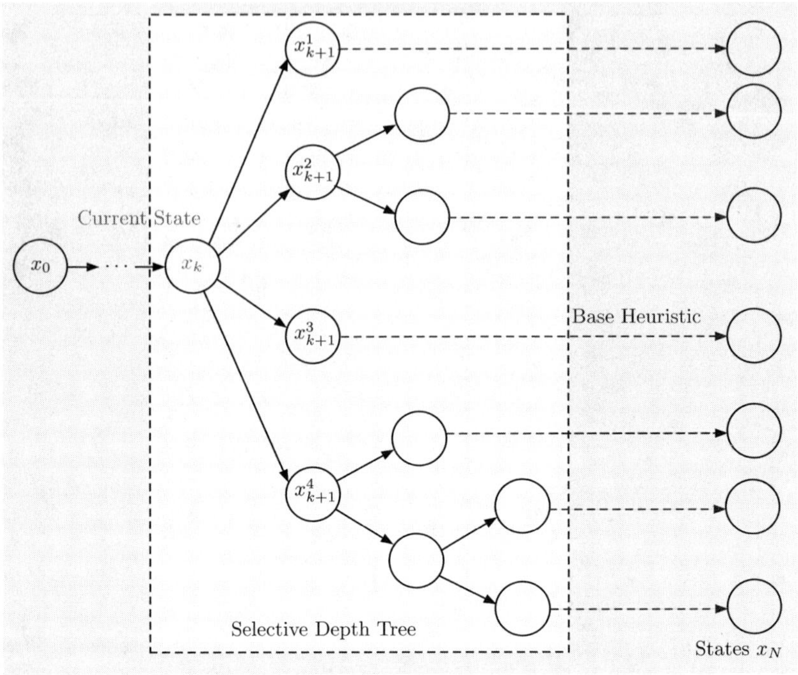

Figure 2.4.6 Illustration of a form of deterministic rollout with selective depth lookahead. After k steps of the algorithm, we have a trajectory that starts at the initial state x_0 and ends at state x_k. We then generate the set of all possible next states (states x_{k+1}^1, x_{k+1}^2, x_{k+1}^3, x_{k+1}^4 in the figure). We "evaluate" these states using the base heuristic, and select some of them for "expansion," i.e., we generate their next states x_{k+2}, evaluate them using the base heuristic, and continue. In the end we have a selective depth tree of next states, and the base heuristic costs from the leaves of the tree. The state x_{k+1} that corresponds to smallest overall cost is chosen by the selective depth lookahead rollout algorithm. For problems with a large number of stages, we can also truncate the rollout trajectories and add a terminal cost function approximation as compensation for the resulting error; cf. Fig. 2.4.5.

lines described earlier for the one-step lookahead case. In still another version of ℓ-step lookahead rollout, we may consider "pruning" the lookahead tree. By this we mean disregarding some of the states that are ℓ steps or less ahead, which are judged less promising according to some criterion (for example the costs of the base heuristic after a one-step lookahead); see Fig. 2.4.6. This may be viewed as *selective depth lookahead*, and aims to limit the number of times that the base heuristic is applied, which can become overwhelming as the length of lookahead is increased. We will encounter again the idea of selective depth lookahead in the context of stochastic rollout and Monte Carlo tree search (see the next section), where in addition to the length of lookahead, the accuracy of the simulation to evaluate the cost of the base heuristic is adapted to the results of earlier computations.

Finally, let us mention a variant of rollout that maintains multiple trajectories, extending from a given state x_k to possibly multiple next states x_{k+1}. These states are the ones considered "most promising" based on the current results of the base heuristic (like being "ϵ-best"), but may be discarded later based on subsequent computations. Such extended forms of rollout are restricted to deterministic problems, and tend to be problem-dependent. We will not consider them further in this book.

2.4.2 Stochastic Rollout and Monte Carlo Tree Search

We will now discuss the extension of the rollout algorithm to stochastic finite-state DP problems. We will restrict ourselves to the case where the base heuristic is a policy $\pi = \{\mu_0, \ldots, \mu_{N-1}\}$ (i.e., is sequentially consistent, in the context of deterministic problems). It is possible to consider more general rollout algorithms that involve base heuristics with a sequential improvement property, but we will not pursue this idea, as it does not seem to have been applied so far in interesting stochastic contexts.

We first note that the cost improvement property that we showed for deterministic problems under the sequential consistency condition carries through for stochastic problems. In particular, let us denote by $J_{k,\pi}(x_k)$ the cost corresponding to starting the base policy at state x_k, and by $J_{k,\tilde{\pi}}(x_k)$ the cost corresponding to starting the rollout algorithm at state x_k. We claim that

$$J_{k,\tilde{\pi}}(x_k) \leq J_{k,\pi}(x_k), \qquad \text{for all } x_k \text{ and } k.$$

We prove this inequality by induction similar to the deterministic case [cf. Eq. (2.27)]. Clearly it holds for $k = N$, since $J_{N,\tilde{\pi}} = J_{N,\pi} = g_N$. Assuming that it holds for index $k+1$, we have for all x_k,

$$\begin{aligned}
J_{k,\tilde{\pi}}(x_k) &= E\Big\{g_k\big(x_k, \tilde{\mu}_k(x_k), w_k\big) + J_{k+1,\tilde{\pi}}\Big(f_k\big(x_k, \tilde{\mu}_k(x_k), w_k\big)\Big)\Big\} \\
&\leq E\Big\{g_k\big(x_k, \tilde{\mu}_k(x_k), w_k\big) + J_{k+1,\pi}\big(f_k(x_k, \tilde{\mu}_k(x_k), w_k)\big)\Big\} \\
&= \min_{u_k \in U_k(x_k)} E\Big\{g_k(x_k, u_k, w_k) + J_{k+1,\pi}\big(f_k(x_k, u_k, w_k)\big)\Big\} \\
&\leq E\Big\{g_k\big(x_k, \mu_k(x_k), w_k\big) + J_{k+1,\pi}\big(f_k(x_k, \mu_k(x_k), w_k)\big)\Big\} \\
&= J_{k,\pi}(x_k),
\end{aligned}$$

where:

(a) The first equality is the DP equation for the rollout policy $\tilde{\pi}$.

(b) The first inequality holds by the induction hypothesis.

(c) The second equality holds by the definition of the rollout algorithm.

(d) The third equality is the DP equation for the policy π that corresponds to the base heuristic.

The induction proof of the cost improvement property is thus complete.

Similar to deterministic problems, it has been observed empirically that for stochastic problems the rollout policy not only does not deteriorate the performance of the base policy, but also typically produces substantial cost improvement; see the case studies referenced at the end of the chapter.

Simulation-Based Implementation of the Rollout Algorithm

A conceptually straightforward way to compute the rollout control at a given state x_k and time k is to consider each possible control $u_k \in U_k(x_k)$ and to generate a "large" number of simulated trajectories of the system starting from (x_k, u_k). Thus a simulated trajectory is obtained from

$$x_{i+1} = f_i\big(x_i, \mu_i(x_i), w_i\big), \qquad i = k+1, \ldots, N-1,$$

where $\{\mu_{k+1}, \ldots, \mu_{N-1}\}$ is the tail portion of the base policy, the starting state of the simulated trajectory is

$$x_{k+1} = f_k(x_k, u_k, w_k),$$

and the disturbance sequence $\{w_k, \ldots, w_{N-1}\}$ is obtained by random sampling. The costs of the trajectories corresponding to a pair (x_k, u_k) can be viewed as samples of the Q-factor

$$Q_k(x_k, u_k) = E\Big\{g_k(x_k, u_k, w_k) + J_{k+1,\bar{\pi}}\big(f_k(x_k, u_k, w_k)\big)\Big\},$$

where $J_{k+1,\bar{\pi}}$ is the cost-to-go function of the base policy, i.e., $J_{k+1,\bar{\pi}}(x_{k+1})$ is the cost of using the base policy starting from x_{k+1}. For problems with a large number of stages, it is also common to truncate the rollout trajectories and add a terminal cost function approximation as compensation for the resulting error.

By Monte Carlo averaging of the costs of the sample trajectories plus the terminal cost (if any), we obtain an approximation to the Q-factor $Q_k(x_k, u_k)$ for each control $u_k \in U_k(x_k)$, which we denote by $\tilde{Q}_k(x_k, u_k)$. We then compute the (approximate) rollout control $\tilde{\mu}_k(x_k)$ with the minimization

$$\tilde{\mu}_k(x_k) \in \arg \min_{u_k \in U_k(x_k)} \tilde{Q}_k(x_k, u_k). \tag{2.30}$$

Example 2.4.3 (Backgammon)

The first impressive application of rollout was given for the ancient two-player game of backgammon, in the paper by Tesauro and Galperin [TeG96]; see Fig. 2.4.7. They implemented a rollout algorithm, which attained a level of play that was better than all computer backgammon programs, and eventually better than the best humans. Tesauro had proposed earlier the use of

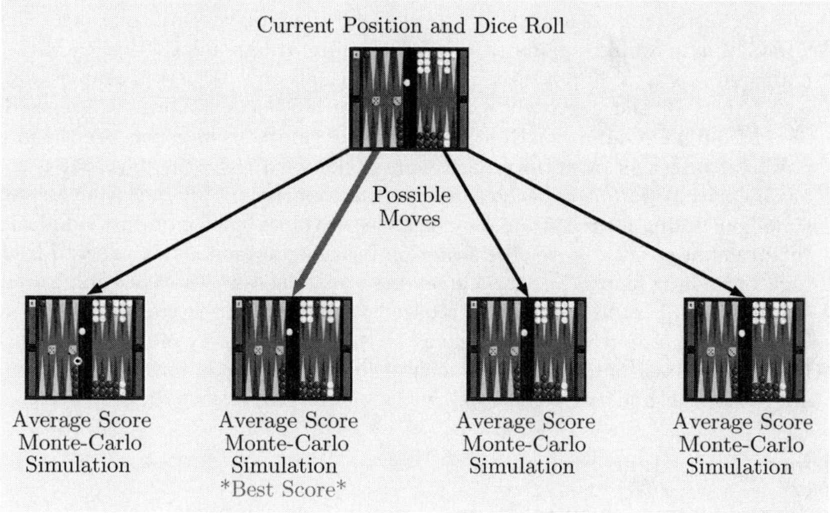

Figure 2.4.7 Illustration of rollout for backgammon. At a given position and roll of the dice, the set of all possible moves is generated, and the outcome of the game for each move is evaluated by "rolling out" (simulating to the end) many games using a suboptimal/heuristic backgammon player (the TD-Gammon player was used for this purpose in [TeG96]), and by Monte-Carlo averaging the scores. The move that results in the best average score is selected for play.

one-step and two-step lookahead with lookahead cost function approximation provided by a neural network, resulting in a backgammon program called TD-Gammon [Tes89a], [Tes89b], [Tes92], [Tes94], [Tes95], [Tes02]. TD-Gammon was trained with the use of the TD(λ) algorithm that will be discussed in Section 5.5, and was used as the base heuristic (for each of the two players) to simulate game trajectories. The rollout algorithm also involved truncation of long game trajectories, using a terminal cost function approximation based on TD-Gammon's position evaluation. Game trajectories are of course random, since they involve the use of dice at each player's turn. Thus the scores of many trajectories have to be generated and Monte Carlo averaged to assess the probability of a win from a given position.

An important issue to consider here is that backgammon is a two-player game and not an optimal control problem that involves a single decision maker. While there is a DP theory for sequential zero-sum games, this theory has not been covered in this book. Thus how are we to interpret rollout algorithms in the context of two-player games? The answer is to treat the two players unequally: one player uses the heuristic policy exclusively (TD-Gammon in the present example). The other player takes the role of the optimizer, and tries to improve on the heuristic policy (TD-Gammon) by using rollout. Thus "policy improvement" in the context of the present example means that when playing against a TD-Gammon opponent, the rollout player achieves a better score on the average than if he/she were to play with the TD-Gammon strategy. In particular, the theory does not guarantee that a rollout player that is trained using TD-Gammon for both players will do better than TD-Gammon would against a non-TD-Gammon opponent. While this is a plausible practical hypothesis, it is one that can only be tested empirically.

Most of the currently existing computer backgammon programs descend from TD-Gammon. Rollout-based backgammon programs are the most powerful in terms of performance, consistent with the principle that a rollout algorithm performs better than its base heuristic. However, they are too time-consuming for real-time play, because of the extensive on-line simulation requirement at each move (the situation in backgammon is exacerbated by its high branching factor, i.e., for a given position, the number of possible successor positions is quite large, as compared for example with chess). They have been used in a limited diagnostic way to assess the quality of neural network-based programs (many articles and empirical works on computer backgammon are posted on-line; see e.g., http://www.bkgm.com/articles/page07.html).

Monte Carlo Tree Search

In the rollout implementation just described, we implicitly assumed that once we reach state x_k, we generate the same large number of trajectories starting from each pair (x_k, u_k), with $u_k \in U(x_k)$, to the end of the horizon. The drawback of this is threefold:

(a) The trajectories may be too long because the horizon length N is large (or infinite, in an infinite horizon context).

(b) Some of the controls u_k may be clearly inferior to others, and may not be worth as much sampling effort.

(c) Some of the controls u_k that appear to be promising, may be worth exploring better through multistep lookahead.

This has motivated lookahead variants, generally referred to as *Monte Carlo tree search* (MCTS for short), which aim to trade off computational economy with a hopefully small risk of degradation in performance. Such variants involve, among others, early discarding of controls deemed to be inferior based on the results of preliminary calculations, and simulation that is limited in scope (either because of a reduced number of simulation samples, or because of a shortened horizon of simulation, or both).

In particular, a simple remedy for (a) above is to use rollout trajectories of reasonably limited length, with some terminal cost approximation at the end (in an extreme case, the rollout may be skipped altogether for some states, i.e., rollout trajectories have zero length). The terminal cost function may be very simple (such as zero) or may be obtained through some auxiliary calculation. In fact the base policy used for rollout may be used to construct the terminal cost function approximation, as noted for the rollout-based backgammon algorithm of Example 2.4.3. In particular, an approximation to the cost function of the base policy may be obtained by training some approximation architecture, such as a neural network (see Chapter 3), and may be used as a terminal cost function.

A simple but less straightforward remedy for (b) is to use some heuristic or statistical test to discard some of the controls u_k, as soon as this is suggested by the early results of simulation. Similarly, to implement (c) one may use some heuristic to increase the length of lookahead selectively for some of the controls u_k. This is similar to the selective depth lookahead procedure for deterministic rollout that we illustrated in Fig. 2.4.6.

The MCTS approach can be based on sophisticated procedures for implementing and combining the ideas just described. The implementation is often adapted to the problem at hand, but the general idea is to use the interim results of the computation and statistical tests to focus the simulation effort along the most promising directions. Thus to implement MCTS one needs to maintain a lookahead tree, which is expanded as the relevant Q-factors are evaluated by simulation, and which balances *the competing desires of exploitation and exploration* (generate and evaluate controls that seem most promising in terms of performance versus assessing the potential of inadequately explored controls). Ideas that were developed in the context of multiarmed bandit problems have played an important role in the construction of this type of MCTS procedures (see the end-of-chapter references).

Example 2.4.4 (Statistical Tests for Adaptive Sampling with a One-Step Lookahead)

Let us consider a typical one-step lookahead selection strategy that is based on adaptive sampling. We are at a state x_k and we try to find a control $\tilde{u}_k$ that minimizes an approximate Q-factor

$$\tilde{Q}_k(x_k,u_k) = E\Big\{g_k(x_k,u_k,w_k) + \tilde{J}_{k+1}\big(f_k(x_k,u_k,w_k)\big)\Big\}$$

over $u_k \in U_k(x_k)$, with $\tilde{Q}_k(x_k,u_k)$ computed by averaging samples of the expression within braces. We assume that $U_k(x_k)$ contains m elements, which for simplicity are denoted $1,\ldots,m$. At the ℓth sampling period, knowing the outcomes of the preceding sampling periods, we select one of the m controls, say i_ℓ, and we draw a sample of $\tilde{Q}_k(x_k,i_\ell)$, whose value is denoted by S_{i_ℓ}. Thus after the nth sampling period we have an estimate $Q_{i,n}$ of the Q-factor of each control $i = 1,\ldots,m$ that has been sampled at least once, given by

$$Q_{i,n} = \frac{\sum_{\ell=1}^{n} \delta(i_\ell = i) S_{i_\ell}}{\sum_{\ell=1}^{n} \delta(i_\ell = i)},$$

where

$$\delta(i_\ell = i) = \begin{cases} 1 & \text{if } i_\ell = i, \\ 0 & \text{if } i_\ell \neq i. \end{cases}$$

Thus $Q_{i,n}$ is the *empirical mean* of the Q-factor of control i (total sample value divided by total number of samples), assuming that i has been sampled at least once.

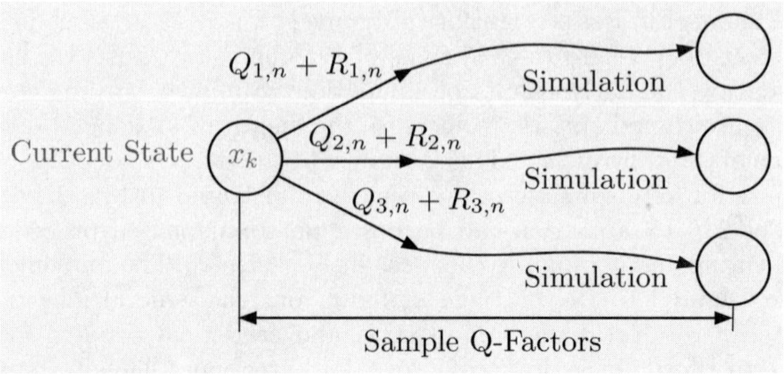

Figure 2.4.8 Illustration of one-step lookahead MCTS at a state x_k. The Q-factor sampled next corresponds to the control i with minimum sum of exploitation index (here taken to be the running average $Q_{i,n}$) and exploration index ($R_{i,n}$, possibly given by the UCB rule).

After n samples have been collected, with each control sampled at least once, we may declare the control i that minimizes $Q_{i,n}$ as the "best" one, i.e., the one that truly minimizes the Q-factor $Q_k(x_k, i)$. However, there is a positive probability that there is an error: the selected control may not minimize the true Q-factor. In adaptive sampling, roughly speaking, we want to design the sample selection strategy and the criterion to stop the sampling, in a way that keeps the probability of error small (by allocating some sampling effort to all controls), and the number of samples limited (by not wasting samples on controls i that appear inferior based on their empirical mean $Q_{i,n}$).

Intuitively, a good sampling policy will balance at time n the desires of exploitation and exploration (i.e., sample controls that seem most promising, in the sense that they have a small empirical mean $Q_{i,n}$, versus assessing the potential of inadequately explored controls, those i that have been sampled a small number of times). Thus it makes sense to sample next the control i that minimizes the sum

$$T_{i,n} + R_{i,n}$$

of two indexes: an *exploitation index* $T_{i,n}$ and an *exploration index* $R_{i,n}$. Usually the exploitation index is chosen to be the empirical mean $Q_{i,n}$. The exploration index is based on a confidence interval formula and depends on the sample count

$$s_i = \sum_{\ell=1}^{n} \delta(i_\ell = i)$$

of control i. A frequently suggested choice is the UCB rule (upper confidence bound), which sets

$$R_{i,n} = -c\sqrt{\frac{\log n}{s_i}},$$

where c is a positive constant that is selected empirically (some analysis suggests values near $c = \sqrt{2}$, assuming that $Q_{i,n}$ is normalized to take values in the range $[-1, 0]$). The UCB rule, first proposed in the paper by Auer, Cesa-Bianchi, and Fischer [ACF02], has been extensively discussed in the literature both for one-step and for multistep lookahead [where it is called UCT (UCB applied to trees; see Kocsis and Szepesvari [KoS06])].†

Its justification is based on probabilistic analyses that relate to the multiarmed bandit problem, and is beyond our scope. Alternatives to the UCT formula have been suggested, and in fact in the AlphaZero program, the exploitation term has a different form from the one above, and depends on the depth of lookahead (see Silver et al. [SHS17]).

Sampling policies for MCTS with multistep lookahead are based on similar sampling ideas to the case of one-step lookahead. A simulated trajectory is run from a node i of the lookahead tree that minimizes the sum $T_{i,n} + R_{i,n}$ of an exploitation index and an exploration index. There are several schemes of this type, but the details are beyond our scope (see the end-of-chapter references).

A major success has been the use of MCTS in two-player game contexts, such as the AlphaGo computer program (Silver et al. [SHM16]), which performs better than the best humans in the game of Go. This program integrates several of the techniques discussed in this chapter, and in Chapters 3 and 4, including MCTS and rollout using a base policy that is trained off-line using a deep neural network. We will discuss neural network training techniques in Chapter 3. The AlphaZero program, which has performed spectacularly well against humans and other programs in the games of Go and chess (Silver et al. [SHS17]), bears some similarity with AlphaGo, and critically relies on MCTS, but does not use rollout.

Randomized Policy Improvement by MCTS

We have described rollout and MCTS as schemes for policy improvement: start with a base policy, and compute an improved policy based on the results of one-step lookahead or multistep lookahead followed by simulation with the base policy. We have implicitly assumed that both the base policy and the rollout policy are deterministic in the sense that they map each state x_k into a unique control $\tilde{\mu}_k(x_k)$ [cf. Eq. (2.30)]. In some contexts, success has been achieved with *randomized policies*, which map a state x_k to a probability distribution over the set of controls $U_k(x_k)$, rather than mapping onto a single control. In particular, the AlphaGo and AlphaZero programs use MCTS to generate and use for training purposes randomized policies, which specify at each board position the probabilities with which the various moves are selected.

A randomized policy can be used as a base policy in a rollout context in exactly the same way as a deterministic policy: for a given state x_k, we

† The paper [ACF02] refers to the rule given here as UCB1 and credits its motivation to the paper by Agrawal [Agr95].

just generate sample trajectories and associated sample Q-factors, starting from each leaf-state of the lookahead tree that is rooted at x_k. We then average the corresponding Q-factor samples. However, the rollout/improved policy, as we have described it, is a deterministic policy, i.e., it applies at x_k the control $\tilde{\mu}_k(x_k)$ that is "best" according to the results of the rollout [cf. Eq. (2.30)]. Still, however, if we wish to generate an improved policy that is randomized, we can simply change the probabilities of different controls in the direction of the deterministic rollout policy. This can be done by increasing by some amount the probability of the "best" control $\tilde{\mu}_k(x_k)$ from its base policy level, while proportionally decreasing the probabilities of the other controls.

The use of MCTS provides a related method to "improve" a randomized policy. In the process of adaptive simulation that is used in MCTS, we generate *frequency counts* of the different controls, i.e., the proportion of rollout trajectories associated with each $u_k \in U_k(x_k)$. We can then obtain the rollout randomized policy by moving the probabilities of the base policy in the direction suggested by the frequency counts, i.e., increase the probability of high-count controls and reduce the probability of the others. This type of policy improvement is reminiscent of gradient-type methods, and has been successful in a number of applications; see Section 5.7 for further discussion, and the end-of-chapter references for such policy improvement implementations in AlphaGo, AlphaZero, and other application contexts.

Variance Reduction in Rollout - Comparing Advantages

When using simulation, sampling is often organized to effect *variance reduction*. By this we mean that for a given problem, the collection and use of samples is structured so that the variance of the simulation error is made smaller, with the same amount of simulation effort. There are several methods of this type for which we refer to textbooks on simulation (see, e.g., Ross [Ros12], and Rubinstein and Kroese [RuK1]).

In this section we discuss a method to reduce the effects of the simulation error in the calculation of the Q-factors in the context of rollout. The key idea is that the selection of the rollout control depends on the values of the Q-factor differences

$$\tilde{Q}_k(x_k, u_k) - \tilde{Q}_k(x_k, \hat{u}_k)$$

for all pairs of controls $(u_k, \hat{u}_k)$. These values must be computed accurately, so that the controls u_k and $\hat{u}_k$ can be accurately compared. On the other hand, the simulation/approximation errors in the computation of the individual Q-factors $\tilde{Q}_k(x_k, u_k)$ may be magnified through the preceding differencing operation.

An alternative approach is to approximate the Q-factor difference $\tilde{Q}_k(x_k, u_k) - \tilde{Q}_k(x_k, \hat{u}_k)$ by sampling the difference

$$C_k(x_k, u_k, \mathbf{w}_k) - C_k(x_k, \hat{u}_k, \mathbf{w}_k), \tag{2.31}$$

where $\mathbf{w}_k = (w_k, w_{k+1}, \ldots, w_{N-1})$ is the same disturbance sequence for the two controls u_k and $\hat{u}_k$, and

$$C_k(x_k, u_k, \mathbf{w}_k) = g_N(x_N) + g_k(x_k, u_k, w_k) + \sum_{i=k+1}^{N-1} g_i\big(x_i, \mu_i(x_i), w_i\big),$$

with $\{\mu_{k+1}, \ldots, \mu_{N-1}\}$ being the tail portion of the base policy. For this to make sense, it is necessary that the probability distribution of each disturbance w_i does not depend on x_i and u_i.

The approximation that is based on the difference (2.31), which involves a common disturbance $\mathbf{w}_k$ for u_k and $\hat{u}_k$, may be far more accurate than the one obtained by differencing samples of $C_k(x_k, u_k, \mathbf{w}_k)$ and $C_k(x_k, \hat{u}_k, \hat{\mathbf{w}}_k)$, which involve two different disturbances $\mathbf{w}_k$ and $\hat{\mathbf{w}}_k$. Indeed, by introducing the zero mean sample errors

$$D_k(x_k, u_k, \mathbf{w}_k) = C_k(x_k, u_k, \mathbf{w}_k) - \tilde{Q}_k(x_k, u_k),$$

it can be seen that the variance of the error in estimating $\tilde{Q}_k(x_k, u_k) - \tilde{Q}_k(x_k, \hat{u}_k)$ with the former method will be no larger than with the latter method if and only if

$$E_{\mathbf{w}_k, \hat{\mathbf{w}}_k}\Big\{\big|D_k(x_k, u_k, \mathbf{w}_k) - D_k(x_k, \hat{u}_k, \hat{\mathbf{w}}_k)\big|^2\Big\} \geq E_{\mathbf{w_k}}\Big\{\big|D_k(x_k, u_k, \mathbf{w}_k) - D_k(x_k, \hat{u}_k, \mathbf{w}_k)\big|^2\Big\}.$$

By expanding the quadratic forms and using the fact $E\big\{D_k(x_k, u_k, \mathbf{w}_k)\big\} = 0$, we see that this condition is equivalent to

$$E\big\{D_k(x_k, u_k, \mathbf{w}_k) D_k(x_k, \hat{u}_k, \mathbf{w}_k)\big\} \geq 0; \tag{2.32}$$

i.e., the errors $D_k(x_k, u_k, \mathbf{w}_k)$ and $D_k(x_k, \hat{u}_k, \mathbf{w}_k)$ being nonnegatively correlated. A little thought should convince the reader that this property is likely to hold in many types of problems.

Roughly speaking, the relation (2.32) holds if changes in the value of u_k (at the first stage) have little effect on the value of the error $D_k(x_k, u_k, \mathbf{w}_k)$ relative to the effect induced by the randomness of $\mathbf{w}_k$. To see this, suppose that there exists a scalar $\gamma < 1$ such that, for all x_k, u_k, and $\hat{u}_k$, there holds

$$E\Big\{\big|D_k(x_k, u_k, \mathbf{w}_k) - D_k(x_k, \hat{u}_k, \mathbf{w}_k)\big|^2\Big\} \leq \gamma E\Big\{\big|D_k(x_k, u_k, \mathbf{w}_k)\big|^2\Big\}. \tag{2.33}$$

Then we have, by using the generic relation $ab \geq a^2 - |a| \cdot |b - a|$ for two scalars a and b,

$$\begin{aligned} D_k(x_k, u_k, \mathbf{w}_k) D_k(x_k, \hat{u}_k, \mathbf{w}_k) &\geq \big|D_k(x_k, u_k, \mathbf{w}_k)\big|^2 \\ &\quad - \big|D_k(x_k, u_k, \mathbf{w}_k)\big| \cdot \big|D_k(x_k, \hat{u}_k, \mathbf{w}_k) - D_k(x_k, u_k, \mathbf{w}_k)\big|, \end{aligned}$$

from which we obtain

$$
\begin{aligned}
E\big\{D_k(x_k,u_k,\mathbf{w}_k)D_k(x_k,\hat{u}_k,\mathbf{w}_k)\big\}
&\geq E\left\{\big|D_k(x_k,u_k,\mathbf{w}_k)\big|^2\right\} \\
&\quad - E\Big\{\big|D_k(x_k,u_k,\mathbf{w}_k)\big|\cdot\big|D_k(x_k,\hat{u}_k,\mathbf{w}_k)-D_k(x_k,u_k,\mathbf{w}_k)\big|\Big\} \\
&\geq E\left\{\big|D_k(x_k,u_k,\mathbf{w}_k)\big|^2\right\} - \frac{1}{2}E\left\{\big|D_k(x_k,u_k,\mathbf{w}_k)\big|^2\right\} \\
&\quad - \frac{1}{2}E\Big\{\big|D_k(x_k,\hat{u}_k,\mathbf{w}_k)-D_k(x_k,u_k,\mathbf{w}_k)\big|^2\Big\} \\
&\geq \frac{1-\gamma}{2}E\left\{\big|D_k(x_k,u_k,\mathbf{w}_k)\big|^2\right\},
\end{aligned}
$$

where for the second inequality we use the generic relation $-|a|\cdot|b|\geq -\frac{1}{2}(a^2+b^2)$ for two scalars a and b, and for the third inequality we use Eq. (2.33).

Thus, under the assumption (2.33), the condition (2.32) holds and guarantees that by averaging cost difference samples rather than differencing (independently obtained) averages of cost samples, the simulation error variance does not increase.

Let us finally note the potential benefit of using Q-factor differences in contexts other than rollout. In particular when approximating Q-factors $Q_k(x_k,u_k)$ using parametric architectures (Section 3.4 in the next chapter), it may be important to approximate and compare instead the differences

$$A_k(x_k,u_k)=Q_k(x_k,u_k)-\min_{u_k\in U_k(x_k)}Q_k(x_k,u_k).$$

The function $A_k(x_k,u_k)$ is also known as the *advantage of the pair* (x_k,u_k), and can serve just as well as $Q_k(x_k,u_k)$ for the purpose of comparing controls, but may work better in the presence of approximation errors. This question will be discussed further in Section 3.4.

2.4.3 Rollout with an Expert

We will now consider a rollout algorithm for a general discrete deterministic optimization context, similar to the one discussed in Section 1.3.2. The distinguishing characteristic of this context is that *we do not know the cost function of the problem.* Instead we have access to a human or software "expert" who can rank any two feasible solutions, without assigning numerical values to them.

More specifically, the problem is to select a sequence $u=(u_1,\ldots,u_N)$, with each u_k belonging to a given finite set U_k, so as to minimize a function $G(u)$ subject to the constraint $u\in U_1\times\cdots\times U_N$.† We assume the following:

† For simplicity we assume that the constraints on the components of u are independent of each other, but it is also possible to deal with more general constraints on u.

(a) The function G is not known. Instead, we can use an "expert" who can compare any two feasible sequences u and $\overline{u}$, in the sense that he/she can determine whether

$$G(u) > G(\overline{u}), \qquad \text{or} \qquad G(u) \leq G(\overline{u}).$$

(b) A heuristic algorithm with the following property is available: Given any $k = 1, \ldots, N-1$, and a partial solution $(u_1, \ldots, u_k)$, it generates a complete feasible solution by concatenating the given partial solution $(u_1, \ldots, u_k)$ with a sequence $(\tilde{u}_{k+1}, \ldots, \tilde{u}_N)$. This complete feasible solution is denoted

$$S_k(u_1, \ldots, u_k) = (u_1, \ldots, u_k, \tilde{u}_{k+1}, \ldots, \tilde{u}_N).$$

It is important to note here that the deterministic rollout algorithm of Section 2.4.1 can be applied to this problem, even though the cost function G is unknown. The reason is that the rollout algorithm uses the cost function only as a means of ranking complete solutions in terms of their cost. Thus, if the ranking of any two solutions can be revealed by the expert, this is all that is needed.† In particular, the rollout algorithm of Section 2.4.1 can be described as follows:

We start with an artificial empty solution, and at the typical step, given the partial solution $(u_1, \ldots, u_k)$, $k < N$, we generate all possible one-step-extended solutions

$$(u_1, \ldots, u_k, u_{k+1}), \qquad u_{k+1} \in U_{k+1},$$

and use the expert to rank the finite set of complete solutions

$$S_{k+1}(u_1, \ldots, u_k, u_{k+1}), \qquad u_{k+1} \in U_{k+1}.$$

We then select the component u_{k+1} that is ranked best by the expert, extend the partial solution $(u_1, \ldots, u_k)$ by adding u_{k+1}, and repeat with the new partial solution $(u_1, \ldots, u_k, u_{k+1})$.

Except for the (mathematically inconsequential) use of an expert rather than a cost function, the preceding rollout algorithm can be viewed as a special case of the one of Section 2.4.1. As a result several of the rollout variants that we have discussed in that section (fortified rollout, rollout with multiple heuristics, and rollout that maintains multiple trajectories) can also be easily adapted to the context of the present section.

---

† Note that for this to be true, it is important that the problem be deterministic, and that the expert ranks solutions using some underlying (though unknown) cost function. In particular, the expert's rankings should have a transitivity property: if u is ranked better than u' and u' is ranked better than u'', then u is ranked better than u''.

Learning to Emulate the Expert

Let us now consider the case where an expert is not available but can be emulated by training with the use of data. In particular, suppose that we are given a set of control sequence pairs $(u^s, \overline{u}^s)$, $s = 1, \ldots, q$, with

$$G(u^s) > G(\overline{u}^s), \qquad s = 1, \ldots, q, \tag{2.34}$$

which we can use for training. Such a set may be obtained in a variety of ways, including querying the expert. We may then train a parametric approximation architecture such as a neural network to produce a function $\tilde{G}(u, r)$, where r is a parameter vector, and use this function in place of the unknown $G(u)$ to implement the preceding rollout algorithm.

A training method, known as *comparison training*, has been suggested for this purpose, and has been used in a variety of game contexts, including backgammon and chess by Tesauro [Tes89b], [Tes01]. Briefly, given the training set of pairs $(u^s, \overline{u}^s)$, $s = 1, \ldots, q$, which satisfy Eq. (2.34), we generate for each $(u^s, \overline{u}^s)$, two solution-cost pairs

$$(u^s, 1), \ (\overline{u}^s, -1), \qquad s = 1, \ldots, q.$$

A parametric architecture $\tilde{G}(\cdot, r)$, involving a parameter vector r, such as a neural network, is then trained by regularized regression with these data to produce an approximation $\tilde{G}(\cdot, \overline{r})$ to be used in place of $G(\cdot)$ in a rollout scheme. We refer to Chapter 3 and to the aforementioned papers by Tesauro for implementation details of the regression procedure. See also Section 3.5 on parametric approximation in policy space through the use of classification methods.

2.5 ON-LINE ROLLOUT FOR DETERMINISTIC INFINITE-SPACES PROBLEMS - OPTIMIZATION HEURISTICS

We have considered so far discrete-spaces applications of rollout, where there is a finite number of relevant Q-factors at each state x_k, which are evaluated by simulation and compared by exhaustive comparison. When the control constraint set is infinite, to implement this approach the constraint set must first be discretized, which is often inconvenient and ineffective. In this section we will discuss an alternative approach for deterministic problems that can deal with an infinite number of controls and Q-factors at x_k without discretization. The idea is to *use a base heuristic that involves a continuous optimization*, and to rely on a nonlinear programming method to solve the corresponding lookahead optimization problem.

To get a sense of the basic idea, consider the one-step lookahead rollout minimization

$$\tilde{\mu}_k(x_k) \in \arg \min_{u_k \in U_k(x_k)} \tilde{Q}_k(x_k, u_k), \tag{2.35}$$

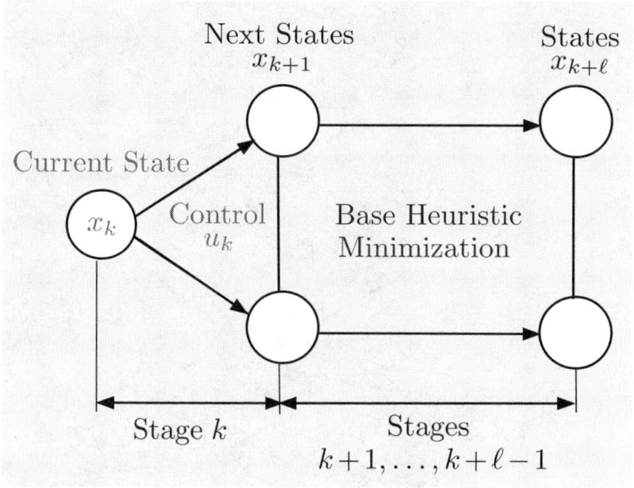

Figure 2.5.1 Schematic illustration of rollout for a deterministic problem with infinite control spaces. The base heuristic is to solve an $(\ell-1)$-stage deterministic optimal control problem, which together with the kth stage minimization over $u_k \in U_k(x_k)$, seamlessly forms an ℓ-stage continuous spaces optimal control/nonlinear programming problem that starts at state x_k.

where $\tilde{Q}_k(x_k,u_k)$ is the approximate Q-factor

$$\tilde{Q}_k(x_k,u_k) = g_k(x_k,u_k) + H_{k+1}\big(f_k(x_k,u_k)\big), \tag{2.36}$$

with $H_{k+1}(x_{k+1})$ being the cost of the base heuristic starting from state x_{k+1} [cf. Eq. (2.24)]. Suppose that we have a differentiable closed-form expression for H_{k+1}, and the functions g_k and f_k are known and are differentiable with respect to u_k. Then the Q-factor $\tilde{Q}_k(x_k,u_k)$ of Eq. (2.36) is also differentiable with respect to u_k, and its minimization (2.35) may be addressed with one of the many gradient-based methods that are available for differentiable unconstrained and constrained optimization.

The preceding approach requires that the heuristic cost $H_{k+1}(x_{k+1})$ be available in closed form, which is highly restrictive, but this difficulty can be circumvented by *using a base heuristic that is itself based on multistep optimization*. In particular, suppose that $H_{k+1}(x_{k+1})$ is the optimal cost of some $(\ell-1)$-stage deterministic optimal control problem that is related to the original problem. Then the rollout algorithm (2.35)-(2.36) can be implemented by solving the *ℓ-stage deterministic optimal control problem, which seamlessly concatenates the first stage minimization over u_k [cf. Eq. (2.35)], with the $(\ell-1)$-stage minimization of the base heuristic*; see Fig. 2.5.1. This ℓ-stage problem may be solvable on-line by standard continuous spaces nonlinear programming or optimal control methods.† An important

† Note, however, that for this to be possible, it is necessary to have a mathematical model of the system; a simulator is not sufficient.

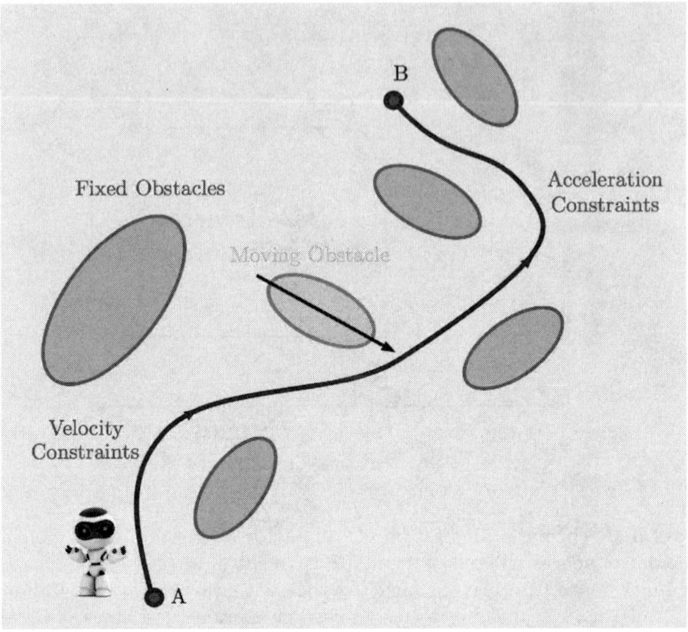

Figure 2.5.2 Illustration of constrained motion of a robot from point A to point B. There are state (position/velocity) constraints, and control (acceleration) constraints. When there are mobile obstacles, the state constraints may change unpredictably, necessitating on-line replanning.

example of methods of this type arises in control system design and is discussed next.

2.5.1 Model Predictive Control

We will consider a classical control problem, where the objective is to keep the state of a deterministic system close to the origin of the state space or close to a given trajectory. This problem has a long history, and has been addressed by a variety of methods. Starting in the late 50s and early 60s, approaches based on state variable system representations and optimal control became popular. The linear-quadratic approach, which we have illustrated by example in Section 1.3.7, was developed during this period, and is still used extensively. Unfortunately, however, linear-quadratic models are often not satisfactory. There are two main reasons for this:

(a) The system may be nonlinear, and it may be inappropriate to use for control purposes a model that is linearized around the desired point or trajectory.

(b) There may be control and/or state constraints, which are not handled adequately through quadratic penalty terms in the cost function. For example, the motion of a robot may be constrained by the presence of obstacles and hardware limitations (see Fig. 2.5.2). The solution obtained from a linear-quadratic model may not be suitable for such a problem, because quadratic penalties treat constraints "softly" and may produce trajectories that violate the constraints.

These inadequacies of the linear-quadratic model have motivated a methodology, called *model predictive control* (MPC for short), which combines elements of several ideas that we have discussed so far: multistep lookahead, rollout with infinite control spaces, and certainty equivalence. Aside from resolving the difficulty with infinitely many Q-factors at x_k, while dealing adequately with state and control constraints, MPC is well-suited for on-line replanning, like all rollout methods.

We will focus primarily on the most common form of MPC, where the system is either deterministic, or else it is stochastic, but it is replaced with a deterministic version by using typical values in place of the uncertain quantities, or state estimates in place of exact state values, similar to the certainty equivalent control approach. Moreover we will consider the case where the objective is to keep the state close to the origin (or more generally some point of interest, called the *set point*); this is called the *regulation problem*. Similar approaches have been developed for the problem of maintaining the state of a nonstationary system along a given state trajectory, and also, with appropriate modifications, to control problems involving disturbances. In particular, in some cases, the trajectory is treated like a sequence of set points, and the subsequently described algorithm is applied repeatedly.

We will consider a deterministic system

$$x_{k+1} = f_k(x_k, u_k),$$

whose state x_k and control u_k are vectors that consist of a finite number of scalar components. The cost per stage is assumed nonnegative

$$g_k(x_k, u_k) \geq 0, \qquad \text{for all } (x_k, u_k),$$

(e.g., a quadratic cost). We impose state and control constraints

$$x_k \in X_k, \qquad u_k \in U_k(x_k), \qquad k = 0, 1, \ldots.$$

We also assume that the system can be kept at the origin at zero cost, i.e.,

$$f_k(0, \overline{u}_k) = 0, \quad g_k(0, \overline{u}_k) = 0 \qquad \text{for some control } \overline{u}_k \in U_k(0).$$

For a given initial state $x_0 \in X_0$, we want to obtain a sequence $\{u_0, u_1, \ldots\}$ such that the states and controls of the system satisfy the state and control constraints with small total cost. In our subsequent development of MPC, we will assume for simplicity an infinite number of stages and no terminal cost, although our discussion can be modified to deal with finite horizon cases.

The MPC Algorithm

Let us describe the MPC algorithm for the deterministic problem just described. At the current state x_k:

(a) MPC solves an ℓ-step lookahead version of the problem, which requires that $x_{k+\ell} = 0$.

(b) If $\{\tilde{u}_k, \ldots, \tilde{u}_{k+\ell-1}\}$ is the optimal control sequence of this problem, MPC applies $\tilde{u}_k$ and discards the other controls $\tilde{u}_{k+1}, \ldots, \tilde{u}_{k+\ell-1}$.

(c) At the next stage, MPC repeats this process, once the next state x_{k+1} is revealed.

In particular, at the typical stage k and state $x_k \in X_k$, the MPC algorithm solves an ℓ-stage optimal control problem involving the same cost function and the requirement $x_{k+\ell} = 0$. This is the problem

$$\min_{u_i,\, i=k,\ldots,k+\ell-1} \sum_{i=k}^{k+\ell-1} g_i(x_i, u_i), \tag{2.37}$$

subject to the system equation constraints

$$x_{i+1} = f_i(x_i, u_i), \qquad i = k, \ldots, k+\ell-1,$$

the state and control constraints

$$x_i \in X_i, \qquad u_i \in U_i(x_i), \qquad i = k, \ldots, k+\ell-1,$$

and the terminal state constraint

$$x_{k+\ell} = 0.$$

Let $\{\tilde{u}_k, \tilde{u}_{k+1}, \ldots, \tilde{u}_{k+\ell-1}\}$ be a corresponding optimal control sequence. The MPC algorithm applies at stage k the first component $\tilde{u}_k$ of this sequence, and discards the remaining components; see Fig. 2.5.3.†

To guarantee that there is an integer ℓ such that the preceding MPC algorithm is feasible, we assume the following.

---

† In the case, where we want the system to follow a given nominal trajectory, rather than stay close to the origin, we should modify the MPC optimization to impose as a terminal constraint that the state $x_{k+\ell}$ should be a point on the nominal trajectory (instead of $x_{k+\ell} = 0$). We should also change the cost function to reflect a penalty for deviating from the given trajectory.

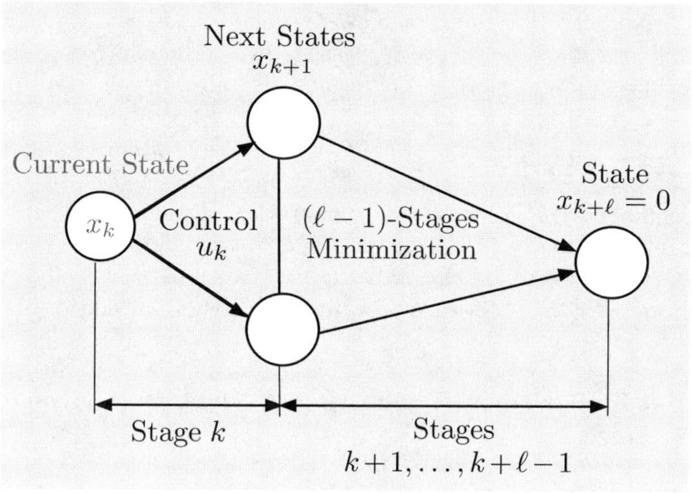

Figure 2.5.3 Illustration of the problem solved by MPC at state x_k. We minimize the cost function over the next ℓ stages while imposing the requirement that $x_{k+\ell}=0$. We then apply the first control of the optimizing sequence.

Constrained Controllability Condition

There exists an integer $\ell > 1$ such that for every initial state $x_k \in X_k$, we can find a sequence of controls $u_k,\ldots,u_{k+\ell-1}$ that drive to 0 the state $x_{k+\ell}$ of the system at time $k+\ell$, while satisfying all the intermediate state and control constraints

$$u_k \in U_k(x_k),\ x_{k+1} \in X_{k+1},\ldots,$$
$$x_{k+\ell-1} \in X_{k+\ell-1},\ u_{k+\ell-1} \in U_{k+\ell-1}(x_{k+\ell-1}).$$

Finding an integer ℓ that satisfies the constrained controllability condition is an important issue to which we will return later.† Generally the constrained controllability condition tends to be satisfied if the control constraints are not too stringent, and the state constraints do not allow a large deviation from the origin. In this case not only can we implement MPC, but also the resulting closed-loop system will tend to be stable; see the following discussion of stability, and the subsequent Example 2.5.2.

Note that the actual state trajectory produced by MPC may never reach the origin (see the subsequent Example 2.5.1). This is because we use

† In the case where we want the system to follow a given nominal trajectory, rather than stay close to the origin, we may want to use a time-dependent lookahead length ℓ_k, to exercise tighter control over critical parts of the nominal trajectory.

only the first control $\tilde{u}_k$ of the kth stage sequence $\{\tilde{u}_k, \tilde{u}_{k+1}, \ldots, \tilde{u}_{k+\ell-1}\}$, which aims at $x_{k+\ell} = 0$. At the next stage $k+1$ the control generated by MPC may be different than $\tilde{u}_{k+1}$, because it will aim one step further to the terminal condition $x_{k+\ell+1} = 0$.

To make the connection of MPC with rollout, we note that *the one-step lookahead function $\tilde{J}$ implicitly used by MPC [cf. Eq. (2.37)] is the cost-to-go function of a certain base heuristic.* This is the heuristic that drives to 0 the state after $\ell-1$ stages (*not ℓ stages*) and keeps the state at 0 thereafter, while observing the state and control constraints, and minimizing the associated $(\ell-1)$-stages cost, in the spirit of our earlier discussion; cf. Fig. 2.5.1.

Sequential Improvement Property and Stability Analysis

It turns out that the base heuristic just described is sequentially improving, so MPC has a cost improvement property, of the type discussed in Section 2.4.1. To see this, let us denote by $\hat{J}_k(x_k)$ the optimal cost of the ℓ-stage problem solved by MPC when at a state $x_k \in X_k$. Let also $H_k(x_k)$ and $H_{k+1}(x_{k+1})$ be the optimal heuristic costs of the corresponding $(\ell-1)$-stage optimization problems that start at x_k and x_{k+1}, and drive the states $x_{k+\ell-1}$ and $x_{k+\ell}$, respectively, to 0. Thus, by the principle of optimality, we have the DP equation

$$\hat{J}_k(x_k) = \min_{u_k \in U_k(x_k)} \big[g_k(x_k, u_k) + H_{k+1}\big(f_k(x_k.u_k)\big)\big].$$

Since having one less stage at our disposal to drive the state to 0 cannot decrease the optimal cost, we have

$$\hat{J}_k(x_k) \leq H_k(x_k).$$

By combining the preceding two relations, we obtain

$$\min_{u_k \in U_k(x_k)} \big[g_k(x_k, u_k) + H_{k+1}\big(f_k(x_k, u_k)\big)\big] \leq H_k(x_k), \tag{2.38}$$

which is the sequential improvement condition for the base heuristic [cf. Eq. (2.28)].†

Often the primary objective in MPC, aside from fulfilling the state and control constraints, is to obtain a stable closed-loop system, i.e., a system that naturally tends to stay close to the origin. This is typically expressed adequately by the requirement of a finite cost over an infinite number of stages:

$$\sum_{k=0}^{\infty} g_k(x_k, u_k) < \infty, \tag{2.39}$$

† Note that the base heuristic is not sequentially consistent, as it fails to satisfy the definition given in Section 2.4.1 (see the subsequent Example 2.5.1).

where $\{x_0, u_0, x_1, u_1, \ldots\}$ is the state and control sequence generated by MPC.

We will now show that the stability condition (2.39) is satisfied by the MPC algorithm. Indeed, from the sequential improvement condition (2.38), we have

$$g_k(x_k, u_k) + H_{k+1}(x_{k+1}) \leq H_k(x_k), \qquad k = 0, 1, \ldots.$$

Adding this relation for all k in a range $[0, K]$, where $K = 0, 1, \ldots$, we obtain

$$H_{K+1}(x_{K+1}) + \sum_{k=0}^{K} g_k(x_k, u_k) \leq H_0(x_0).$$

Since (in view of the nonnegativity of the cost per stage) we have

$$0 \leq H_{K+1}(x_{K+1}),$$

it follows that

$$\sum_{k=0}^{K} g_k(x_k, u_k) \leq H_0(x_0), \qquad K = 0, 1, \ldots,$$

and taking the limit as $K \to \infty$, we obtain

$$\sum_{k=0}^{\infty} g_k(x_k, u_k) \leq H_0(x_0) < \infty, \tag{2.40}$$

[$H_0(x_0)$ is finite because the transfer from x_0 to $x_\ell = 0$ is feasible by the constrained controllability condition]. We have thus verified the stability condition (2.39).

Example 2.5.1

Consider a scalar linear system and a quadratic cost

$$x_{k+1} = x_k + u_k, \qquad g_k(x_k, u_k) = x_k^2 + u_k^2,$$

where the state and control constraints are

$$x_k \in X_k = \big\{x \mid |x| \leq 1.5\big\}, \qquad u_k \in U_k(x_k) = \big\{u \mid |u| \leq 1\big\}.$$

We apply the MPC algorithm with $\ell = 2$. For this value of ℓ, the constrained controllability assumption is satisfied, since the 2-step sequence of controls

$$u_0 = -\text{sgn}(x_0), \qquad u_1 = -x_1 = -x_0 + \text{sgn}(x_0)$$

drives the state x_2 to 0, for any x_0 with $|x_0| \leq 1.5$.

At state $x_k \in X_k$, MPC minimizes the two-stage cost

$$x_k^2 + u_k^2 + (x_k + u_k)^2 + u_{k+1}^2,$$

subject to the control constraints

$$|u_k| \leq 1, \qquad |u_{k+1}| \leq 1,$$

and the state constraints

$$|x_{k+1}| \leq 1.5, \qquad x_{k+2} = x_k + u_k + u_{k+1} = 0.$$

This is a quadratic programming problem, which can be solved with available software, and in this case analytically, because of its simplicity. In particular, it can be verified that the minimization yields

$$\tilde{u}_k = -\frac{2}{3}x_k, \qquad \tilde{u}_{k+1} = -(x_k + \tilde{u}_k).$$

Thus the MPC algorithm selects $\tilde{u}_k = -\frac{2}{3}x_k$, which results in the closed-loop system

$$x_{k+1} = \frac{1}{3}x_k, \qquad k = 0, 1, \ldots.$$

Note that while this closed-loop system is stable, its state is never driven to 0 if started from $x_0 \neq 0$. Moreover, it is easily verified that the base heuristic is not sequentially consistent. For example, starting from $x_k = 1$, the base heuristic generates the sequence

$$\left\{x_k = 1, u_k = -\frac{2}{3}, x_{k+1} = \frac{1}{3}, u_{k+1} = -\frac{1}{3}, x_{k+2} = 0, u_{k+2} = 0, \ldots\right\},$$

while starting from the next state $x_{k+1} = \frac{1}{3}$ it generates the sequence

$$\left\{x_{k+1} = \frac{1}{3}, u_{k+1} = -\frac{2}{9}, x_{k+2} = \frac{1}{9}, u_{k+2} = -\frac{1}{9}, x_{k+3} = 0, u_{k+3} = 0, \ldots\right\},$$

so the sequential consistency condition of Section 2.4.1 is violated.

Regarding the choice of the horizon length ℓ for the MPC calculations, note that if the constrained controllability assumption is satisfied for some value of ℓ, it is also satisfied for all larger values of ℓ. This argues for a large value of ℓ. On the other hand, the optimal control problem solved at each stage by MPC becomes larger and hence more difficult as ℓ increases. Thus, the horizon length is usually chosen on the basis of some experimentation: first ensure that ℓ is large enough for the constrained controllability assumption to hold, and then by further experimentation to ensure overall satisfactory performance.

2.5.2 Target Tubes and the Constrained Controllability Condition

We now return to the constrained controllability condition, which asserts that the state constraint sets X_k are such that starting from anywhere within X_k, it is possible to drive to 0 the state of the system within some number of steps ℓ, while staying within X_m at each intermediate step $m = k+1, \ldots, m = k+\ell-1$. Unfortunately, this assumption masks some major complications. In particular, the control constraint set may not be sufficiently rich to compensate for natural instability tendencies of the system. As a result it may be impossible to keep the state within X_k over a sufficiently long period of time, something that may be viewed as a form of instability. Here is an example.

Example 2.5.2

Consider the scalar linear system

$$x_{k+1} = 2x_k + u_k,$$

which is unstable, and the control constraint

$$|u_k| \leq 1.$$

Then if $0 \leq x_0 < 1$, it can be seen that by using the control $u_0 = -1$, the next state satisfies,

$$x_1 = 2x_0 - 1 < x_0,$$

and is closer to 0 than the preceding state x_0. Similarly, using controls $u_k = -1$, every subsequent state x_{k+1} will get closer to 0 than x_k. Eventually, after a sufficient number of steps $\bar{k}$, with controls $u_k = -1$ for $k < \bar{k}$, the state $x_{\bar{k}}$ will satisfy

$$0 \leq x_{\bar{k}} \leq \frac{1}{2}.$$

Once this happens, the feasible control $u_{\bar{k}} = -2x_{\bar{k}}$ will drive the state $x_{\bar{k}+1}$ to 0.

Similarly, when $-1 < x_0 \leq 0$, by applying control $u_k = 1$ for a sufficiently large number of steps $\bar{k}$, the state $x_{\bar{k}}$ will be driven into the region $[-1/2, 0]$, and then the feasible control $u_{\bar{k}} = -2x_{\bar{k}}$ will drive the state $x_{\bar{k}+1}$ to 0.

Suppose now that the state constraint is of the form $X_k = [-\beta, \beta]$ for all k, and let us explore what happens for different values of β. The preceding discussion shows that if $0 < \beta < 1$ the constrained controllability assumption is satisfied, and for every initial state $x_0 \in X_0$, the states x_k can be kept within X_k and can be driven to 0 in a finite number ℓ of steps. The number ℓ depends on β, and in particular if $0 < \beta < 1/2$, ℓ can be taken equal to 1.

On the other hand, if $\beta \geq 1$, it is impossible to drive the state to 0 from every initial state $x_0 \in [1, \beta]$ without violating the constraint $|u_k| \leq 1$, so the constrained controllability assumption is violated. In fact if the initial state

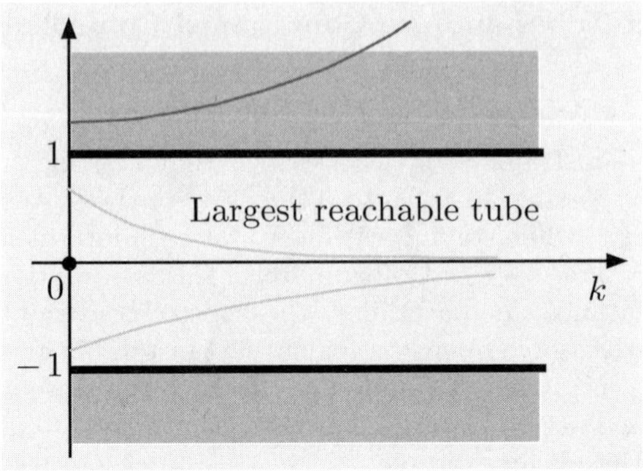

Figure 2.5.4 Schematic illustration of state trajectories under MPC for Example 2.5.2. If the initial state lies within the set $(-1, 1)$ the constrained controllability condition is satisfied for sufficiently large ℓ, and the MPC algorithm yields a stable controller. If the initial state lies outside this set, MPC cannot be implemented because the constrained controllability condition fails to holds. Moreover, the system is unstable starting from such an initial state. In this example, the largest reachable tube is $\{X, X, \ldots\}$ with $X = \big\{x \mid |x| \leq 1\big\}$.

satisfies $|x_0| > 1$, the state trajectory diverges in the sense that $|x_k| \to \infty$ for any control sequence that satisfies the constraint $|u_k| \leq 1$; see Fig. 2.5.4. Thus if $\beta \geq 1$, either a larger control constraint set or an initial condition that is close to 0, or both, are needed to satisfy the constrained controllability condition.

The critical construction in the preceding example is to identify sets of states $\overline{X}_k \subset X_k$ such that starting from within $\overline{X}_k$ we are guaranteed to stay within the "tube" $\{\overline{X}_{k+1}, \overline{X}_{k+2}, \ldots, \overline{X}_N\}$ for all subsequent times with appropriate choices of control. Tubes of this type can serve as state constraints in MPC. Moreover, when the sets $\overline{X}_k$ are bounded, *a guarantee that the state can stay within the tube amounts to a form of closed-loop stability guarantee*. In the remainder of this section, we will address the issue of how such tubes can be constructed.

Formally, a tube $\{\overline{X}_0, \overline{X}_1, \ldots, \overline{X}_N\}$ is just a sequence of subsets with $\overline{X}_k \subset X_k$ for all $k = 0, \ldots, N$. The tube is called *reachable* if it has the property that for every k and $x_k \in \overline{X}_k$ there exists a $u_k \in U_k(x_k)$ such that $f_k(x_k, u_k) \in \overline{X}_{k+1}$. A reachable tube was also called an *effective target tube* in [Ber71], and for simplicity it will be called a *target tube* in this section; the latter name is widely used in the current literature.† If the original tube

† The concept of target tubes and reachability in discrete-time systems was introduced in the author's 1971 Ph.D. thesis ([Ber71], available on-line) and associated papers [BeR71], [Ber72], [BeR73], where various methods for constructing target tubes were given, for both deterministic and minimax/game problems,

of state constraints $\{X_0, X_1, \ldots, X_N\}$ is not reachable, the constrained controllability condition cannot be satisfied. In this case, it is necessary to compute a reachable tube to use as a set of state constraints in place of the original tube $\{X_0, X_1, \ldots, X_N\}$. Thus obtaining a reachable tube is a prerequisite towards satisfying the constrained controllability assumption, and serves as the first step in the analysis and design of MPC schemes with state constraints.

Given an N-stage deterministic problem with state constraints $x_k \in X_k$, $k = 0, \ldots, N$, we can obtain a reachable tube $\{\overline{X}_0, \overline{X}_1, \ldots, \overline{X}_N\}$ by a recursive algorithm that starts with

$$\overline{X}_N = X_N,$$

and generates $\overline{X}_k$, $k = 0, \ldots, N-1$, going backwards,

$$\overline{X}_k = \bigl\{x_k \in X_k \mid \text{for some } u_k \in U_k(x_k) \text{ we have } f_k(x_k, u_k) \in \overline{X}_{k+1}\bigr\}.$$

Generally, it is difficult to compute the sets $\overline{X}_k$ of the reachable tube, but algorithms that produce inner approximations have been constructed. The author's thesis [Ber71] and subsequent papers [BeR71], [Ber72], [BeR73], [Ber07], gave inner ellipsoidal approximations for both finite and infinite horizon problems with perfect and partial state information, which involve linear systems and ellipsoidal constraints. Other authors have developed polyhedral approximations; see the textbook by Borelli, Bemporad, and Morari [BBM17].

Example 2.5.2 (continued)

In this example, if the state constraints are

$$X_k = \bigl\{x_k \mid |x_k| \le 1\bigr\}, \qquad k = 0, 1, \ldots, N, \tag{2.41}$$

the tube $\{X_k \mid k = 0, \ldots, N\}$ is reachable for any N. However, this is not true for the constraint sets

$$X_k = \bigl\{x_k \mid |x_k| \le 2\bigr\}, \qquad k = 0, 1, \ldots, N.$$

For example for $x_0 = 2$ the next state $x_1 = 2x_0 + u_k = 4 + u_k$ will not satisfy $|x_1| \le 2$ for any of the feasible controls u_k with $|u_k| \le 1$. Thus it is necessary to replace the tube of original constraints $\{X_k \mid k = 0, \ldots, N\}$ with

---

including easily characterized ellipsoids for linear systems. Reachability can be defined for finite horizon as well as infinite horizon problems. Reachability concepts were studied by several authors much later, particularly for MPC and related contexts, and have been generalized to continuous-time systems (see the end-of-chapter references).

a reachable tube $\{\overline{X}_k \mid k = 0, \ldots, N\}$. It can be verified that the largest tube that is reachable for some value of N is the one with

$$\overline{X}_k = \big\{x_k \mid |x_k| \leq 1\big\}, \qquad k = 0, 1, \ldots, N.$$

Calculating a reachable tube is relatively easy for one-dimensional problems, but becomes complicated for multidimensional problems, where approximations are required in general. On the other hand, it can be seen that in order for the MPC minimizations to be feasible for $\ell = 2$, the initial condition must satisfy $|x_0| \leq 3/4$.

There is a subtle (and relatively insignificant) difference between reachability of the tube $\{X_k \mid k = 0, \ldots, N\}$ and satisfying the constrained controllability condition. The latter implies the former, but the converse is not true, as the preceding example illustrates. In particular, if X_k is given by Eq. (2.41), the tube is reachable, but the constrained controllability assumption is not satisfied because starting at the boundary of X_k, we cannot drive the state to 0 in any number of steps, using controls u_k with $|u_k| \leq 1$. By contrast, if we remove from X_k the boundary points 1 and -1, the tube is still reachable while the constrained controllability condition is satisfied. Generally, except for boundary effects of this type, tube reachability typically implies the constrained controllability condition.

On the other hand, it is important to note that the constrained controllability condition implies just that there exists a horizon threshold $\overline{\ell} > 1$ for which the MPC scheme with $\ell \geq \overline{\ell}$ is valid in the sense that it guarantees satisfaction of the state constraints. If ℓ is below the threshold, the MPC scheme may not work, as the preceding example illustrates.

2.5.3 Variants of Model Predictive Control

The MPC scheme that we have described is just the starting point for a broad methodology with many variations, which often relate to the suboptimal control methods that we have discussed so far in this chapter. For example, in the problem solved by MPC at each stage, instead of the requirement of driving the system state to 0 in ℓ steps, one may use a large penalty for the state being nonzero after ℓ steps. Then, the preceding stability analysis goes through, as long as the terminal penalty is chosen so that the sequential improvement condition (2.38) is satisfied and Eq. (2.40) can be shown. We can view the terminal penalty as a terminal cost function approximation in the context of rollout.

In another variant, instead of aiming to drive the state to 0 after ℓ steps, one aims to reach a sufficiently small neighborhood of the origin, within which a stabilizing controller, designed by other methods, may be used.

We finally mention variants of MPC methods, which combine rollout and terminal cost function approximation, and which can deal with uncertainty and system disturbances; see the end-of-chapter references. As an illustration, let us provide an example of a scheme that combines MPC with certainty equivalent control ideas (cf. Section 2.3.2).

Example 2.5.3 (MPC for Stochastic Systems and Certainty Equivalent Approximation)

Consider the stochastic system

$$x_{k+1} = f_k(x_k, u_k, w_k),$$

with the expected cost of a policy $\pi = \{\mu_0, \dots, \mu_{N-1}\}$ starting at x_0 defined by

$$J_\pi(x_0) = E\left\{ g_N(x_N) + \sum_{k=0}^{N-1} g\big(x_k, \mu_k(x_k), w_k\big) \right\};$$

cf. the framework of Section 1.2. We assume that for all k, there are state and control constraints of the form

$$x_k \in X_k, \qquad u_k \in U_k(x_k),$$

and that the stochastic disturbances w_k take values within some known set W_k.

An important characteristic of this problem is that a policy must maintain reachability of the tube $\{X_0, X_1, \dots\}$, even under worst-case disturbance values. For this it is necessary that for each state $x_k \in X_k$, the control u_k is chosen from within the subset $\tilde{U}_k(x_k)$ given by

$$\tilde{U}_k(x_k) = \big\{u_k \in U_k(x_k) \mid f(x_k, u_k, w_k) \in X_{k+1}, \text{ for all } w_k \in W_k\big\}.$$

We assume that $\tilde{U}_k(x_k)$ is nonempty for all $x_k \in X_k$ and is somehow available. This is not automatically satisfied; similar to the deterministic case discussed earlier, the target tube $\{X_0, X_1, \dots\}$ must be properly constructed using reachability methods, the sets $U_k(x_k)$ must be sufficiently "rich" to ensure that this is possible, and the sets $\tilde{U}_k(x_k)$ must be computed.

We will now describe a rollout/MPC method that generalizes the one given earlier for deterministic problems. It satisfies the state and control constraints, and uses assumed certainty equivalence to define the base policy over $\ell - 1$ steps, where $\ell > 1$ is some integer. In particular, at a given state $x_k \in X_k$, this method first fixes the disturbances $w_{k+1}, \dots, w_{k+\ell-1}$ to some typical values. It then applies the control $\tilde{u}_k$ that minimizes over $u_k \in \tilde{U}_k(x_k)$ the Q-factor

$$\tilde{Q}_k(x_k, u_k) = E\Big\{g(x_k, u_k, w_k) + H_{k+1}\big(f(x_k, u_k, w_k)\big)\Big\}, \tag{2.42}$$

where $H_{k+1}(x_{k+1})$ is the optimal cost of the deterministic transfer from x_{k+1} to 0 in $\ell - 1$ steps with controls $\tilde{u}_m$ from the sets $\tilde{U}_m(x_m)$, $m = k+1, \dots, k+\ell-1$, and with the disturbances fixed at their typical values. Here we require a constrained controllability condition that guarantees that this transfer is possible.

Note that the minimization over $u_k \in \tilde{U}_k(x_k)$ of the Q-factor (2.42) can be implemented by optimizing over sequences $\{u_k, u_{k+1}, \ldots, u_{k+\ell-1}\}$ an ℓ-stage trajectory of a deterministic system. This is the problem that seamlessly concatenates the first stage minimization over u_k [cf. Eq. (2.42)] with the $(\ell-1)$-stage minimization of the base heuristic. Consistent with the general rollout approach of this section, it may be possible to address this problem with gradient-based optimization methods.

A drawback of the MPC method of the preceding example is that it may not be well suited for on-line implementation because of the substantial amount of computation required at each state x_k. An alternative is to introduce *approximation in policy space on top of approximation in value space*; cf. Section 2.1.5. To this end, we can generate a large number of sample states x_k^s, $s = 1, \ldots, q$, and calculate the corresponding controls u_k^s using the Q-factor minimization

$$u_k^s \in \arg \min_{u_k \in \tilde{U}(x_k^s)} E\Big\{ g(x_k^s, u_k, w_k) + H_{k+1}\big(f(x_k^s, u_k, w_k)\big)\Big\},$$

[cf. Eq. (2.42)]. We can then use the pairs (x_k^s, u_k^s), $s = 1, \ldots, q$, and some form of regression to train a Q-factor parametric architecture $\tilde{Q}_k(x_k, u_k, \bar{r}_k)$ such as a neural network [cf. the approximation in policy space approach of Eq. (2.5)]. Once this is done, the MPC controls can be generated on-line using the minimization

$$\tilde{\mu}_k(x_k) \in \arg \min_{u_k \in \tilde{U}_k(x_k)} \tilde{Q}_k(x_k, u_k, \bar{r}_k);$$

cf. Eq. (2.6). This type of approximation in policy space approach may be applied more broadly in MPC methods where the on-line computational requirements are excessive.

2.6 NOTES AND SOURCES

When considering approximations within a large and challenging problem setting, a primary question is what to approximate. In this chapter, and indeed in the entire book, we argue that *in RL the key objects to approximate are values and policies*, leading to a broad division between approximation in value space and approximation in policy space approaches.

The structure of approximation in value space involves limited lookahead minimization, and is depicted in the key Figs. 2.1.1 and 2.2.1. There we have argued that there are three candidate areas for approximation to consider:

(a) *Cost-to-go function approximation*, which defines the approximate Q-factors at a given state.

(b) *Simplification of expected values* involved in the calculation of the approximate Q-factors.

(c) *Simplification of the Q-factor minimization* over all admissible controls.

There are several candidate approaches for each of these three approximations. Indeed most of our subsequent account revolves around the development of the relevant algorithmic machinery and the insight needed to bring to bear the right combination of approximations on a given problem. Significantly, *the three approximations are largely decoupled from each other*, allowing a broad spectrum of mixtures of algorithmic choices.

One more choice in approximation in value space is important, namely *whether to use one-step or multistep lookahead minimization* (also whether to try MCTS or not). A simple practical guideline is to use the longest lookahead that the computational budget restrictions will allow. The reason is that, except for unusual circumstances, longer lookahead will typically yield at least some improvement in performance. Still, however, it may be important to use selectively long lookahead; this may be used to properly allocate computational resources during on-line lookahead optimization. Some insight about this issue may be drawn from the game of chess, where the use of long lookahead is very beneficial; generally, it is widely recognized that chess positions that are dynamic and involve principal lines of play call for longer lookahead. This is consistent with MCTS and suggests to extend the lookahead at states that seem promising (are likely to occur under an optimal policy, based on the interim results of computation), and also involve critical decisions (committing decisions whose consequences are "significant" and/or "irreversible"). Thus, we may try selectively to extend the lookahead and/or improve the quality of the cost function approximation at such states, assuming of course that we can recognize them in some way (which will be likely problem-dependent).

Approximation in policy space is based on somewhat different algorithmic ideas than approximation in value space: optimization within a restricted class of parametrized policies. It requires a methodological viewpoint that is less connected with DP, and it will receive less attention in this book. Still, approximation in policy space is an important approach that may be favored in circumstances to be explained in Section 5.7. Moreover it may be combined with approximation in value space, as we have already noted in Section 2.1.5, and we will discuss further in Section 5.7.

In Section 1.5 we provided a list of many exact DP and approximate DP/RL textbooks and survey papers. In what follows in this section, and in other similar end-of-chapter sections, we will aim at a more targeted if inevitably incomplete set of citations to specific research topics.

Sections 2.1, 2.2: Approximation in value space has been considered in an ad hoc manner since the early days of DP, motivated by the curse of dimensionality. The idea was reframed and coupled with model-free simulation methods that originated in the 1980s in artificial intelligence.

Section 2.3: The problem approximation approach has a long history in optimal control and operations research. The author's book [Ber17], Section 6.2.1, provides some examples of enforced decomposition in flexible manufacturing and multiarmed bandit problems; see also the thesis by Kimemia [Kim82], and the papers by Kimemia, Gershwin, and Bertsekas [KGB82], and Whittle [Whi88]. The author's paper [Ber07] describes a few additional schemes based on constraint relaxation.

Section 2.4: The main idea of rollout algorithms, obtaining an improved policy starting from some other suboptimal policy, has appeared in several DP application contexts. The name "rollout" was coined by Tesauro in specific reference to rolling the dice in the game of backgammon [TeG96]. In Tesauro's proposal, a given backgammon position is evaluated by "rolling out" many games starting from that position, using a simulator, and the results are averaged to provide a "score" for the position; see Example 2.4.3. The use of the name "rollout" has gradually expanded beyond its original context; for example the samples collected through trajectory simulation are referred to as "rollouts" by some authors. In this book, we will adopt the original intended meaning: policy improvement starting from a base policy, which is evaluated with some form of Monte Carlo simulation.

The application of rollout algorithms to discrete deterministic optimization problems, the notions of sequential consistency, sequential improvement, fortified rollout, and the use of multiple heuristics (also called "parallel rollout") were first given in the paper by Bertsekas, Tsitsiklis, and Wu [BTW97], and also in the books by Bertsekas and Tsitsiklis [BeT96], and Bertsekas [Ber98]. Rollout algorithms for stochastic problems were further formalized in the papers by Bertsekas [Ber97b], [Ber05a], [Ber05b], and Bertsekas and Castanon [BeC99]. A survey of rollout in discrete optimization is given by the author in [Ber13a]. Variants of rollout algorithms for the case where the control u consists of m independently constrained components, $u = (u^1, \ldots, u^m)$, as is common in multiagent systems, can be implemented far more economically than unstructured problems of comparable size, as shown in the author's paper [Ber19c].

The expert-based rollout scheme of Section 2.4.3 seems new in the form given here, but is inspired by the method of *comparison training* proposed by Tesauro [Tes89a], [Tes89b], [Tes01], and subsequently used by several other authors (for some recent references, see [DNW16], [TCW19]). This is a general method for training an approximation architecture to choose between two alternatives, using a dataset of expert choices in place of an explicit cost function.

There have been many works and applications relating to rollout algorithms. See Secomandi [Sec00], [Sec01], [Sec03], Ferris and Voelker [FeV02], [FeV04], McGovern, Moss, and Barto [MMB02], Savagaonkar, Givan, and Chong [SGC02], Bertsimas and Popescu [BeP03], Guerriero and Mancini [GuM03], Tu and Pattipati [TuP03], Wu, Chong, and Givan [WCG03], Chang, Givan, and Chong [CGC04], Meloni, Pacciarelli, and Pranzo [MPP04], Yan, Diaconis, Rusmevichientong, and Van Roy [YDR04], Besse and Chaib-draa [BeC08], Sun et al. [SZL08], Bertazzi et al. [BBG13], Sun et al. [SLJ13], Tesauro et al. [TGL13], Antunes and Heemels [AnH14], Beyme and Leung [BeL14], Goodson, Thomas, and Ohlmann [GTO15], Khashooei, Antunes, and Heemels [KAH15], Li and Womer [LiW15], Mastin and Jaillet [MaJ15], Huang, Jia, and Guan [HJG16], Simroth, Holfeld, and Brunsch [SHB15], Lan, Guan, and Wu [LGW16], Ulmer [Ulm17], Bertazzi and Secomandi [BeS18], Guerriero, Di Puglia, and Macrina [GDP18], Sarkale et al. [SNC18], Ulmer at al. [UGM18], Chu, Xu, and Li [CXL19]. These works discuss variants and problem-specific adaptations of rollout algorithms for a broad variety of practical problems, and consistently report positive computational experience.

The idea of rollout that uses limited lookahead, adaptive pruning of the lookahead tree, and rollout truncation with cost function approximation at the end of the rollout was suggested by Tesauro and Galperin [TeG96] in the context of backgammon. Related ideas appeared earlier in the paper by Abramson [Abr90], in a game playing context.

The paper and the 2007 1st edition of the monograph by Chang, Hu, Fu, and Marcus [CFH05], [CFH13] proposed and analyzed adaptive sampling in connection with DP, and early forms of Monte Carlo tree search, including statistical tests to control the sampling process. The name "Monte Carlo tree search" (Section 2.4.2) has become popular, and in its current use, it encompasses a broad range of methods that involve adaptive sampling, rollout, extensions to sequential games, and the use of statistical tests. We refer to the papers by Coulom [Cou06], the survey by Browne et al. [BPW12], and the discussions by Chang et al. [CFH16] and Fu [Fu17]. Statistical tests for adaptive sampling has been inspired by works on multiarmed bandit problems; see Lai and Robbins [LaR85], Agrawal [Agr95], Burnetas and Katehakis [BuK97], Meuleau and Bourgine [MeB99], Auer, Cesa-Bianchi, and Fischer [ACF02], Peret and Garcia [PeG04], Kocsis and Szepesvari [KoS06], Dimitrakakis and Lagoudakis [DiL08], Audibert, Munos, and Szepesvari [AMS09], and the monograph by Munos [Mun14]. The technique for variance reduction in the calculation of Q-factor differences (Section 2.4.2) was given in the author's paper [Ber97b].

Rollout can be applied to problems with partial state information, reformulated in belief space. The computational requirements increase substantially over the perfect state information case, but not prohibitively so. We discuss some of the possibilities in Section 5.7.3, in the context of infinite horizon problems and policy iteration, including the use of rollout truncation, cost function approximation, and parallelization.

Section 2.5: The MPC approach is popular in a variety of control system design contexts, and particularly in chemical process control and robotics, where meeting explicit control and state constraints is an important practical issue. However, it requires a mathematical model and it is better suited for deterministic than for stochastic problems.

The literature on MPC is voluminous. For a survey, which gives many of the early references, see Morari and Lee [MoL99], and for a more recent survey, see Mayne [May14]. For related textbooks, see Maciejowski [Mac02], Camacho and Bordons [CaB04], Kouvaritakis and Cannon [KoC15], and Borelli, Bemporad, and Morari [BBM17].

The view of MPC as a rollout algorithm, first suggested in the author's review paper [Ber05a], provides a suboptimal control perspective and a connection with the RL methodology. The stability analysis given here is based on the work of Keerthi and Gilbert [KeG88]. The recent paper by Krener [Kre19] discusses methods to estimate the optimal cost function for use as terminal cost function approximation, aiming to achieve stabilization with MPC lookahead that is as small as possible. Generally, any problem approximation method that yields a "good" estimate of the optimal cost-to-go function is a promising candidate for terminal cost function approximation in the context of MPC and rollout. Proving stability of the overall scheme, however, depends on whether this cost function approximation satisfies a sequential improvement condition (sometimes also referred to as a "Lyapunov condition" in the control literature).

In our account of MPC, we have restricted ourselves to deterministic problems possibly involving tight state constraints as well as control constraints. Problems with stochastic uncertainty and state constraints are more challenging because of the difficulty of guaranteeing that the constraints are satisfied; see the survey by Mayne [May14] for a review of various approaches that have been used in this context. The textbook [Ber17], Section 6.4, describes MPC for problems with set membership uncertainty and state constraints, using target tube/reachability concepts, which originated in the author's PhD thesis and related papers [Ber71], [Ber72], [BeR71], [BeR73]. Target tubes were also used subsequently in MPC and other contexts by several authors; see the surveys by Blanchini [Bla99] and Mayne [May14]. Reachability for continuous-time games has been studied by Mitchell, Bayen, and Tomlin [MBT05]. For an alternative recent approach to reachability, which is based on the notion of Conditional Value-at-Risk (CVaR), see Chapman et al. [CLT19].

Problems with partial state information pose considerable challenges for MPC. One possibility for problems with a continuous control space is to combine MPC with certainty equivalence, whereby states are replaced with estimates, as discussed in Section 2.3.2. For POMDP with finite state and control spaces, MPC cannot be applied, at least in the form given here. However, rollout is generally applicable and may provide a viable option for suboptimal control. We discuss this approach in Section 5.7.3.

3

Parametric Approximation

Contents

Clearly, for the success of approximation in value space, it is important to select a class of lookahead functions $\tilde{J}_k$ that is suitable for the problem at hand. In the preceding chapter we discussed several methods for choosing $\tilde{J}_k$ based mostly on problem approximation and rollout. In this chapter we discuss an alternative approach, whereby $\tilde{J}_k$ is chosen to be a member of a parametric class of functions, including neural networks, with the parameters "optimized" or "trained" by using some algorithm. The training methods used for parametric approximation in value space can also be used for approximation in policy space, as we will discuss in Section 3.5.

3.1 APPROXIMATION ARCHITECTURES

The starting point for the schemes of this chapter is a class of functions $\tilde{J}_k(x_k, r_k)$ that for each k, depend on the current state x_k and a vector $r_k = (r_{1,k}, \ldots, r_{m_k,k})$ of m_k "tunable" scalar parameters, also called *weights*. By adjusting the weights, one can change the "shape" of $\tilde{J}_k$ so that it is a reasonably good approximation to the true optimal cost-to-go function J_k^*. The class of functions $\tilde{J}_k(x_k, r_k)$ is called an *approximation architecture*, and the process of choosing the parameter vectors r_k is commonly called *training* or *tuning* the architecture.

The simplest training approach is to do some form of semi-exhaustive or semi-random search in the space of parameter vectors and adopt the parameters that result in best performance of the associated one-step lookahead controller (according to some criterion). More systematic approaches are based on numerical optimization, such as for example a least squares fit that aims to match the cost approximation produced by the architecture to a "training set," i.e., a large number of pairs of state and cost values that are obtained through some form of sampling process. Throughout this chapter we will focus on this latter approach.

3.1.1 Linear and Nonlinear Feature-Based Architectures

There is a large variety of approximation architectures, based for example on polynomials, wavelets, radial basis functions, discretization/interpolation schemes, neural networks, and others. A particularly interesting type of cost approximation involves *feature extraction*, a process that maps the state x_k into some vector $\phi_k(x_k)$, called the *feature vector* associated with x_k at time k. The vector $\phi_k(x_k)$ consists of scalar components

$$\phi_{1,k}(x_k), \ldots, \phi_{m_k,k}(x_k),$$

called *features*. A feature-based cost approximation has the form

$$\tilde{J}_k(x_k, r_k) = \hat{J}_k\big(\phi_k(x_k), r_k\big),$$

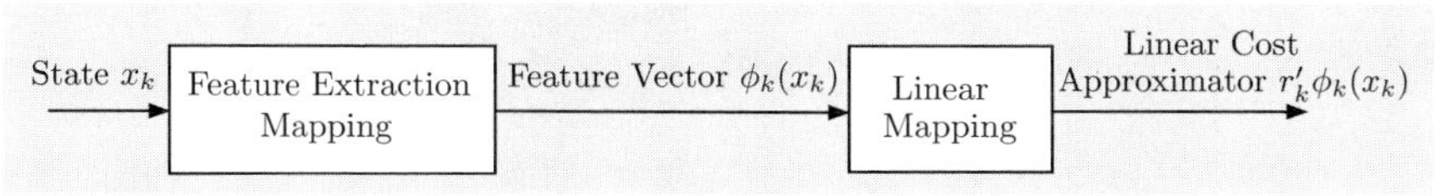

Figure 3.1.1 The structure of a linear feature-based architecture. We use a feature extraction mapping to generate an input $\phi_k(x_k)$ to a linear mapping defined by a weight vector r_k.

where r_k is a parameter vector and $\hat{J}_k$ is some function. Thus, the cost approximation depends on the state x_k through its feature vector $\phi_k(x_k)$.

Note that we are allowing for different features $\phi_k(x_k)$ and different parameter vectors r_k for each stage k. This is necessary for nonstationary problems (e.g., if the state space changes over time), and also to capture the effect of proximity to the end of the horizon. On the other hand, for stationary problems with a long or infinite horizon, where the state space does not change with k, it is common to use the same features and parameters for all stages. The subsequent discussion can easily be adapted to infinite horizon methods, as we will discuss later.

Features are often handcrafted, based on whatever human intelligence, insight, or experience is available, and are meant to capture the most important characteristics of the current state. There are also systematic ways to construct features, including the use of data and neural networks, which we will discuss shortly. In this section, we provide a brief and selective discussion of architectures, and we refer to the specialized literature (e.g., Bertsekas and Tsitsiklis [BeT96], Bishop [Bis95], Haykin [Hay08], Sutton and Barto [SuB18]), and the author's [Ber12], Section 6.1.1, for more detailed presentations.

One idea behind using features is that the optimal cost-to-go functions J_k^* may be complicated nonlinear mappings, so it is sensible to try to break their complexity into smaller, less complex pieces. In particular, if the features encode much of the nonlinearity of J_k^*, we may be able to use a relatively simple architecture $\hat{J}_k$ to approximate J_k^*. For example, with a well-chosen feature vector $\phi_k(x_k)$, a good approximation to the cost-to-go is often provided by *linearly* weighting the features, i.e.,

$$\tilde{J}_k(x_k, r_k) = \hat{J}_k\big(\phi_k(x_k), r_k\big) = \sum_{\ell=1}^{m_k} r_{\ell,k}\phi_{\ell,k}(x_k) = r_k'\phi_k(x_k), \tag{3.1}$$

where $r_{\ell,k}$ and $\phi_{\ell,k}(x_k)$ are the ℓth components of r_k and $\phi_k(x_k)$, respectively, and $r_k'\phi_k(x_k)$ denotes the inner product of r_k and $\phi_k(x_k)$, viewed as column vectors of $\Re^{m_k}$ (a prime denotes transposition, so r_k' is a row vector); see Fig. 3.1.1.

This is called a *linear feature-based architecture*, and the scalar parameters $r_{\ell,k}$ are also called *weights*. Among other advantages, these architectures admit simpler training algorithms than their nonlinear counterparts.

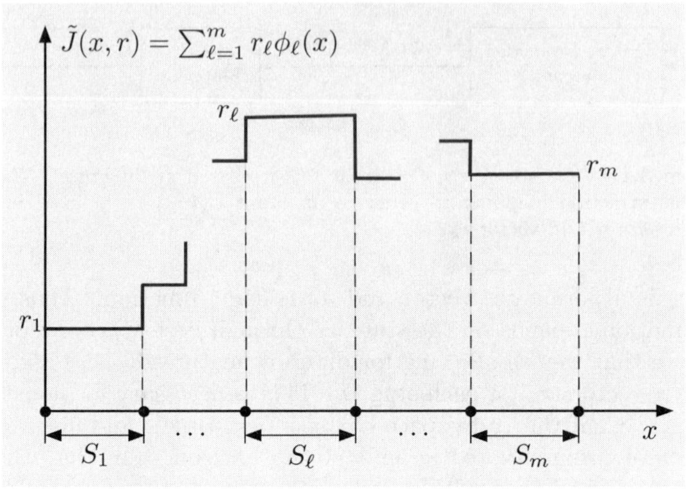

Figure 3.1.2 Illustration of a piecewise constant architecture. The state space is partitioned into subsets $S_1,\ldots,S_m$, with each subset S_ℓ defining the feature

$$\phi_\ell(x)=\begin{cases}1 & \text{if } x\in S_\ell,\\ 0 & \text{if } x\notin S_\ell,\end{cases}$$

with its own weight r_ℓ.

Mathematically, the approximating function $\tilde J_k(x_k,r_k)$ can be viewed as a member of the subspace spanned by the features $\phi_{\ell,k}(x_k)$, $\ell=1,\ldots,m_k$, which for this reason are also referred to *basis functions*. We provide a few examples, where for simplicity we drop the index k.

Example 3.1.1 (Piecewise Constant Approximation)

Suppose that the state space is partitioned into subsets $S_1,\ldots,S_m$, so that every state belongs to one and only one subset. Let the ℓth feature be defined by membership to the set S_ℓ, i.e.,

$$\phi_\ell(x)=\begin{cases}1 & \text{if } x\in S_\ell,\\ 0 & \text{if } x\notin S_\ell.\end{cases}$$

Consider the architecture

$$\tilde J(x,r)=\sum_{\ell=1}^m r_\ell\phi_\ell(x),$$

where r is the vector consists of the m scalar parameters $r_1,\ldots,r_m$. It can be seen that $\tilde J(x,r)$ is the piecewise constant function that has value r_ℓ for all states within the set S_ℓ; see Fig. 3.1.2.

The piecewise constant approximation is an example of a linear feature-based architecture that involves exclusively *local features*. These are features that take a nonzero value only for a relatively small subset of states. Thus a change of a single weight causes a change of the value of $\tilde{J}(x,r)$ for relatively few states x. At the opposite end we have linear feature-based architectures that involve *global features*. These are features that take nonzero values for a large number of states. The following is an example.

Example 3.1.2 (Polynomial Approximation)

An important case of linear architecture is one that uses polynomial basis functions. Suppose that the state consists of n components $x^1,\ldots,x^n$, each taking values within some range of integers. For example, in a queueing system, x^i may represent the number of customers in the ith queue, where $i = 1,\ldots,n$. Suppose that we want to use an approximating function that is quadratic in the components x^i. Then we can define a total of $1+n+n^2$ basis functions that depend on the state $x = (x^1,\ldots,x^n)$ via

$$\phi_0(x) = 1, \qquad \phi_i(x) = x^i, \qquad \phi_{ij}(x) = x^i x^j, \quad i,j = 1,\ldots,n.$$

A linear approximation architecture that uses these functions is given by

$$\tilde{J}(x,r) = r_0 + \sum_{i=1}^{n} r_i x^i + \sum_{i=1}^{n}\sum_{j=1}^{n} r_{ij} x^i x^j,$$

where the parameter vector r has components r_0, r_i, and r_{ij}, with $i,j = 1,\ldots,n$. Indeed, any kind of approximating function that is polynomial in the components $x^1,\ldots,x^n$ can be constructed similarly.

A more general polynomial approximation may be based on some other known features of the state. For example, we may start with a feature vector

$$\phi(x) = \big(\phi_1(x),\ldots,\phi_m(x)\big)',$$

and transform it with a quadratic polynomial mapping. In this way we obtain approximating functions of the form

$$\tilde{J}(x,r) = r_0 + \sum_{i=1}^{m} r_i \phi_i(x) + \sum_{i=1}^{m}\sum_{j=1}^{m} r_{ij}\phi_i(x)\phi_j(x),$$

where the parameter r has components r_0, r_i, and r_{ij}, with $i,j = 1,\ldots,m$. This can also be viewed as a linear architecture that uses the basis functions

$$w_0(x) = 1, \qquad w_i(x) = \phi_i(x), \qquad w_{ij}(x) = \phi_i(x)\phi_j(x), \quad i,j = 1,\ldots,m.$$

The preceding example architectures are generic in the sense that they can be applied to many different types of problems. Other architectures rely on problem-specific insight to construct features, which are then combined into a relatively simple architecture. The following are two examples involving games.

Example 3.1.3 (Tetris)

Let us revisit the game of tetris, which we discussed in Example 1.3.4. We can model the problem of finding an optimal playing strategy as a finite horizon problem with a very large horizon.

In Example 1.3.4 we viewed as state the pair of the board position x and the shape of the current falling block y. We viewed as control, the horizontal positioning and rotation applied to the falling block. However, the DP algorithm can be executed over the space of x only, since y is an uncontrollable state component; cf. Section 1.3.5. The optimal cost-to-go function is a vector of huge dimension (there are 2^{200} board positions in a "standard" tetris board of width 10 and height 20). However, it has been successfully approximated in practice by low-dimensional linear architectures.

In particular, the following features have been proposed in [BeI96]: the heights of the columns, the height differentials of adjacent columns, the wall height (the maximum column height), the number of holes of the board, and the constant 1 (the unit is often included as a feature in cost approximation architectures, as it allows for a constant shift in the approximating function). These features are readily recognized by tetris players as capturing important aspects of the board position.† There are a total of 22 features for a "standard" board with 10 columns. Of course the $2^{200} \times 22$ matrix of feature values cannot be stored in a computer, but for any board position, the corresponding row of features can be easily generated, and this is sufficient for implementation of the associated approximate DP algorithms. For recent works involving approximate DP methods and the preceding 22 features, see [Sch13], [GGS13], and [SGG15], which reference several other related papers.

Example 3.1.4 (Computer Chess)

Computer chess programs that involve feature-based architectures have been available for many years, and are still used widely (they have been upstaged in the mid-2010s by alternative types of chess programs, which use neural network techniques that will be discussed later). These programs are based on approximate DP for minimax problems,‡ a feature-based parametric architecture, and multistep lookahead. For the most part, however, the computer chess training methodology has been qualitatively different from the parametric approximation methods that we consider in this book.

In particular, with few exceptions, the training of chess architectures has been done with ad hoc hand-tuning techniques (as opposed to some form of optimization). Moreover, the features have traditionally been hand-crafted

---

† The use of feature-based approximate DP methods for the game of tetris was first suggested in the paper [TsV96], which introduced just two features (in addition to the constant 1): the wall height and the number of holes of the board. Most studies have used the set of features of [BeI96] described here, but other sets of features have also been used; see [ThS09] and the discussion in [GGS13].

‡ We have not discussed DP for minimax problems and two-player games, but the ideas of approximation in value space apply to these contexts as well; see [Ber17] for a discussion that is focused on computer chess.

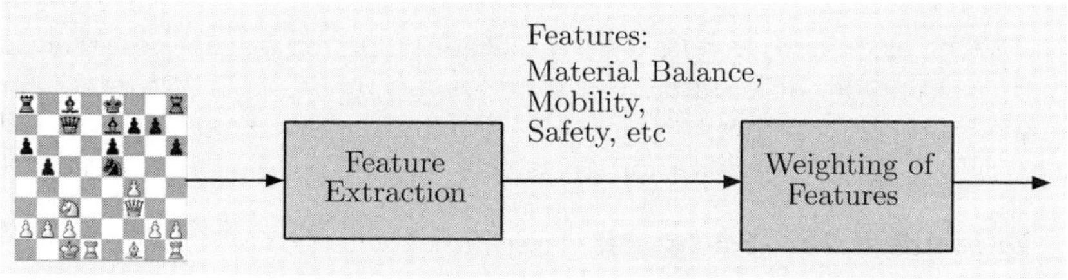

Figure 3.1.3 A feature-based architecture for computer chess.

based on chess-specific knowledge (as opposed to automatically generated through a neural network or some other method). Indeed, it has been argued for a long time that the success of chess programs in outperforming the best humans can be more properly attributed to their use of long lookahead and the brute-force calculating power of modern computers, than to algorithmic approaches that can learn powerful playing strategies, which humans have difficulty conceiving or executing. Thus computer chess was viewed as an approximate DP context where a lot of computational power is available, but innovative and sophisticated algorithms are hard to apply. However, this assessment has changed radically following the impressive success of the AlphaZero chess program (Silver et al. [SHS17]).

The fundamental principles on which all computer chess programs are based were laid out by Shannon [Sha50]. He proposed limited lookahead and evaluation of the end positions by means of a "scoring function" (in our terminology this plays the role of a cost function approximation). This function may involve, for example, the calculation of a numerical value for each of a set of major features of a position that chess players easily recognize (such as material balance, mobility, pawn structure, and other positional factors), together with a method to combine these numerical values into a single score. Shannon then went on to describe various strategies of exhaustive and selective search over a multistep lookahead tree of moves.

We may view the scoring function as a feature-based architecture for evaluating a chess position/state (cf. Fig. 3.1.3). In most computer chess programs, the features are weighted linearly, i.e., the architecture $\tilde J(x,r)$ that is used for limited lookahead is linear [cf. Eq. (3.1)]. In many cases, the weights are determined manually, by trial and error based on experience. However, in some programs, the weights are determined with supervised learning techniques that use examples of grandmaster play, i.e., by adjustment to bring the play of the program as close as possible to the play of chess grandmasters. This is a technique that applies more broadly in artificial intelligence; see Tesauro [Tes89b], [Tes01].

In a recent computer chess breakthrough, the entire idea of extracting features of a position through human expertise was abandoned in favor of feature discovery through self-play and the use of neural networks. The first program of this type to attain supremacy over humans, as well as over the best computer programs that use human expertise-based features, was AlphaZero (Silver et al. [SHS17]). This program is based on DP principles of policy iteration and Monte Carlo tree search. It will be discussed further later.

Our next example relates to a methodology for feature construction, where the number of features may increase as more data is collected. For a simple example, consider the piecewise constant approximation of Example 3.1.1, where more pieces are progressively added based on new data.

Example 3.1.5 (Feature Extraction from Data)

We have viewed so far feature vectors $\phi(x)$ as functions of x, obtained through some unspecified process that is based on prior knowledge about the cost function being approximated. On the other hand, features may also be extracted from data. For example suppose that with some preliminary calculation using data, we have identified some suitable states $x(\ell)$, $\ell = 1, \ldots, m$, that can serve as "anchors" for the construction of Gaussian basis functions of the form

$$\phi_\ell(x) = e^{-\frac{\|x - x(\ell)\|^2}{2\sigma^2}}, \qquad \ell = 1, \ldots, m, \tag{3.2}$$

where σ is a scalar "variance" parameter. This type of function is known as a *radial basis function*. It is concentrated around the state $x(\ell)$, and it is weighed with a scalar weight r_ℓ to form a parametric linear feature-based architecture, which can be trained using additional data. Several other types of data-dependent basis functions, such as support vector machines, are used in machine learning, where they are often referred to as *kernels*.

While it is possible to use a preliminary calculation to obtain the anchors $x(\ell)$ in Eq. (3.2), and then use additional data for training, one may also consider enrichment of the set of basis functions simultaneously with training. In this case the number of the basis functions increases as the training data is collected. A motivation here is that the quality of the approximation may increase with additional basis functions. This idea underlies a field of machine learning, known as *kernel methods* or sometimes *nonparametric methods*.

A further discussion is outside our scope. We refer to the literature; see e.g., books such as Cristianini and Shawe-Tylor [ChS00], [ShC04], Scholkopf and Smola [ScS02], Bishop [Bis06], Kung [Kun14], surveys such as Hofmann, Scholkopf, and Smola [HSS08], Pillonetto et al. [PDC14], and RL-related discussions such as Dietterich and Wang [DiW02], Ormoneit and Sen [OrS02], Engel, Mannor, and Meir [EMM05], Jung and Polani [JuP07], Reisinger, Stone, and Miikkulainen [RSM08], Busoniu et al. [BBS10], Bethke [Bet10]. In what follows, we will focus on parametric architectures with a fixed and given feature vector.

The next example considers a feature extraction strategy that is particularly relevant to problems of partial state information.

Example 3.1.6 (Feature Extraction from Sufficient Statistics)

The concept of a sufficient statistic, which originated in inference methodologies, plays an important role in DP. As discussed in Section 1.3, it refers to quantities that summarize all the essential content of the state x_k for optimal control selection at time k.

In particular, consider a partial information context where at time k we have accumulated an *information* record (also called the *past history*)

$$I_k = (z_0, \dots, z_k, u_0, \dots, u_{k-1}),$$

which consists of the past controls $u_0, \dots, u_{k-1}$ and the state-related measurements $z_0, \dots, z_k$ obtained at the times $0, \dots, k$. The control u_k is allowed to depend only on I_k, and the optimal policy is a sequence of the form $\{\mu_0^*(I_0), \dots, \mu_{N-1}^*(I_{N-1})\}$. We say that a function $S_k(I_k)$ is a *sufficient statistic at time* k if the control function μ_k^* depends on I_k only through $S_k(I_k)$, i.e., for some function $\hat{\mu}_k$, we have

$$\mu_k^*(I_k) = \hat{\mu}_k\big(S_k(I_k)\big),$$

where μ_k^* is optimal.

There are several examples of sufficient statistics, and they are typically problem-dependent. A trivial possibility is to view I_k itself as a sufficient statistic, and a more sophisticated possibility is to view the *belief state* b_k as a sufficient statistic (this is the conditional probability distribution of x_k given I_k; cf. Section 1.3.6). For a proof that b_k is indeed a sufficient statistic and for a more detailed discussion of other possible sufficient statistics, see [Ber17], Chapter 4. For a mathematically more advanced discussion, see [BeS78], Chapter 10.

Since a sufficient statistic contains all the relevant information for optimal control purposes, an idea that suggests itself is to introduce features of a given sufficient statistic and to train a corresponding approximation architecture accordingly. As examples of potentially good features, one may consider some special characteristic of I_k (such as whether some alarm-like "special" event has been observed), or a partial history (such as the last m measurements and controls in I_k, or more sophisticated versions based on the concept of a *finite-state controller* proposed by White [Whi91], and White and Scherer [WhS94], and further discussed by Hansen [Han98], Kaelbling, Littman, and Cassandra [KLC98], Meuleau et al. [MPK99], Poupart and Boutilier [PoB04], Yu and Bertsekas [YuB06], Saldi, Yuksel, and Linder [SYL17]). In the case where the belief state b_k is used as a sufficient statistic, examples of good features may be a point estimate based on b_k, the variance of this estimate, and other quantities that can be simply extracted from b_k.

Of course it is possible to supplement a sufficient statistic with features of other sufficient statistics, and thus obtain an enlarged/richer sufficient statistic. In problem-specific contexts, and in the presence of approximations, this may yield improved results.

Unfortunately, in many situations an adequate set of features is not known, so it is important to have methods that construct features automatically, to supplement whatever features may already be available. Indeed, there are architectures that do not rely on the knowledge of good features. We have noted the kernel methods of Example 3.1.5 in this connection. Another very popular possibility is *neural networks*, which we will describe in Section 3.2. Some of these architectures involve training that constructs simultaneously both the feature vectors $\phi_k(x_k)$ and the parameter vectors r_k that weigh them.

3.1.2 Training of Linear and Nonlinear Architectures

In what follows in this section we discuss the process of choosing the parameter vector r of a parametric architecture $\tilde{J}(x, r)$. This is generally referred to as *training*. The most common type of training is based on a least squares optimization, also known as *least squares regression*. Here a set of state-cost training pairs (x^s, β^s), $s = 1, \ldots, q$, called the *training set*, is collected and r is determined by solving the problem

$$\min_r \sum_{s=1}^{q} \big(\tilde{J}(x^s, r) - \beta^s\big)^2. \tag{3.3}$$

Thus r is chosen to minimize the sum of squared errors between the sample costs β^s and the architecture-predicted costs $\tilde{J}(x^s, r)$. Here there is some "target" cost function J that we aim to approximate with $\tilde{J}(\cdot, r)$, and the sample cost β^s is the value $J(x^s)$ plus perhaps some error or "noise."

The cost function of the training problem (3.3) is generally nonconvex, which may pose challenges, since there may exist multiple local minima. However, for a linear architecture the cost function is convex quadratic, and the training problem admits a closed-form solution. In particular, for the linear architecture $\tilde{J}(x, r) = r'\phi(x)$, the problem becomes

$$\min_r \sum_{s=1}^{q} \big(r'\phi(x^s) - \beta^s\big)^2.$$

By setting the gradient of the quadratic objective to 0, we obtain

$$\sum_{s=1}^{q} \phi(x^s)\big(r'\phi(x^s) - \beta^s\big) = 0,$$

or

$$\sum_{s=1}^{q} \phi(x^s)\phi(x^s)'r = \sum_{s=1}^{q} \phi(x^s)\beta^s.$$

Thus by matrix inversion we obtain the minimizing parameter vector

$$\hat{r} = \left(\sum_{s=1}^{q} \phi(x^s)\phi(x^s)'\right)^{-1} \sum_{s=1}^{q} \phi(x^s)\beta^s. \tag{3.4}$$

If the inverse above does not exist, an additional quadratic in r, called a *regularization* function, is added to the least squares objective to deal with this, and also to help with other issues to be discussed later. A singular value decomposition approach may also be used to deal with the matrix inversion issue; see [BeT96], Section 3.2.2.

Thus a linear architecture has the important advantage that the training problem can be solved exactly and conveniently with the formula (3.4) (of course it may be solved by any other algorithm that is suitable for linear least squares problems). By contrast, if we use a nonlinear architecture, such as a neural network, the associated least squares problem is nonquadratic and also nonconvex. Despite this fact, through a combination of sophisticated implementation of special gradient algorithms, called *incremental*, and powerful computational resources, neural network methods have been successful in practice as we will discuss in Section 3.2.

3.1.3 Incremental Gradient and Newton Methods

We will now digress to discuss special methods for solution of the least squares training problem (3.3), assuming a parametric architecture that is differentiable in the parameter vector. This methodology can be properly viewed as a subject in nonlinear programming and iterative algorithms, and as such it can be studied independently of the approximate DP methods of this book. Thus the reader who has already some exposure to the subject may skip to the next section, and return later as needed.

Incremental methods have a rich theory, and our presentation in this section is brief, focusing primarily on implementation and intuition. Some surveys and book references are provided at the end of the chapter, which include a more detailed treatment. In particular, the neuro-dynamic programming book [BeT96] contains convergence analysis for both deterministic and stochastic training problems, while the author's nonlinear programming book [Ber16a] contains an extensive discussion of incremental methods. Since, we want to cover problems that are more general than the specific least squares training problem (3.3), we will adopt a broader formulation and notation that are standard in nonlinear programming.

Incremental Gradient Methods

We view the training problem (3.3) as a special case of the minimization of a sum of component functions

$$f(y) = \sum_{i=1}^{m} f_i(y), \tag{3.5}$$

where each f_i is a differentiable scalar function of the n-dimensional column vector y (this is the parameter vector). Thus we use the more common symbols y and m in place of r and q, respectively, and we replace the squared error terms $\big(\tilde{J}(x^s, r) - \beta^s\big)^2$ in the training problem (3.3) with the generic terms $f_i(y)$.

The (ordinary) gradient method for problem (3.5) generates a sequence $\{y^k\}$ of iterates, starting from some initial guess y^0 for the minimum of the cost function f. It has the form†

$$y^{k+1} = y^k - \gamma^k \nabla f(y^k) = y^k - \gamma^k \sum_{i=1}^{m} \nabla f_i(y^k), \tag{3.6}$$

where γ^k is a positive stepsize parameter. The incremental gradient method is similar to the ordinary gradient method, but uses the gradient of a single component of f at each iteration. It has the general form

$$y^{k+1} = y^k - \gamma^k \nabla f_{i_k}(y^k), \tag{3.7}$$

where i_k is some index from the set $\{1, \ldots, m\}$, chosen by some deterministic or randomized rule. Thus a single component function f_{i_k} is used at iteration k, with great economies in gradient calculation cost over the ordinary gradient method (3.6), particularly when m is large.

The method for selecting the index i_k of the component to be iterated on at iteration k is important for the performance of the method. Three common rules are:

(1) A *cyclic order*, the simplest rule, whereby the indexes are taken up in the fixed deterministic order $1, \ldots, m$, so that i_k is equal to (k modulo m) plus 1. A contiguous block of iterations involving the components

$$f_1, \ldots, f_m$$

in this order and exactly once is called a *cycle*.

(2) A *uniform random order*, whereby the index i_k chosen randomly by sampling over all indexes with a uniform distribution, independently of the past history of the algorithm. This rule may perform better than the cyclic rule in some circumstances.

(3) A *cyclic order with random reshuffling*, whereby the indexes are taken up one by one within each cycle, but their order after each cycle is reshuffled randomly (and independently of the past). This rule is used widely in practice, particularly when the number of components m is modest, for reasons to be discussed later.

Note that in the cyclic cases, it is essential to include all components in a cycle, for otherwise some components will be sampled more often than others, leading to a bias in the convergence process. Similarly, it is necessary to sample according to the uniform distribution in the random order case.

---

† We use standard calculus notation for gradients; see, e.g., [Ber16a], Appendix A. In particular, $\nabla f(y)$ denotes the n-dimensional column vector whose components are the first partial derivatives $\partial f(y)/\partial y_i$ of f with respect to the components $y_1, \ldots, y_n$ of the column vector y.

Focusing for the moment on the cyclic rule (with or without reshuffling), we note that the motivation for the incremental gradient method is faster convergence: we hope that far from the solution, a single cycle of the method will be as effective as several (as many as m) iterations of the ordinary gradient method (think of the case where the components f_i are similar in structure). Near a solution, however, the incremental method may not be as effective.

To be more specific, we note that there are two complementary performance issues to consider in comparing incremental and nonincremental methods:

(a) *Progress when far from convergence.* Here the incremental method can be much faster. For an extreme case take m large and all components f_i identical to each other. Then an incremental iteration requires m times less computation than a classical gradient iteration, but gives exactly the same result, when the stepsize is scaled to be m times larger. While this is an extreme example, it reflects the essential mechanism by which incremental methods can be much superior: far from the minimum a single component gradient will point to "more or less" the right direction, at least most of the time; see the following example.

(b) *Progress when close to convergence.* Here the incremental method can be inferior. In particular, the ordinary gradient method (3.6) can be shown to converge with a constant stepsize under reasonable assumptions, see e.g., [Ber16a], Chapter 1. However, the incremental method requires a diminishing stepsize, and its ultimate rate of convergence can be much slower.

This type of behavior is illustrated in the following example.

Example 3.1.7

Assume that y is a scalar, and that the problem is

$$\begin{aligned} &\text{minimize} \quad f(y) = \tfrac{1}{2}\sum_{i=1}^{m}(c_i y - b_i)^2 \\ &\text{subject to} \quad y \in \Re, \end{aligned}$$

where c_i and b_i are given scalars with $c_i \neq 0$ for all i. The minimum of each of the components $f_i(y) = \frac{1}{2}(c_i y - b_i)^2$ is

$$y_i^* = \frac{b_i}{c_i},$$

while the minimum of the least squares cost function f is

$$y^* = \frac{\sum_{i=1}^{m} c_i b_i}{\sum_{i=1}^{m} c_i^2}.$$

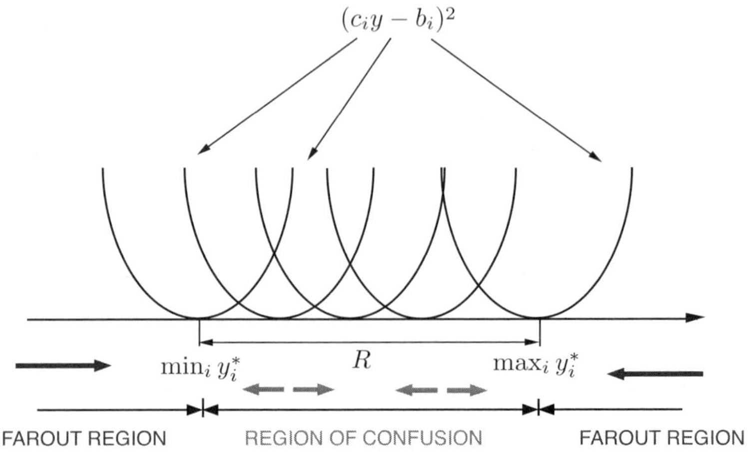

Figure 3.1.4. Illustrating the advantage of incrementalism when far from the optimal solution. The region of component minima

$$R = \left[\min_i y_i^*, \max_i y_i^*\right],$$

is labeled as the "region of confusion." It is the region where the method does not have a clear direction towards the optimum. The ith step in an incremental gradient cycle is a gradient step for minimizing $(c_i y - b_i)^2$, so if y lies outside the region of component minima $R = \left[\min_i y_i^*, \max_i y_i^*\right]$, (labeled as the "farout region") and the stepsize is small enough, progress towards the solution y^* is made.

It can be seen that y^* lies within the range of the component minima

$$R = \left[\min_i y_i^*, \max_i y_i^*\right],$$

and that for all y *outside* the range R, the gradient

$$\nabla f_i(y) = c_i(c_i y - b_i)$$

has the same sign as $\nabla f(y)$ (see Fig. 3.1.4). As a result, when outside the region R, the incremental gradient method

$$y^{k+1} = y^k - \gamma^k c_{i_k}(c_{i_k} y^k - b_{i_k})$$

approaches y^* at each step, provided the stepsize γ^k is small enough. In fact it can be verified that it is sufficient that

$$\gamma^k \leq \min_i \frac{1}{c_i^2}.$$

However, for y *inside* the region R, the ith step of a cycle of the incremental gradient method need not make progress. It will approach y^* (for small enough stepsize γ^k) only if the current point y^k does not lie in the interval connecting y_i^* and y^*. This induces an oscillatory behavior within the region R, and as a result, the incremental gradient method will typically not converge to y^* unless $\gamma^k \to 0$. By contrast, the ordinary gradient method, which takes the form

$$y^{k+1} = y^k - \gamma \sum_{i=1}^{m} c_i(c_i y^k - b_i),$$

can be verified to converge to y^* for any constant stepsize γ with

$$0 < \gamma \leq \frac{1}{\sum_{i=1}^{m} c_i^2}.$$

However, for y outside the region R, a full iteration of the ordinary gradient method need not make more progress towards the solution than a single step of the incremental gradient method. In other words, with comparably intelligent stepsize choices, *far from the solution (outside R), a single pass through the entire set of cost components by incremental gradient is roughly as effective as m passes by ordinary gradient.*

The preceding example assumes that each component function f_i has a minimum, so that the range of component minima is defined. In cases where the components f_i have no minima, a similar phenomenon may occur, as illustrated by the following example (the idea here is that we may combine several components into a single component that has a minimum).

Example 3.1.8:

Consider the case where f is the sum of increasing and decreasing convex exponentials, i.e.,

$$f_i(y) = a_i e^{b_i y}, \qquad y \in \Re,$$

where a_i and b_i are scalars with $a_i > 0$ and $b_i \neq 0$. Let

$$I^+ = \{i \mid b_i > 0\}, \qquad I^- = \{i \mid b_i < 0\},$$

and assume that I^+ and I^- have roughly equal numbers of components. Let also y^* be the minimum of $\sum_{i=1}^{m} f_i$.

Consider the incremental gradient method that given the current point, call it y^k, chooses some component f_{i_k} and iterates according to the incremental gradient iteration

$$y^{k+1} = y^k - \alpha^k \nabla f_{i_k}(y^k).$$

Then it can be seen that if $y^k >> y^*$, y^{k+1} will be substantially closer to y^* if $i \in I^+$, and negligibly further away than y^* if $i \in I^-$. The net effect, averaged over many incremental iterations, is that if $y^k >> y^*$, an incremental gradient iteration makes roughly one half the progress of a full gradient iteration, with m times less overhead for calculating gradients. The same is true if $y^k << y^*$. On the other hand as y^k gets closer to y^* the advantage of incrementalism is reduced, similar to the preceding example. In fact in order for the incremental method to converge, a diminishing stepsize is necessary, which will ultimately make the convergence slower than the one of the nonincremental gradient method with a constant stepsize.

The discussion of the preceding examples relies on y being one-dimensional, but in many multidimensional problems the same qualitative behavior can be observed. In particular, a pass through the ith component f_i by the incremental gradient method can make progress towards the solution in the region where the component gradient $\nabla f_{i_k}(y^k)$ makes an angle less than 90 degrees with the cost function gradient $\nabla f(y^k)$. If the components f_i are not "too dissimilar", this is likely to happen in a region of points that are not too close to the optimal solution set. This behavior has been verified in many practical contexts, including the training of neural networks (cf. the next section), where incremental gradient methods have been used extensively, frequently under the name *backpropagation methods*.

Stepsize Choice and Diagonal Scaling

The choice of the stepsize γ^k plays an important role in the performance of incremental gradient methods. On close examination, it turns out that the direction used by the method differs from the gradient direction by an error that is proportional to the stepsize, and for this reason a diminishing stepsize is essential for convergence to a local minimum of f (this is the best we can hope for since f may not be convex).

However, it turns out that for the cyclic rule without reshuffling a peculiar form of convergence also typically occurs for a constant but sufficiently small stepsize. In this case, the iterates converge to a "limit cycle", whereby the ith iterates within the cycles converge to a different limit than the jth iterates for $i \neq j$. The sequence $\{y^k\}$ of the iterates obtained at the end of cycles converges, except that the limit obtained *need not* be a minimum of f, even when f is convex. The limit tends to be close to the minimum point when the constant stepsize is small (see Section 2.4 of [Ber16a] for analysis and examples). In practice, it is common to use a constant stepsize for a (possibly prespecified) number of iterations, then decrease the stepsize by a certain factor, and repeat, up to the point where the stepsize reaches a prespecified floor value.

An alternative possibility is to use a diminishing stepsize rule of the form

$$\gamma^k = \min\left\{\gamma, \frac{\beta_1}{k + \beta_2}\right\},$$

where γ, β_1, and β_2 are some positive scalars. There are also variants of the method that use a constant stepsize throughout, and generically converge to a stationary point of f under reasonable assumptions. In one type of such method the degree of incrementalism gradually diminishes as the method progresses (see [Ber97a]). Another incremental approach with similar aims, is the aggregated incremental gradient method, which will be discussed later in this section.

Regardless of whether a constant or a diminishing stepsize is ultimately used, to maintain the advantage of faster convergence when far from the solution, the incremental method must use a much larger stepsize than the corresponding nonincremental gradient method (as much as m times larger so that the size of the incremental gradient step is comparable to the size of the nonincremental gradient step).

One possibility is to use an adaptive stepsize rule, whereby, roughly speaking, the stepsize is reduced (or increased) when the progress of the method indicates that the algorithm is (or is not) oscillating. There are formal ways to implement such stepsize rules with sound convergence properties (see [Tse98], [MYF03]).

The difficulty with stepsize selection may also be addressed with *diagonal scaling*, i.e., using a stepsize γ_j^k that is different for each of the components y_j of y. Second derivatives can be very useful for this purpose. In generic nonlinear programming problems of unconstrained minimization of a function f, it is common to use diagonal scaling with stepsizes

$$\gamma_j^k = \gamma \left(\frac{\partial^2 f(y^k)}{\partial^2 y_j} \right)^{-1}, \qquad j = 1, \ldots, n,$$

where γ is a constant that is nearly equal 1 (the second derivatives may also be approximated by gradient difference approximations). However, in least squares training problems, this type of scaling is inconvenient because of the additive form of f as a sum of a large number of component functions:

$$f(y) = \sum_{i=1}^{m} f_i(y),$$

cf. Eq. (3.5). The neural network literature includes a number of practical scaling schemes, some of which have been incorporated in publicly and commercially available software. Also, later in this section, when we discuss incremental Newton methods, we will provide another type of diagonal scaling that uses second derivatives and is well suited to the additive character of the cost function f.

Stochastic Gradient Descent

Incremental gradient methods are related to methods that aim to minimize an expected value

$$f(y) = E\big\{F(y, w)\big\},$$

where w is a random variable, and $F(\cdot, w) : \Re^n \mapsto \Re$ is a differentiable function for each possible value of w. The *stochastic gradient method* for minimizing f is given by

$$y^{k+1} = y^k - \gamma^k \nabla_y F(y^k, w^k), \tag{3.8}$$

where w^k is a sample of w and $\nabla_y F$ denotes gradient of F with respect to y. This method has a rich theory and a long history, and it is strongly related to the classical algorithmic field of *stochastic approximation*; see the books [BeT96], [KuY03], [Spa03], [Mey07], [Bor08], [BPP13]. The method is also often referred to as *stochastic gradient descent*, particularly in the context of machine learning applications.

If we view the expected value cost $E\{F(y, w)\}$ as a weighted sum of cost function components, we see that the stochastic gradient method (3.8) is related to the incremental gradient method

$$y^{k+1} = y^k - \gamma^k \nabla f_{i_k}(y^k) \tag{3.9}$$

for minimizing a finite sum $\sum_{i=1}^m f_i$, when randomization is used for component selection. An important difference is that the former method involves stochastic sampling of cost components $F(y, w)$ from a possibly infinite population, under some probabilistic assumptions, while in the latter the set of cost components f_i is predetermined and finite. However, it is possible to view the incremental gradient method (3.9), with uniform randomized selection of the component function f_i (i.e., with i_k chosen to be any one of the indexes $1, \ldots, m$, with equal probability $1/m$, and independently of preceding choices), as a stochastic gradient method.

Despite the apparent similarity of the incremental and the stochastic gradient methods, the view that the problem

$$\begin{aligned} &\text{minimize} \quad f(y) = \sum_{i=1}^m f_i(y) \\ &\text{subject to} \quad y \in \Re^n, \end{aligned} \tag{3.10}$$

can simply be treated as a special case of the problem

$$\begin{aligned} &\text{minimize} \quad f(y) = E\{F(y, w)\} \\ &\text{subject to} \quad y \in \Re^n, \end{aligned}$$

is questionable.

One reason is that once we convert the finite sum problem to a stochastic problem, we preclude the use of methods that exploit the finite sum structure, such as the incremental aggregated gradient methods to be discussed in the next subsection. Another reason is that the finite-component problem (3.10) is often genuinely deterministic, and to view it as a stochastic problem at the outset may mask some of its important characteristics, such as the number m of cost components, or the sequence in which the components are ordered and processed. These characteristics may potentially be algorithmically exploited. For example, with insight into the problem's structure, one may be able to discover a special deterministic or partially randomized order of processing the component functions that is superior to a uniform randomized order. On the other hand analysis indicates that in the absence of problem-specific knowledge that can be exploited to select a favorable deterministic order, a uniform randomized order (each component f_i chosen with equal probability $1/m$ at each iteration, independently of preceding choices) sometimes has superior worst-case complexity; see [NeB00], [NeB01], [BNO03], [Ber15a], [WaB16].

Finally, let us note the popular hybrid technique, which reshuffles randomly the order of the cost components after each cycle. Practical experience indicates that it has somewhat better performance to the uniform randomized order when m is large. One possible reason is that random reshuffling allocates exactly one computation slot to each component in an m-slot cycle, while uniform random sampling allocates one computation slot to each component *on the average*. A nonzero variance in the number of slots that any fixed component gets within a cycle, may be detrimental to performance. While it seems difficult to establish this fact analytically, a justification is suggested by the view of the incremental gradient method as a gradient method with error in the computation of the gradient. The error has apparently greater variance in the uniform random sampling method than in the random reshuffling method, and heuristically, if the variance of the error is larger, the direction of descent deteriorates, suggesting slower convergence.

Incremental Aggregated Gradient Methods

Another algorithm that is well suited for least squares training problems is the *incremental aggregated gradient method*, which has the form

$$y^{k+1} = y^k - \gamma^k \sum_{\ell=0}^{m-1} \nabla f_{i_{k-\ell}}(y^{k-\ell}), \tag{3.11}$$

where f_{i_k} is the new component function selected for iteration k.† In the most common version of the method the component indexes i_k are selected in a cyclic order $[i_k = (k \text{ modulo } m) + 1]$. Random selection of the index i_k has also been suggested.

From Eq. (3.11) it can be seen that the method computes the gradient incrementally, one component per iteration. However, in place of the single

† In the case where $k < m$, the summation in Eq. (3.11) should go up to $\ell = k$, and the stepsize should be replaced by a corresponding larger value.

component gradient $\nabla f_{i_k}(y^k)$, used in the incremental gradient method (3.7), it uses the sum of the component gradients computed in the past m iterations, which is an approximation to the total cost gradient $\nabla f(y^k)$.

The idea of the method is that by aggregating the component gradients one may be able to reduce the error between the true gradient $\nabla f(y^k)$ and the incrementally computed approximation used in Eq. (3.11), and thus attain a faster asymptotic convergence rate. Indeed, it turns out that under certain conditions the method exhibits a linear convergence rate, just like in the nonincremental gradient method, without incurring the cost of a full gradient evaluation at each iteration (a strongly convex cost function and with a sufficiently small constant stepsize are required for this; see [Ber16a], Section 2.4.2, and the references quoted there). This is in contrast with the incremental gradient method (3.7), for which a linear convergence rate can be achieved only at the expense of asymptotic error, as discussed earlier.

A disadvantage of the aggregated gradient method (3.11) is that it requires that the most recent component gradients be kept in memory, so that when a component gradient is reevaluated at a new point, the preceding gradient of the same component is discarded from the sum of gradients of Eq. (3.11). There have been alternative implementations that ameliorate this memory issue, by recalculating the full gradient periodically and replacing an old component gradient by a new one; we refer to the specialized literature on the subject. Another potential drawback of the aggregated gradient method is that for a large number of terms m, one hopes to converge within the first cycle through the components f_i, thereby reducing the effect of aggregating the gradients of the components.

Incremental Newton Methods

We will now consider an incremental version of Newton's method for unconstrained minimization of an additive cost function of the form

$$f(y) = \sum_{i=1}^{m} f_i(y),$$

where the functions f_i are convex and twice continuously differentiable.†

The ordinary Newton's method, at the current iterate y^k, obtains the next iterate y^{k+1} by minimizing over y the quadratic approximation/second

† We will denote by $\nabla^2 f(y)$ the $n \times n$ Hessian matrix of f at y, i.e., the matrix whose (i,j)th component is the second partial derivative $\partial^2 f(y)/\partial y^i \partial y^j$. A beneficial consequence of assuming convexity of f_i is that the Hessian matrices $\nabla^2 f_i(y)$ are positive semidefinite, which facilitates the implementation of the algorithms to be described. On the other hand, the algorithmic ideas of this section may also be adapted for the case where the functions f_i are nonconvex.

order expansion

$$\tilde{f}(y; y^k) = \nabla f(y^k)'(y - y^k) + \tfrac{1}{2}(y - y^k)'\nabla^2 f(y^k)(y - y^k),$$

of f at y^k. Similarly, the incremental form of Newton's method minimizes a sum of quadratic approximations of components of the general form

$$\tilde{f}_i(y; \psi) = \nabla f_i(\psi)'(y - \psi) + \tfrac{1}{2}(y - \psi)'\nabla^2 f_i(\psi)(y - \psi), \tag{3.12}$$

as we will now explain.

As in the case of the incremental gradient method, we view an iteration as a cycle of m subiterations, each involving a single additional component f_i, and its gradient and Hessian at the current point within the cycle. In particular, if y^k is the vector obtained after k cycles, the vector y^{k+1} obtained after one more cycle is

$$y^{k+1} = \psi_{m,k},$$

where starting with $\psi_{0,k} = y^k$, we obtain $\psi_{m,k}$ after the m steps

$$\psi_{i,k} \in \arg\min_{y \in \Re^n} \sum_{\ell=1}^{i} \tilde{f}_\ell(y; \psi_{\ell-1,k}), \qquad i = 1, \ldots, m, \tag{3.13}$$

where $\tilde{f}_\ell$ is defined as the quadratic approximation (3.12). If all the functions f_i are quadratic, it can be seen that the method finds the solution in a single cycle.† The reason is that when f_i is quadratic, each $f_i(y)$ differs from $\tilde{f}_i(y; \psi)$ by a constant, which does not depend on y. Thus the difference

$$\sum_{i=1}^{m} f_i(y) - \sum_{i=1}^{m} \tilde{f}_i(y; \psi_{i-1,k})$$

is a constant that is independent of y, and minimization of either sum in the above expression gives the same result.

It is important to note that the computations of Eq. (3.13) can be carried out efficiently. For simplicity, let as assume that $\tilde{f}_1(y; \psi)$ is a positive definite quadratic, so that for all i, $\psi_{i,k}$ is well defined as the unique

† Here we assume that the m quadratic minimizations (3.13) to generate $\psi_{m,k}$ have a solution. For this it is sufficient that the first Hessian matrix $\nabla^2 f_1(y^0)$ be positive definite, in which case there is a unique solution at every iteration. A simple possibility to deal with this requirement is to add to f_1 a small positive regularization term, such as $\epsilon\|y - y^0\|^2$. A more sound possibility is to lump together several of the component functions (enough to ensure that the sum of their quadratic approximations at y^0 is positive definite), and to use them in place of f_1. This is generally a good idea and leads to smoother initialization, as it ensures a relatively stable behavior of the algorithm for the initial iterations.

solution of the minimization problem in Eq. (3.13). We will show that the incremental Newton method (3.13) can be implemented with the incremental update formula

$$\psi_{i,k} = \psi_{i-1,k} - D_{i,k}\nabla f_i(\psi_{i-1,k}), \tag{3.14}$$

where $D_{i,k}$ is given by

$$D_{i,k} = \left(\sum_{\ell=1}^{i} \nabla^2 f_\ell(\psi_{\ell-1,k})\right)^{-1}, \tag{3.15}$$

and (assuming existence of the required inverses) is generated iteratively as

$$D_{i,k} = \left(D_{i-1,k}^{-1} + \nabla^2 f_i(\psi_{i-1,k})\right)^{-1}. \tag{3.16}$$

Indeed, from the definition of the method (3.13), the quadratic function $\sum_{\ell=1}^{i-1} \tilde f_\ell(y;\psi_{\ell-1,k})$ is minimized by $\psi_{i-1,k}$ and its Hessian matrix is $D_{i-1,k}^{-1}$, so we have

$$\sum_{\ell=1}^{i-1} \tilde f_\ell(y;\psi_{\ell-1,k}) = \tfrac{1}{2}(y-\psi_{i-1,k})' D_{i-1,k}^{-1}(y-\psi_{i-1,k}) + \text{constant}.$$

Thus, by adding $\tilde f_i(y;\psi_{i-1,k})$ to both sides of this expression, we obtain

$$\begin{aligned}\sum_{\ell=1}^{i} \tilde f_\ell(y;\psi_{\ell-1,k}) &= \tfrac{1}{2}(y-\psi_{i-1,k})' D_{i-1,k}^{-1}(y-\psi_{i-1,k}) + \text{constant}\\ &\quad + \tfrac{1}{2}(y-\psi_{i-1,k})'\nabla^2 f_i(\psi_{i-1,k})(y-\psi_{i-1,k}) + \nabla f_i(\psi_{i-1,k})'(y-\psi_{i-1,k}).\end{aligned}$$

Since by definition $\psi_{i,k}$ minimizes this function, we obtain Eqs. (3.14)-(3.16).

The recursion (3.16) for the matrix $D_{i,k}$ can often be efficiently implemented by using convenient formulas for the inverse of the sum of two matrices. In particular, if f_i is given by

$$f_i(y) = h_i(a_i'y - b_i),$$

for some twice differentiable convex function $h_i : \Re \mapsto \Re$, vector a_i, and scalar b_i, we have

$$\nabla^2 f_i(\psi_{i-1,k}) = \nabla^2 h_i(a_i'\psi_{i-1,k} - b_i)\, a_i a_i',$$

and the recursion (3.16) can be written as

$$D_{i,k} = D_{i-1,k} - \frac{D_{i-1,k}a_i a_i' D_{i-1,k}}{\nabla^2 h_i(a_i'\psi_{i-1,k} - b_i)^{-1} + a_i' D_{i-1,k} a_i}.$$

This is the well-known Sherman-Morrison formula for the inverse of the sum of an invertible matrix and a rank-one matrix.

We have considered so far a single cycle of the incremental Newton method. Similar to the case of the incremental gradient method, we may cycle through the component functions multiple times. In particular, we may apply the incremental Newton method to the extended set of components

$$f_1, f_2, \ldots, f_m, \; f_1, f_2, \ldots, f_m, \; f_1, f_2, \ldots. \tag{3.17}$$

The resulting method asymptotically resembles an incremental gradient method with diminishing stepsize of the type described earlier. Indeed, from Eq. (3.15), the matrix $D_{i,k}$ diminishes roughly in proportion to $1/k$. From this it follows that the asymptotic convergence properties of the incremental Newton method are similar to those of an incremental gradient method with diminishing stepsize of order $O(1/k)$. Thus its convergence rate is slower than linear.

To accelerate the convergence of the method one may employ a form of restart, so that $D_{i,k}$ does not converge to 0. For example $D_{i,k}$ may be reinitialized and increased in size at the beginning of each cycle. For problems where f has a unique nonsingular minimum y^* [one for which $\nabla^2 f(y^*)$ is nonsingular], one may design incremental Newton schemes with restart that converge linearly to within a neighborhood of y^* (and even superlinearly if y^* is also a minimum of all the functions f_i, so there is no region of confusion); see [Ber96a]. Alternatively, the update formula (3.16) may be modified to

$$D_{i,k} = \left(\beta_k D_{i-1,k}^{-1} + \nabla^2 f_\ell(\psi_{i,k})\right)^{-1}, \tag{3.18}$$

by introducing a parameter $\beta_k \in (0, 1)$, which can be used to accelerate the practical convergence rate of the method.

Incremental Newton Method with Diagonal Approximation

Generally, with proper implementation, the incremental Newton method is often substantially faster than the incremental gradient method, in terms of numbers of iterations (there are theoretical results suggesting this property for stochastic versions of the two methods; see the end-of-chapter references). However, in addition to computation of second derivatives, the incremental Newton method involves greater overhead per iteration due to the matrix-vector calculations in Eqs. (3.14), (3.16), and (3.18), so it is suitable only for problems where n, the dimension of y, is relatively small.

A way to remedy in part this difficulty is to approximate $\nabla^2 f_i(\psi_{i,k})$ by a diagonal matrix, and recursively update a diagonal approximation of $D_{i,k}$ using Eqs. (3.16) or (3.18). In particular, one may set to 0 the off-diagonal components of $\nabla^2 f_i(\psi_{i,k})$. In this case, the iteration (3.14) becomes a diagonally scaled version of the incremental gradient method, and involves comparable overhead per iteration (assuming the required diagonal second derivatives are easily computed or approximated). As an additional scaling option, one may multiply the diagonal components with a stepsize parameter that is close to 1 and add a small positive constant (to bound them away from 0). Ordinarily this method is easily implementable, and requires little experimentation with stepsize selection.

Example 3.1.9: (Diagonalized Newton Method for Linear Least Squares)

Consider the problem of minimizing the sum of squares of linear scalar functions:

$$\begin{aligned} &\text{minimize} \quad f(y) = \tfrac{1}{2}\sum_{i=1}^{m}(c_i'y - b_i)^2 \\ &\text{subject to} \quad y \in \Re^n, \end{aligned}$$

where for all i, c_i is a given nonzero vector in $\Re^n$ and b_i is a given scalar. In this example, we will assume that we apply the incremental Newton method to the extended set of components

$$f_1, f_2, \ldots, f_m, f_1, f_2, \ldots, f_m, f_1, f_2, \ldots,$$

c.f. Eq. (3.17).

The inverse Hessian matrix for the kth cycle is given by

$$D_{i,k} = \left((k-1)\sum_{\ell=1}^{m} c_\ell c_\ell' + \sum_{\ell=1}^{i} c_\ell c_\ell'\right)^{-1}, \qquad i = 1, \ldots, m,$$

and its jth diagonal term (the one to multiply the partial derivative $\partial f_i/\partial x^j$ within cycle $k+1$) is

$$\gamma_{i,j}^k = \frac{1}{(k-1)\sum_{\ell=1}^{m}(c_\ell^j)^2 + \sum_{\ell=1}^{i}(c_\ell^j)^2}. \tag{3.19}$$

The diagonalized version of the incremental Newton method is given for each of the n components of $\psi_{i,k}$ by

$$\psi_{i,k}^j = \psi_{i-1,k}^j - \gamma_{i,j}^k c_i^j (c_i'\psi_{i-1,k} - b_i), \qquad j = 1, \ldots, n,$$

where $\psi_{i,k}^j$ and c_i^j are the jth components of the vectors $\psi_{i,k}$ and c_i, respectively. Contrary to the incremental Newton iteration, it does not involve matrix inversion. Note that the diagonal scaling term (3.19) tends to 0 as k increases, for each component j. As noted earlier, this diagonal term may be multiplied by a constant that is less or equal to 1 for a better guarantee of asymptotic convergence.

In an alternative implementation of the diagonal Hessian approximation idea, we reinitialize the inverse Hessian matrix to 0 (or to small multiple of the identity) at the beginning of each cycle. Then instead of the scalar $\gamma_{i,j}^k$ of Eq. (3.19), the jth diagonal term of the inverse Hessian will be

$$\frac{1}{\sum_{\ell=1}^{i} (c_\ell^j)^2},$$

and will not converge to 0. It may be modified by multiplication with a suitable diminishing scalar for practical purposes.

3.2 NEURAL NETWORKS

There are several different types of neural networks that can be used for a variety of tasks, such as pattern recognition, classification, image and speech recognition, and others. In this section, we focus on our finite horizon DP context, and the role that neural networks can play in approximating the optimal cost-to-go functions J_k^*. As an example within this context, we may first use a neural network to construct an approximation to J_{N-1}^*. Then we may use this approximation to approximate J_{N-2}^*, and continue this process backwards in time, to obtain approximations to all the optimal cost-to-go functions J_k^*, $k = 1, \ldots, N-1$, as we will discuss in more detail in Section 3.3.

To describe the use of neural networks in finite horizon DP, let us consider the typical stage k, and for convenience drop the index k; the subsequent discussion applies to each value of k separately. We consider parametric architectures $\tilde J(x, v, r)$ of the form

$$\tilde J(x, v, r) = r'\phi(x, v) \tag{3.20}$$

that depend on two parameter vectors v and r. Our objective is to select v and r so that $\tilde J(x, v, r)$ approximates some cost function that can be sampled (possibly with some error). The process is to collect a training set that consists of a large number of state-cost pairs (x^s, β^s), $s = 1, \ldots, q$, and to find a function $\tilde J(x, v, r)$ of the form (3.20) that matches the training set in a least squares sense, i.e., (v, r) minimizes

$$\sum_{s=1}^{q} \big(\tilde J(x^s, v, r) - \beta^s\big)^2.$$

We postpone for later the question of how the training pairs (x^s, β^s) are generated, and how the training problem is solved.† Notice the different

† The least squares training problem used here is based on *nonlinear regression*. This is a classical method for approximating the expected value of a function with a parametric architecture, and involves a least squares fit of the architecture to simulation-generated samples of the expected value. We refer to machine learning and statistics textbooks for more discussion.

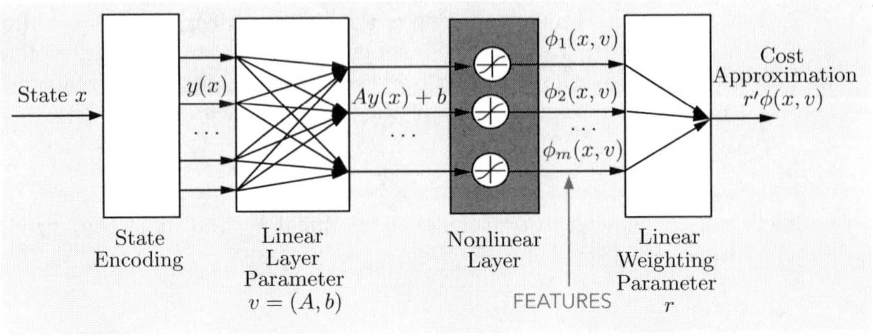

Figure 3.2.1 A perceptron consisting of a linear layer and a nonlinear layer. It provides a way to compute features of the state, which can be used for cost function approximation. The state x is encoded as a vector of numerical values $y(x)$, which is then transformed linearly as $Ay(x)+b$ in the linear layer. The m scalar output components of the linear layer, become the inputs to nonlinear functions that produce the m scalars $\phi_\ell(x,v)=\sigma\big((Ay(x)+b)_\ell\big)$, which can be viewed as features that are in turn linearly weighted with parameters r_ℓ.

roles of the two parameter vectors here: v *parametrizes* $\phi(x,v)$, which in some interpretation may be viewed as a feature vector, and r *is a vector of linear weighting parameters for the components of* $\phi(x,v)$.

A neural network architecture provides a parametric class of functions $\tilde J(x,v,r)$ of the form (3.20) that can be used in the optimization framework just described. The simplest type of neural network is the *single layer perceptron*; see Fig. 3.2.1. Here the state x is encoded as a vector of numerical values $y(x)$ with components $y_1(x),\ldots,y_n(x)$, which is then transformed linearly as

$$Ay(x)+b,$$

where A is an $m\times n$ matrix and b is a vector in $\Re^m$.† This transformation is called the *linear layer* of the neural network. We view the components of A and b as parameters to be determined, and we group them together into the parameter vector $v=(A,b)$.

Each of the m scalar output components of the linear layer,

$$\big(Ay(x)+b\big)_\ell, \qquad \ell=1,\ldots,m,$$

becomes the input to a nonlinear differentiable function σ that maps scalars to scalars. Typically σ is differentiable and monotonically increasing. A simple and popular possibility is the *rectified linear unit* (ReLU for short),

† The method of encoding x into the numerical vector $y(x)$ is problem-dependent, but it is important to note that some of the components of $y(x)$ could be some known interesting features of x that can be designed based on problem-specific knowledge.

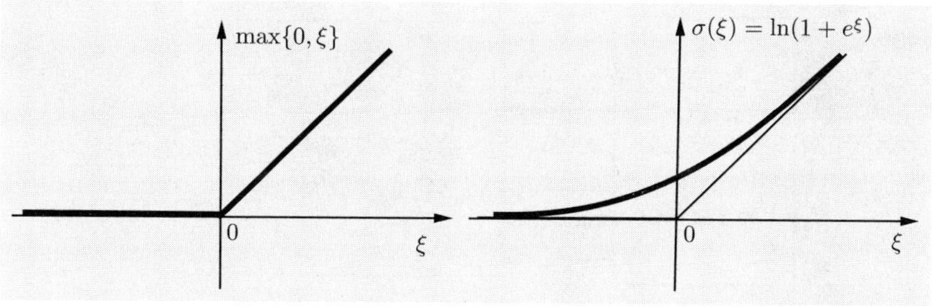

Figure 3.2.2 The rectified linear unit $\sigma(\xi) = \ln(1 + e^{\xi})$. It is the function $\max\{0, \xi\}$ with its corner "smoothed out." Its derivative is $\sigma'(\xi) = e^{\xi}/(1 + e^{\xi})$, and approaches 0 and 1 as $\xi \to -\infty$ and $\xi \to \infty$, respectively.

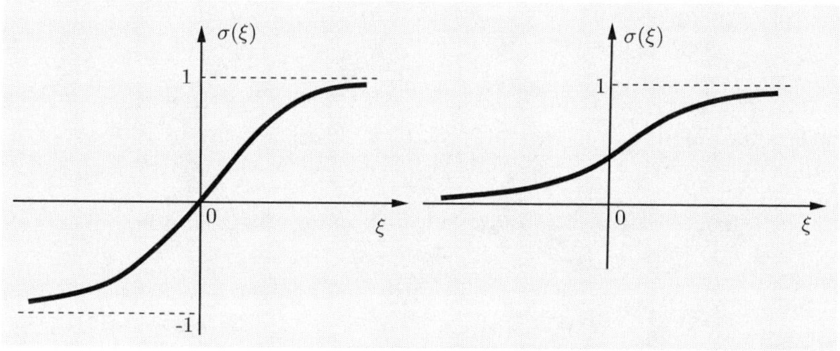

Figure 3.2.3 Some examples of sigmoid functions. The hyperbolic tangent function is on the left, while the logistic function is on the right.

which is simply the function $\max\{0, \xi\}$, approximated by a differentiable function σ by some form of smoothing operation; for example $\sigma(\xi) = \ln(1+e^{\xi})$, which illustrated in Fig. 3.2.2. Other functions, used since the early days of neural networks, have the property

$$-\infty < \lim_{\xi \to -\infty} \sigma(\xi) < \lim_{\xi \to \infty} \sigma(\xi) < \infty;$$

see Fig. 3.2.3. Such functions are called *sigmoids*, and some common choices are the *hyperbolic tangent* function

$$\sigma(\xi) = \tanh(\xi) = \frac{e^{\xi} - e^{-\xi}}{e^{\xi} + e^{-\xi}},$$

and the *logistic* function

$$\sigma(\xi) = \frac{1}{1 + e^{-\xi}}.$$

In what follows, we will ignore the character of the function σ (except for differentiability), and simply refer to it as a "nonlinear unit" and to the corresponding layer as a "nonlinear layer."

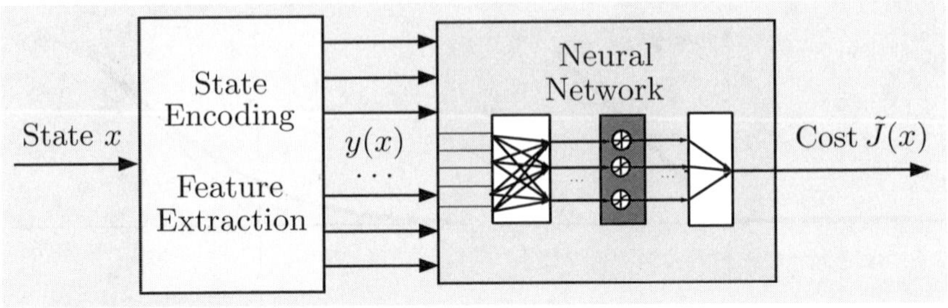

Figure 3.2.4 Nonlinear architecture with a view of the state encoding process as a feature extraction mapping preceding the neural network.

At the outputs of the nonlinear units, we obtain the scalars

$$\phi_\ell(x, v) = \sigma\big((Ay(x) + b)_\ell\big), \qquad \ell = 1, \ldots, m.$$

One possible interpretation is to view $\phi_\ell(x, v)$ as features of x, which are linearly combined using weights r_ℓ, $\ell = 1, \ldots, m$, to produce the final output

$$\tilde{J}(x, v, r) = \sum_{\ell=1}^{m} r_\ell \phi_\ell(x, v) = \sum_{\ell=1}^{m} r_\ell \sigma\big((Ay(x) + b)_\ell\big).$$

Note that each value $\phi_\ell(x, v)$ depends on just the ℓth row of A and the ℓth component of b, not on the entire vector v. In some cases this motivates placing some constraints on individual components of A and b to achieve special problem-dependent "handcrafted" effects.

State Encoding and Direct Feature Extraction

The state encoding operation that transforms x into the neural network input $y(x)$ can be instrumental in the success of the approximation scheme. Examples of possible state encodings are components of the state x, numerical representations of qualitative characteristics of x, and more generally features of x, i.e., functions of x that aim to capture "important nonlinearities" of the optimal cost-to-go function. With the latter view of state encoding, we may consider the approximation process as consisting of a feature extraction mapping, followed by a neural network with input the extracted features of x, and output the cost-to-go approximation; see Fig. 3.2.4.

The idea here is that with a good feature extraction mapping, the neural network need not be very complicated (may involve few nonlinear units and corresponding parameters), and may be trained more easily. This intuition is borne out by simple examples and practical experience. However, as is often the case with neural networks, it is hard to support it with a quantitative analysis.

Figure 3.2.5 Illustration of the graph of the cost function of a two-dimensional neural network training problem. The cost function is nonconvex, and it becomes flat for large values of the weights. In particular, the cost tends asymptotically to a constant as the vector of weights is changed towards ∞ along rays emanating from the origin. The graph in the figure corresponds to an extremely simple training problem: a single input, a single nonlinear unit, two weights, and five input-output data pairs (corresponding to the five "petals" of the graph).

3.2.1 Training of Neural Networks

Given a set of state-cost training pairs (x^s, β^s), $s = 1, \ldots, q$, the parameters of the neural network A, b, and r are obtained by solving the training problem

$$\min_{A,b,r} \sum_{s=1}^{q} \left(\sum_{\ell=1}^{m} r_\ell \sigma\left(\left(Ay(x^s) + b\right)_\ell \right) - \beta^s \right)^2 . \tag{3.21}$$

Note that the cost function of this problem is generally nonconvex, so there may exist multiple local minima. In fact the graph of cost function of the above training problem can be very complicated; see Fig. 3.2.5, which illustrates a very simple special case.

In practice it is common to augment the cost function of this problem with a *regularization* function, such as a quadratic in the parameters A, b, and r. This is customary in least squares problems in order to make the problem easier to solve algorithmically. However, in the context of neural network training, regularization is primarily important for a different reason: it helps to avoid *overfitting*, which occurs when the number of parameters of the neural network is relatively large (comparable to the size of the training set). In this case a neural network model matches the training data very well but may not do as well on new data. This is a well known difficulty in machine learning, which is the subject of much current research, particularly in the context of deep neural networks.

We refer to machine learning and neural network textbooks for a discussion of algorithmic questions regarding regularization and other issues that relate to the practical implementation of the training process. In any case, the training problem (3.21) is an unconstrained nonconvex differentiable optimization problem that can in principle be addressed with any of the standard gradient-type methods. Significantly, it is well-suited for the incremental methods discussed in Section 3.1.3 (a sense of this can be obtained by comparing the cost structure illustrated in Fig. 3.2.5 with the one of Fig. 3.1.4, which explains the advantage of incrementalism).

Let us now consider a few issues regarding the neural network formulation and training process just described:

(a) The first issue is to select a method to solve the training problem (3.21). While we can use any unconstrained optimization method that is based on gradients, in practice it is important to take into account the cost function structure of problem (3.21). The salient characteristic of this cost function is that it is the sum of a potentially very large number q of component functions. This makes the computation of the cost function value of the training problem and/or its gradient very costly. For this reason the incremental methods of Section 3.1.3 are universally used for training.† Experience has shown that these methods can be vastly superior to their nonincremental counterparts in the context of neural network training.

    The implementation of the training process has benefited from experience that has been accumulated over time, and has provided guidelines for scaling, regularization, initial parameter selection, and other practical issues; we refer to books on neural networks such as Bishop [Bis95], Goodfellow, Bengio, and Courville [GBC16], and Haykin [Hay08] for related accounts. Still, incremental methods can be quite slow, and training may be a time-consuming process. Fortunately, the training is ordinarily done off-line, in which case computation time may not be a serious issue. Moreover, in practice the neural network training problem typically need not be solved with great accuracy.

---

† The incremental methods are valid for an arbitrary order of component selection within the cycle, but it is common to randomize the order at the beginning of each cycle. Also, in a variation of the basic method, we may operate on a batch of several components at each iteration, called a *minibatch*, rather than a single component. This has an averaging effect, which reduces the tendency of the method to oscillate and allows for the use of a larger stepsize; see the end-of-chapter references.

(b) Another important question is how well we can approximate the optimal cost-to-go functions J_k^* with a neural network architecture, assuming we can choose the number of the nonlinear units m to be as large as we want. The answer to this question is quite favorable and is provided by the so-called *universal approximation theorem*. Roughly, the theorem says that assuming that x is an element of a Euclidean space X and $y(x) \equiv x$, a neural network of the form described can approximate arbitrarily closely (in an appropriate mathematical sense), over a closed and bounded subset $S \subset X$, any piecewise continuous function $J : S \mapsto \Re$, provided the number m of nonlinear units is sufficiently large. For proofs of the theorem at different levels of generality, we refer to Cybenko [Cyb89], Funahashi [Fun89], Hornik, Stinchcombe, and White [HSW89], and Leshno et al. [LLP93]. For intuitive explanations we refer to Bishop ([Bis95], pp. 129-130) and Jones [Jon90].

(c) While the universal approximation theorem provides some assurance about the adequacy of the neural network structure, it does not predict how many nonlinear units we may need for "good" performance in a given problem. Unfortunately, this is a difficult question to even pose precisely, let alone to answer adequately. In practice, one is reduced to trying increasingly larger values of m until one is convinced that satisfactory performance has been obtained for the task at hand. Experience has shown that in many cases the number of required nonlinear units and corresponding dimension of A can be very large, adding significantly to the difficulty of solving the training problem. This has given rise to many suggestions for modifications of the neural network structure. One possibility is to concatenate multiple single layer perceptrons so that the output of the nonlinear layer of one perceptron becomes the input to the linear layer of the next, giving rise to deep neural networks, which we will discuss in Section 3.2.2.

What are the Features that can be Produced by Neural Networks?

The short answer is that just about any feature that can be of practical interest can be produced or be closely approximated by a neural network. What is needed is a single layer that consists of a sufficiently large number of nonlinear units, preceded and followed by a linear layer. This is a consequence of the universal approximation theorem. In particular it is not necessary to have more than one nonlinear layer (although it is possible that fewer nonlinear units may be needed with a neural network that involves multiple nonlinear layers).

As an illustration, we will consider features of a scalar state variable x, and a neural network that uses the rectifier function

$$\sigma(\xi) = \max\{0, \xi\}$$

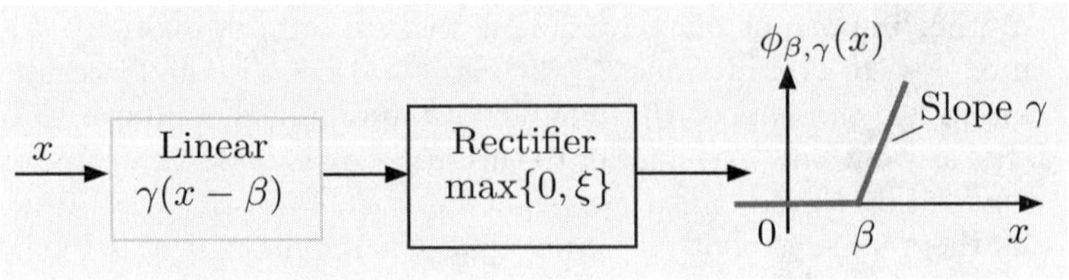

Figure 3.2.6 The feature

$$\phi_{\beta,\gamma}(x) = \max\{0, \gamma(x-\beta)\},$$

produced by a rectifier preceded by the linear function $L(x) = \gamma(x-\beta)$.

as the basic nonlinear unit. Thus a single rectifier preceded by a linear function

$$L(x) = \gamma(x-\beta),$$

where β and γ are scalars, produces the feature

$$\phi_{\beta,\gamma}(x) = \max\{0, \gamma(x-\beta)\}, \tag{3.22}$$

illustrated in Fig. 3.2.6.

We can now construct more complex features by adding or subtracting single rectifier features of the form (3.22). In particular, subtracting two rectifier functions with the same slope but two different horizontal shifts β_1 and β_2, we obtain the feature

$$\phi_{\beta_1,\beta_2,\gamma}(x) = \phi_{\beta_1,\gamma}(x) - \phi_{\beta_2,\gamma}(x)$$

shown in Fig. 3.2.7(a). By subtracting again two features of the preceding form, we obtain the "pulse" feature

$$\phi_{\beta_1,\beta_2,\beta_3,\beta_4,\gamma}(x) = \phi_{\beta_1,\beta_2,\gamma}(x) - \phi_{\beta_3,\beta_4,\gamma}(x),$$

shown in Fig. 3.2.7(b). The "pulse" feature can be used in turn as a fundamental block to approximate any desired feature by a linear combination of "pulses." This explains why neural networks are capable of producing features of any shape by using linear layers to precede and follow nonlinear layers, at least in the case of a scalar state x. The mechanism for feature formation just described can be extended to the case of a multidimensional x, and lies at the heart of the universal approximation theorem and its proof (see Cybenko [Cyb89]).

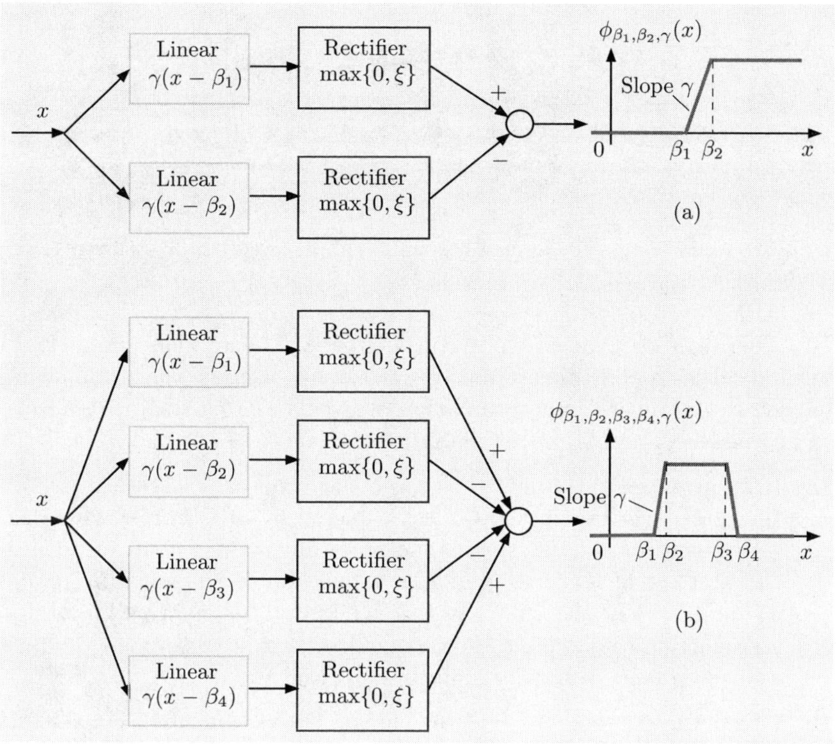

Figure 3.2.7 (a) Illustration of how the feature

$$\phi_{\beta_1,\beta_2,\gamma}(x) = \phi_{\beta_1,\gamma}(x) - \phi_{\beta_2,\gamma}(x)$$

is formed by a neural network of a two-rectifier nonlinear layer, preceded by a linear layer.

(b) Illustration of how the "pulse" feature

$$\phi_{\beta_1,\beta_2,\beta_3,\beta_4,\gamma}(x) = \phi_{\beta_1,\beta_2,\gamma}(x) - \phi_{\beta_3,\beta_4,\gamma}(x)$$

is formed by a neural network of a four-rectifier nonlinear layer, preceded by a linear layer.

3.2.2 Multilayer and Deep Neural Networks

An important generalization of the single layer perceptron architecture involves a concatenation of multiple layers of linear and nonlinear functions; see Fig. 3.2.8. In particular the outputs of each nonlinear layer become the inputs of the next linear layer. In some cases it may make sense to add as additional inputs some of the components of the state x or the state encoding $y(x)$.

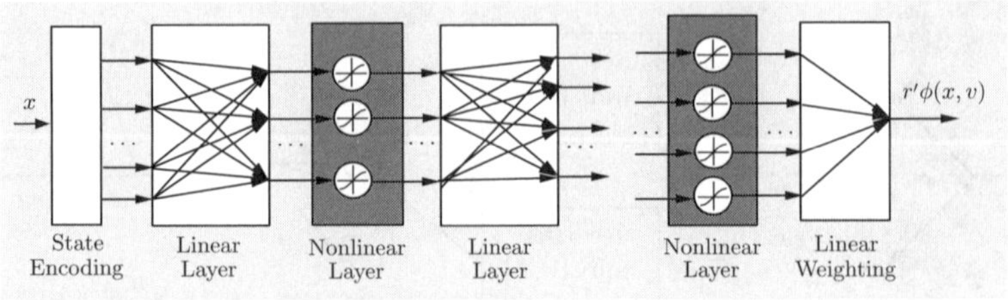

Figure 3.2.8 A deep neural network, with multiple layers. Each nonlinear layer constructs the inputs of the next linear layer.

There are a few questions to consider here. The first has to do with the reason for having multiple nonlinear layers, when a single one is sufficient to guarantee the universal approximation property. Here are some qualitative (and somewhat speculative) explanations:

(a) If we view the outputs of each nonlinear layer as features, we see that the multilayer network produces a hierarchy of features, where each set of features is a function of the preceding set of features [except for the first set of features, which is a function of the encoding $y(x)$ of the state x]. In the context of specific applications, this hierarchical structure can be exploited to specialize the role of some of the layers and to enhance some characteristics of the state.

(b) Given the presence of multiple linear layers, one may consider the possibility of using matrices A with a particular sparsity pattern, or other structure that embodies special linear operations such as convolution. When such structures are used, the training problem often becomes easier, because the number of parameters in the linear layers is drastically decreased.

(c) Overparametrization (more weights than data, as in a multilayer neural network) may help to mitigate the detrimental effects of overfitting, and the attendant need for regularization. The explanation for this fascinating phenomenon (observed as early as the late 90s) is the subject of much current research; see [ZBH16], [BMM18], [BRT18], [SJL18], [BLL19], [HMR19], [MVS19] for some representative works.

Note that while in the early days of neural networks practitioners tended to use few nonlinear layers (say one to three), more recently a lot of success in certain problem domains (including image and speech processing, as well as approximate DP) has been achieved with so called *deep neural networks*, which involve a considerably larger number of layers. In particular, the use of deep neural networks has been an important factor for the success of the AlphaGo and AlphaZero programs that play Go and chess, respectively; see [SHM16], [SHS17]. By contrast, Tesauro's backgammon program and its descendants (cf. Section 2.4.2) have performed well with one or two nonlinear layers [PaR12].

Training and Backpropagation

Let us now consider the training problem for multilayer networks. It has the form

$$\min_{v,r} \sum_{s=1}^{q} \left(\sum_{\ell=1}^{m} r_\ell \phi_\ell(x^s, v) - \beta^s \right)^2,$$

where v represents the collection of all the parameters of the linear layers, and $\phi_\ell(x,v)$ is the ℓth feature produced at the output of the final nonlinear layer. Various types of incremental gradient methods, which modify the weight vector in the direction opposite to the gradient of a single sample term

$$\left(\sum_{\ell=1}^{m} r_\ell \phi_\ell(x^s, v) - \beta^s \right)^2,$$

can also be applied here. They are the methods most commonly used in practice. An important fact is that the gradient with respect to v of each feature $\phi_\ell(x,v)$ can be efficiently calculated, as we will describe shortly.

Multilayer perceptrons can be represented compactly by introducing certain mappings to describe the linear and the nonlinear layers. In particular, let $L_1, \ldots, L_{m+1}$ denote the matrices representing the linear layers; that is, at the output of the 1st linear layer we obtain the vector $L_1 x$ and at the output of the kth linear layer ($k > 1$) we obtain $L_k \xi$, where ξ is the output of the preceding nonlinear layer. Similarly, let $\Sigma_1, \ldots, \Sigma_m$ denote the mappings representing the nonlinear layers; that is, when the input of the kth nonlinear layer ($k > 1$) is the vector y with components $y(j)$, we obtain at the output the vector $\Sigma_k y$ with components $\sigma\big(y(j)\big)$. The output of the multilayer perceptron is

$$F(L_1, \ldots, L_{m+1}, x) = L_{m+1} \Sigma_m L_m \cdots \Sigma_1 L_1 x.$$

The special nature of this formula has an important computational consequence: the gradient (with respect to the weights) of the squared error between the output and a desired output y,

$$E(L_1, \ldots, L_{m+1}) = \frac{1}{2}\big(y - F(L_1, \ldots, L_{m+1}, x)\big)^2,$$

can be efficiently calculated using a special procedure known as *backpropagation*, which is just an intelligent way of using the chain rule.† In particular, the partial derivative of the cost function $E(L_1, \ldots, L_{m+1})$ with

† The name *backpropagation* is used in several different ways in the neural networks literature. For example feedforward neural networks of the type shown in Fig. 3.2.8 are sometimes referred to as *backpropagation networks*. The somewhat abstract derivation of the backpropagation formulas given here comes from Section 3.1.1 of the book [BeT96].

respect to $L_k(i,j)$, the ijth component of the matrix L_k, is given by

$$\frac{\partial E(L_1,\ldots,L_{m+1})}{\partial L_k(i,j)} = -e' L_{m+1}\overline{\Sigma}_m L_m \cdots L_{k+1}\overline{\Sigma}_k I_{ij} \Sigma_{k-1} L_{k-1} \cdots \Sigma_1 L_1 x, \tag{3.23}$$

where e is the error vector

$$e = y - F(L_1,\ldots,L_{m+1},x),$$

$\overline{\Sigma}_n$, $n = 1,\ldots,m$, is the diagonal matrix with diagonal terms equal to the derivatives of the nonlinear functions σ of the nth hidden layer evaluated at the appropriate points, and I_{ij} is the matrix obtained from L_k by setting all of its components to 0 except for the ijth component which is set to 1. This formula can be used to obtain efficiently all of the terms needed in the partial derivatives (3.23) of E using a two-step calculation:

(a) Use a forward pass through the network to calculate sequentially the outputs of the linear layers

$$L_1 x, L_2\Sigma_1 L_1 x, \ldots, L_{m+1}\Sigma_m L_m \cdots \Sigma_1 L_1 x.$$

This is needed in order to obtain the points at which the derivatives in the matrices $\overline{\Sigma}_n$ are evaluated, and also in order to obtain the error vector $e = y - F(L_1,\ldots,L_{m+1},x)$.

(b) Use a backward pass through the network to calculate sequentially the terms

$$e' L_{m+1}\overline{\Sigma}_m L_m \cdots L_{k+1}\overline{\Sigma}_k$$

in the derivative formulas (3.23), starting with $e' L_{m+1}\overline{\Sigma}_m$, proceeding to $e' L_{m+1}\overline{\Sigma}_m L_m \overline{\Sigma}_{m-1}$, and continuing to $e' L_{m+1}\overline{\Sigma}_m \cdots L_2 \overline{\Sigma}_1$.

As a final remark, we mention that the ability to simultaneously extract features and optimize their linear combination is not unique to neural networks. Other approaches that use a multilayer architecture have been proposed (see the survey by Schmidhuber [Sch15]), and they admit similar training procedures based on appropriately modified forms of backpropagation. An example of an alternative multilayer architecture approach is the Group Method for Data Handling (GMDH), which is principally based on the use of polynomial (rather than sigmoidal) nonlinearities. The GMDH approach was investigated extensively in the Soviet Union starting with the work of Ivakhnenko in the late 60s; see e.g., [Iva68]. It has been used in a large variety of applications, and its similarities with the neural network methodology have been noted (see the survey by Ivakhnenko [Iva71], and the large literature summary at the web site http://www.gmdh.net). Most of the GMDH research relates to inference-type problems, and apparently there have not been any applications to approximate DP to date. However, this may be a fruitful area of investigation, since in some applications it may turn out that polynomial nonlinearities are more suitable than sigmoidal or rectified linear unit nonlinearities.

3.3 SEQUENTIAL DYNAMIC PROGRAMMING APPROXIMATION

Let us describe a popular approach for training an approximation architecture $\tilde{J}_k(x_k, r_k)$ for a finite horizon DP problem. The parameter vectors r_k are determined sequentially, with an algorithm known as *fitted value iteration*, starting from the end of the horizon, and proceeding backwards as in the DP algorithm: first r_{N-1} then r_{N-2}, and so on. The algorithm samples the state space for each stage k, and generates a large number of states x_k^s, $s = 1, \ldots, q$. It then determines sequentially the parameter vectors r_k to obtain a good "least squares fit" to the DP algorithm.

In particular, each r_k is determined by generating a large number of sample states and solving a least squares problem that aims to minimize the error in satisfying the DP equation for these states at time k. At the typical stage k, having obtained r_{k+1}, we determine r_k from the least squares problem

$$r_k \in \arg\min_r \sum_{s=1}^{q} \Bigg(\tilde{J}_k(x_k^s, r) \\ - \min_{u \in U_k(x_k^s)} E\Big\{ g(x_k^s, u, w_k) + \tilde{J}_{k+1}\big(f_k(x_k^s, u, w_k), r_{k+1}\big)\Big\}\Bigg)^2$$

where x_k^s, $i = 1, \ldots, q$, are the sample states that have been generated for the kth stage. Since r_{k+1} is assumed to be already known, the complicated minimization term in the right side of this equation is the known scalar

$$\beta_k^s = \min_{u \in U_k(x_k^s)} E\Big\{ g(x_k^s, u, w_k) + \tilde{J}_{k+1}\big(f_k(x_k^s, u, w_k), r_{k+1}\big)\Big\}, \tag{3.24}$$

so that r_k is obtained as

$$r_k \in \arg\min_r \sum_{s=1}^{q} \big(\tilde{J}_k(x_k^s, r) - \beta_k^s\big)^2. \tag{3.25}$$

The algorithm starts at stage $N - 1$ with the minimization

$$r_{N-1} \in \arg\min_r \sum_{s=1}^{q} \Bigg(\tilde{J}_{N-1}(x_{N-1}^s, r) \\ - \min_{u \in U_{N-1}(x_{N-1}^s)} E\Big\{ g_{N-1}(x_{N-1}^s, u, w_{N-1}) + g_N\big(f_{N-1}(x_{N-1}^s, u, w_{N-1})\big)\Big\}\Bigg)^2$$

and ends with the calculation of r_0 at $k = 0$.

In the case of a linear architecture, where the approximate cost-to-go functions are

$$\tilde{J}_k(x_k, r_k) = r_k' \phi_k(x_k), \qquad k = 0, \ldots, N-1,$$

the least squares problem (3.25) greatly simplifies, and admits the closed form solution

$$r_k = \left(\sum_{s=1}^{q} \phi_k(x_k^s) \phi_k(x_k^s)' \right)^{-1} \sum_{s=1}^{q} \beta_k^s \phi_k(x_k^s);$$

cf. Eq. (3.4). For a nonlinear architecture such as a neural network, incremental gradient algorithms may be used.

An important implementation issue is how to select the sample states x_k^s, $s = 1, \ldots, q$, $k = 0, \ldots, N-1$. In practice, they are typically obtained by some form of Monte Carlo simulation, but the distribution by which they are generated is important for the success of the method. In particular, it is important that the sample states are "representative" in the sense that they are visited often under a nearly optimal policy. More precisely, the frequencies with which various states appear in the sample should be roughly proportional to the probabilities of their occurrence under an optimal policy. This point will be discussed later, in the context of infinite horizon problems, and the notion of "representative" state will be better quantified in probabilistic terms.

Aside from the issue of selection of the sampling distribution that we have just described, a difficulty with fitted value iteration arises when the horizon is very long, since then the number of parameters may become excessive. In this case, however, the problem is often stationary, in the sense that the system and cost per stage do not change as time progresses. Then it may be possible to treat the problem as one with an infinite horizon and bring to bear additional methods for training approximation architectures; see the relevant discussions in Chapters 5 and 6.

3.4 Q-FACTOR PARAMETRIC APPROXIMATION

We will now consider an alternative form of approximation in value space and fitted value iteration. It involves approximation of the optimal Q-factors of state-control pairs (x_k, u_k) at time k, with no intermediate approximation of cost-to-go functions. The optimal Q-factors are defined by

$$Q_k^*(x_k, u_k) = E\Big\{ g_k(x_k, u_k, w_k) + J_{k+1}^*\big(f_k(x_k, u_k, w_k)\big) \Big\}, \quad k = 0, \ldots, N-1, \tag{3.26}$$

where J_{k+1}^* is the optimal cost-to-go function for stage $k+1$. Thus $Q_k^*(x_k, u_k)$ is the cost attained by using u_k at state x_k, and subsequently using an optimal policy.

As noted in Section 1.2, the DP algorithm can be written as

$$J_k^*(x_k) = \min_{u \in U_k(x_k)} Q_k^*(x_k, u_k),$$

and by using this equation, we can write Eq. (3.26) in the following equivalent form that relates Q_k^* with Q_{k+1}^*:

$$Q_k^*(x_k, u_k) = E\Big\{ g_k(x_k, u_k, w_k) + \min_{u \in U_{k+1}(f_k(x_k,u_k,w_k))} Q_{k+1}^*\big(f_k(x_k, u_k, w_k), u\big)\Big\}. \tag{3.27}$$

This suggests that in place of the Q-factors $Q_k^*(x_k, u_k)$, we may use Q-factor approximations and Eq. (3.27) as the basis for suboptimal control.

We can obtain such approximations by using methods that are similar to the ones we have considered so far (parametric approximation, enforced decomposition, certainty equivalent control, etc). Parametric Q-factor approximations $\tilde{Q}_k(x_k, u_k, r_k)$ may involve a neural network, or a feature-based linear architecture. The feature vector may depend on just the state, or on both the state and the control. In the former case, the architecture has the form

$$\tilde{Q}_k(x_k, u_k, r_k) = r_k(u_k)'\phi_k(x_k), \tag{3.28}$$

where $r_k(u_k)$ is a separate weight vector for each control u_k. In the latter case, the architecture has the form

$$\tilde{Q}_k(x_k, u_k, r_k) = r_k'\phi_k(x_k, u_k), \tag{3.29}$$

where r_k is a weight vector that is independent of u_k. The architecture (3.28) is suitable for problems with a relatively small number of control options at each stage. In what follows, we will focus on the architecture (3.29), but the discussion with few modifications, also applies to the architecture (3.28) and to nonlinear architectures as well.

We may adapt the fitted value iteration approach of the preceding section to compute sequentially the parameter vectors r_k in Q-factor parametric approximations, starting from $k = N - 1$. This algorithm is based on Eq. (3.27), with r_k obtained by solving least squares problems similar to the ones of the cost function approximation case [cf. Eq. (3.25)]. As an example, the parameters r_k of the architecture (3.29) are computed sequentially by collecting sample state-control pairs (x_k^s, u_k^s), $s = 1, \ldots, q$, and solving the linear least squares problems

$$r_k \in \arg\min_r \sum_{s=1}^{q} \big(r'\phi_k(x_k^s, u_k^s) - \beta_k^s\big)^2, \tag{3.30}$$

where

$$\beta_k^s = E\Big\{g_k(x_k^s, u_k^s, w_k) + \min_{u\in U_{k+1}(f_k(x_k^s,u_k^s,w_k))} r_{k+1}'\phi_{k+1}\big(f_k(x_k^s, u_k^s, w_k), u\big)\Big\}. \tag{3.31}$$

Thus, having obtained r_{k+1}, we obtain r_k through a least squares fit that aims to minimize the sum of the squared errors in satisfying Eq. (3.27). Note that the solution of the least squares problem (3.30) can be obtained in closed form as

$$r_k = \left(\sum_{s=1}^{q} \phi_k(x_k^s, u_k^s)\phi_k(x_k^s, u_k^s)'\right)^{-1} \sum_{s=1}^{q} \beta_k^s \phi_k(x_k^s, u_k^s);$$

[cf. Eq. (3.4)]. Once r_k has been computed, the one-step lookahead control $\tilde{\mu}_k(x_k)$ is obtained on-line as

$$\tilde{\mu}_k(x_k) \in \arg \min_{u\in U_k(x_k)} \tilde{Q}_k(x_k, u, r_k), \tag{3.32}$$

without the need to calculate any expected value. This latter property is a primary incentive for using Q-factors in approximate DP, particularly when there are tight constraints on the amount of on-line computation that is possible in the given practical setting.

The samples β_k^s of Eq. (3.31) involve the computation of an expected value. In an alternative implementation, we may replace β_k^s with an average of just a few samples (even a single sample) of the random variable

$$g_k(x_k^s, u_k^s, w_k) + \min_{u\in U_{k+1}(f_k(x_k^s,u_k^s,w_k))} r_{k+1}'\phi_{k+1}\big(f_k(x_k^s, u_k^s, w_k), u\big), \tag{3.33}$$

collected according to the probability distribution of w_k. This distribution may either be known explicitly, or in a model-free situation, through a computer simulator; cf. the discussion of Section 2.1.4. In particular, to implement this scheme, we only need a simulator that for any pair (x_k, u_k) generates a sample of the stage cost $g_k(x_k, u_k, w_k)$ and the next state $f_k(x_k, u_k, w_k)$ according to the distribution of w_k.

Note that the samples of the random variable (3.33) do not require the computation of an expected value like the samples (3.24) in the cost approximation method of the preceding chapter. Moreover the samples of (3.33) involve a simpler minimization than the samples (3.24). This is an important advantage of working with Q-factors rather than state costs.

Having obtained the weight vectors $r_0, \ldots, r_{N-1}$, and hence the one-step lookahead policy $\tilde{\pi} = \{\tilde{\mu}_0, \ldots, \tilde{\mu}_{N-1}\}$ through Eq. (3.32), a further possibility is to approximate this policy with a parametric architecture. This is *approximation in policy space built on top of approximation in value space*; see the discussion of Section 2.1.5. The idea here is to simplify even further the on-line computation of the suboptimal controls by avoiding the minimization of Eq. (3.32).

Advantage Updating

Let us finally note an alternative to computing Q-factor approximations. It is motivated by the potential benefit of approximating Q-factor differences rather than Q-factors. In this method, called *advantage updating*, instead of computing and comparing $Q_k^*(x_k,u_k)$ for all $u_k \in U_k(x_k)$, we compute

$$A_k(x_k,u_k) = Q_k^*(x_k,u_k) - \min_{u_k \in U_k(x_k)} Q_k^*(x_k,u_k).$$

The function $A_k(x_k,u_k)$ can serve just as well as $Q_k^*(x_k,u_k)$ for the purpose of comparing controls, but may have a much smaller range of values than $Q_k^*(x_k,u_k)$.

In the absence of approximations, selecting controls by advantage updating is clearly equivalent to selecting controls by comparing their Q-factors. However, when approximation is involved, using A_k can be important, because Q_k^* may embody sizable quantities that are independent of u, and which may interfere with algorithms such as the fitted value iteration (3.30)-(3.31). In particular, when training an architecture to approximate Q_k^*, the training algorithm may naturally try to capture the large scale behavior of Q_k^*, which may be irrelevant because it may not be reflected in the Q-factor differences A_k. However, with advantage updating, we may instead focus the training process on finer scale variations of Q_k^*, which may be all that matters. This question is discussed further and is illustrated with an example in the neuro-dynamic programming book [BeT96], Section 6.6.2.

The technique of subtracting a suitable constant (often called a *baseline*) from a quantity that is estimated by simulation is a useful idea; see Fig. 3.4.1 (in the case of advantage updating, the constants depend on x_k, but the same general idea applies). It can also be used in the context of the sequential DP approximation method of Section 3.3, as well as in conjunction with other simulation-based methods in RL.

3.5 PARAMETRIC APPROXIMATION IN POLICY SPACE BY CLASSIFICATION

We have focused so far on approximation in value space using parametric architectures. In this section we will discuss briefly how the cost function approximation methods of this chapter can be suitably adapted for the purpose of approximation in policy space, whereby we select the policy by using optimization over a parametric family of some form.

In particular, suppose that for a given stage k, we have access to dataset of "good" sample state-control pairs (x_k^s, u_k^s), $s = 1,\ldots,q$, obtained through some unspecified process, such as rollout or problem approximation. We may then wish to "learn" this process by training the parameter

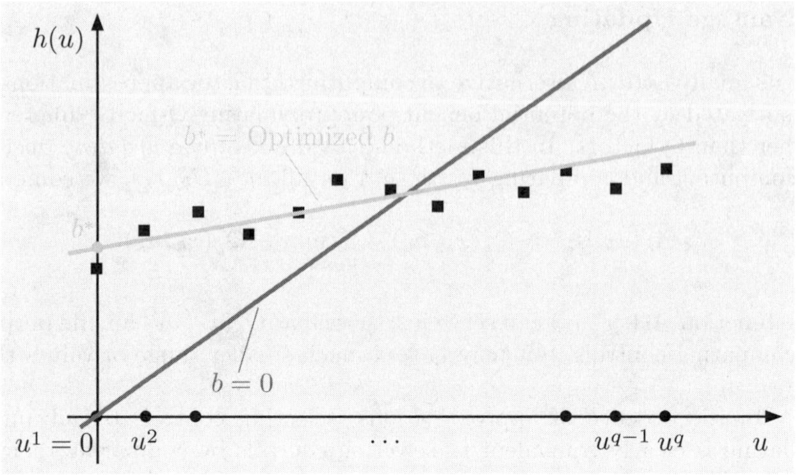

Figure 3.4.1. Illustration of the idea of subtracting a baseline constant from a cost or Q-factor approximation. Here we have samples $h(u^1),\ldots,h(u^q)$ of a scalar function $h(u)$ at sample points $u^1,\ldots,u^q$, and we want to approximate $h(u)$ with a linear function $\tilde h(u,r)=ru$, where r is a scalar tunable weight. We subtract a baseline constant b from the samples, and we solve the problem

$$\bar r \in \arg\min_r \sum_{s=1}^q \Big(\big(h(u^s)-b\big)-ru^s\Big)^2.$$

By properly adjusting b, we can improve the quality of the approximation, which after subtracting b from all the sample values, takes the form $\tilde h(u,b,r)=b+ru$. Conceptually, b serves as an additional weight (multiplying the basis function 1), which enriches the approximation architecture.

vector r_k of a parametric family of policies $\tilde\mu_k(x_k,r_k)$, using least squares minimization/regression:

$$\bar r_k \in \arg\min_{r_k} \sum_{s=1}^q \big\|u_k^s-\tilde\mu_k(x_k^s,r_k)\big\|^2; \tag{3.34}$$

cf. our discussion of approximation in policy space in Section 2.1.5.

It is useful to make the connection of this regression approach with *classification*, an important problem in machine learning. This is the problem of constructing an algorithm, called a *classifier*, which assigns a given "object" to one of a finite number of "categories" based on its "characteristics." Here we use the term "object" generically. In some cases, the classification may relate to persons or situations. In other cases, an object may represent a hypothesis, and the problem is to decide which of the hypotheses is true, based on some data. In the context of approximation in policy space, objects correspond to states, and categories correspond to controls to be applied at the different states. Thus in this case, we view each sample $\big(x_k^s, u_k^s\big)$ as an object-category pair.

Generally, in (multiclass) classification we assume that we have a population of objects, each belonging to one of m categories $c = 1, \ldots, m$. We want to be able to assign a category to any object that is presented to us. Mathematically, we represent an object with a vector x (e.g., some raw description or a vector of features of the object), and we aim to construct a rule that assigns to every possible object x a unique category c.

To illustrate a popular classification method, let us assume that if we draw an object x at random from this population, the conditional probability of the object being of category c is $p(c \mid x)$. If we know the probabilities $p(c \mid x)$, we can use a classical statistical approach, whereby we assign x to the category $c^*(x)$ that has maximal posterior probability, i.e.,

$$c^*(x) \in \arg \max_{c=1,\ldots,m} p(c \mid x). \tag{3.35}$$

This is called the Maximum a Posteriori rule (or MAP rule for short; see for example the book [BeT08], Section 8.2, for a discussion).

When the probabilities $p(c \mid x)$ are unknown, we may try to estimate them using a least squares optimization, based on the following property.

Proposition 3.5.1: (Least Squares Property of Conditional Probabilities) Let $\xi(x)$ be any prior distribution of x, so that the joint distribution of (c, x) is

$$\zeta(c, x) = \sum_x \xi(x) \sum_{c=1}^m p(c \mid x).$$

Let $z(c, x)$ be the function of (c, x) defined by

$$z(c, x) = \begin{cases} 1 & \text{if } x \text{ is of category } c, \\ 0 & \text{otherwise.} \end{cases}$$

For any function $h(c, x)$ of (c, x), consider

$$E\Big\{\big(z(c, x) - h(c, x)\big)^2\Big\},$$

the expected value with respect to the distribution $\zeta(c, x)$ of the random variable $\big(z(c, x) - h(c, x)\big)^2$. Then $p(c \mid x)$ minimizes this expected value over all functions $h(c, x)$, i.e., for all functions h, we have

$$E\Big\{\big(z(c, x) - p(c \mid x)\big)^2\Big\} \leq E\Big\{\big(z(c, x) - h(c, x)\big)^2\Big\}. \tag{3.36}$$

The proof of the proposition may be found in textbooks that deal with Bayesian least squares estimation (see, e.g., [BeT08], Section 8.3).†

The proposition states that $p(c\,|\,x)$ is the function of (c,x) that minimizes

$$E\Big\{\big(z(c,x)-h(c,x)\big)^2\Big\} \tag{3.37}$$

over all functions h of (c,x), *independently of the prior distribution of* x. This suggests that we can obtain approximations to the probabilities $p(c\,|\,x)$, $c=1,\ldots,m$, by minimizing an empirical/simulation based approximation of the expected value (3.37).

More specifically, let us assume that we have a training set consisting of q object-category pairs (x^s,c^s), $s=1,\ldots,q$, and corresponding vectors

$$z^s(c)=\begin{cases}1 & \text{if } c^s=c,\\ 0 & \text{otherwise,}\end{cases} \qquad c=1,\ldots,m,$$

and adopt a parametric approach. In particular, for each category $c=1,\ldots,m$, we approximate the probability $p(c\,|\,x)$ with a function $\tilde h(c,x,r)$

---

† For a quick argument, consider for any pair (c,x) the conditional expected value $E\Big\{\big(z(c,x)-y\big)^2 \;\big|\; c,x\Big\}$, where y is any scalar. Given (c,x), the random variable $z(c,x)$ takes the value $z(c,x)=1$ with probability $p(c\,|\,x)$ and the value $z(c,x)=0$ with probability $1-p(c\,|\,x)$, so we have

$$E\Big\{\big(z(c,x)-y\big)^2 \;\big|\; c,x\Big\}=p(c\,|\,x)(y-1)^2+\big(1-p(c\,|\,x)\big)y^2.$$

We minimize this expression with respect to y, by setting to 0 its derivative, i.e.,

$$0=2p(c\,|\,x)(y-1)+2\big(1-p(c\,|\,x)\big)y=2\big(-p(c\,|\,x)+y\big).$$

We thus obtain the minimizing value of y, namely $p(c\,|\,x)$, so that

$$E\Big\{\big(z(c,x)-p(c\,|\,x)\big)^2 \,|\, c,x\Big\}\le E\Big\{\big(z(c,x)-y\big)^2 \,|\, c,x\Big\}, \qquad \text{for all scalars } y.$$

For any function h of (c,x) we set $y=h(c,x)$ in the above expression and obtain

$$E\Big\{\big(z(c,x)-p(c\,|\,x)\big)^2 \,|\, c,x\Big\}\le E\Big\{\big(z(c,x)-h(c,x)\big)^2 \,|\, c,x\Big\}.$$

Since this is true for all (c,x), we also have

$$\sum_{(c,x)}\zeta(c,x)E\Big\{\big(z(c,x)-p(c\,|\,x)\big)^2 \,|\, c,x\Big\}\le\sum_{(c,x)}\zeta(c,x)E\Big\{\big(z(c,x)-h(c,x)\big)^2 \,|\, c,x\Big\},$$

for all functions h, i.e., Eq. (3.36) holds.

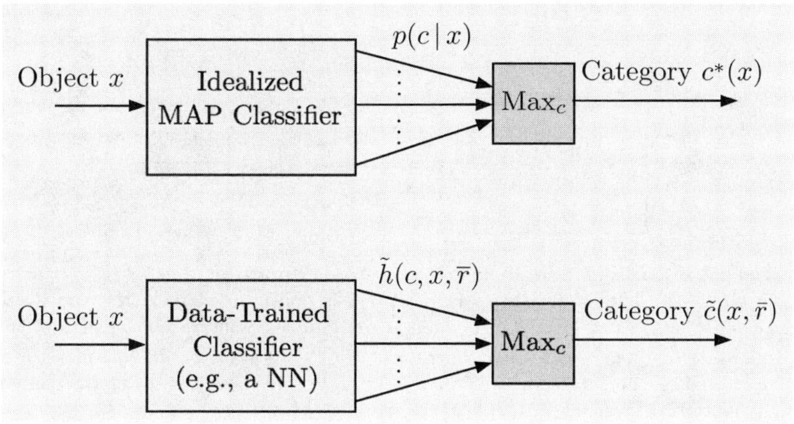

Figure 3.5.1. Illustration of the MAP classifier $c^*(x)$ for the case where the probabilities $p(c \,|\, x)$ are known [cf. Eq. (3.35)], and its data-trained version $\tilde{c}(x, \bar{r})$ [cf. Eq. (3.39)]. The classifier may be obtained by using the data set (x_k^s, u_k^s), $s = 1, \ldots, q$, and an approximation architecture such as a feature-based architecture or a neural network.

that is parametrized by a vector r, and optimize over r the empirical approximation to the expected squared error of Eq. (3.37). In particular, we obtain r by the least squares regression:

$$\bar{r} \in \arg\min_r \sum_{s=1}^{q} \sum_{c=1}^{m} \big(z^s(c) - \tilde{h}(c, x^s, r) \big)^2. \tag{3.38}$$

The functions $\tilde{h}(c, x, r)$ may be provided for example by a feature-based architecture or a neural network.

Note that each training pair (x^s, c^s) is used to generate m examples for use in the regression problem (3.38): $m - 1$ "negative" examples of the form $(x^s, 0)$, corresponding to the $m - 1$ categories $c \neq c^s$, and one "positive" example of the form $(x^s, 1)$, corresponding to $c = c^s$. Note also that the incremental training methods described in Sections 3.1.3 and 3.2.1 can be applied to the solution of this problem.

The regression problem (3.38) approximates the minimization of the expected value (3.37), so we conclude that its solution $\tilde{h}(c, x, \bar{r})$, $c = 1, \ldots, m$, approximates the probabilities $p(c \,|\, x)$. Once this solution is obtained, we may use it to classify a new object x according to the rule

$$\text{Estimated Object Category} = \tilde{c}(x, \bar{r}) \in \arg\max_{c=1,\ldots,m} \tilde{h}(c, x, \bar{r}), \tag{3.39}$$

which approximates the MAP rule (3.35); cf. Fig. 3.5.1.

Returning to approximation in policy space, for a given training set (x_k^s, u_k^s), $s = 1, \ldots, q$, the classifier just described provides (approximations

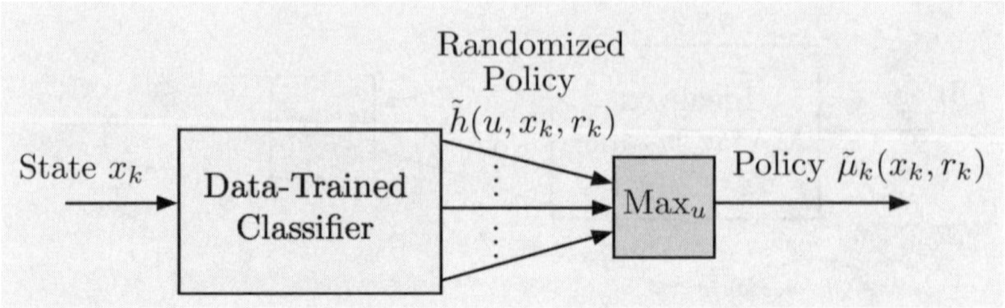

Figure 3.5.2 Illustration of classification-based approximation in policy space. The classifier, defined by the parameter r_k, is constructed by using the training set (x_k^s, u_k^s), $s = 1, \ldots, q$. It yields a randomized policy that consists of the probability $\tilde{h}(u, x_k, r_k)$ of using control $u \in U_k(x_k)$ at state x_k. This policy is approximated by the deterministic policy $\tilde{\mu}_k(x_k, r_k)$ that uses at state x_k the control that maximizes over $u \in U_k(x_k)$ the probability $\tilde{h}(u, x_k, r_k)$ [cf. Eq. (3.39)].

to) the "probabilities" of using the controls $u_k \in U_k(x_k)$ at the states x_k, so it yields a "randomized" policy $\tilde{h}(u, x_k, r_k)$ for stage k [once the values $\tilde{h}(u, x_k, r_k)$ are normalized so that, for any given x_k, they add to 1]; cf. Fig. 3.5.2. In practice, this policy is usually approximated by the deterministic policy $\tilde{\mu}_k(x_k, r_k)$ that uses at state x_k the control of maximal probability at that state; cf. Eq. (3.39).

For the simpler case of a classification problem with just two categories, say A and B, a similar formulation is to hypothesize a relation of the following form between object x and its category:

$$\text{Object Category} = \begin{cases} A & \text{if } \tilde{h}(x, r) = 1, \\ B & \text{if } \tilde{h}(x, r) = -1, \end{cases}$$

where $\tilde{h}$ is a given function and r is the unknown parameter vector. Given a set of q object-category pairs $(x^1, z^1), \ldots, (x^q, z^q)$ where

$$z^s = \begin{cases} 1 & \text{if } x \text{ is of category } A, \\ -1 & \text{if } x \text{ is of category } B, \end{cases}$$

we obtain r by the least squares regression:

$$\bar{r} \in \arg\min_r \sum_{s=1}^{q} \big(z^s - \tilde{h}(x^s, r)\big)^2.$$

The optimal parameter vector $\bar{r}$ is used to classify a new object with data vector x according to the rule

$$\text{Estimated Object Category} = \begin{cases} A & \text{if } \tilde{h}(x, \bar{r}) > 0, \\ B & \text{if } \tilde{h}(x, \bar{r}) < 0. \end{cases}$$

In the context of DP, this classifier may be used, among others, in stopping problems where there are just two controls available at each state: stopping (i.e., moving to a termination state) and continuing (i.e., moving to some nontermination state).

There are several variations of the preceding classification schemes, for which we refer to the specialized literature. Moreover, there are several commercially and publicly available software packages for solving the associated regression problems and their variants. They can be brought to bear on the problem of parametric approximation in policy space using any training set of state-control pairs, regardless of how it was obtained.

3.6 NOTES AND SOURCES

Section 3.1: Our discussion of approximation architectures, neural networks, and training has been limited, and aimed just to provide the connection with approximate DP and a starting point for further exploration. The literature on the subject is vast, and the textbooks mentioned in the references to Chapter 1 provide detailed accounts and many sources, in addition to the ones given in Sections 3.1.3 and 3.2.1.

There are two broad directions of inquiry in parametric architectures:

(1) The design of architectures, either in a general or a problem-specific context. Research in this area has intensified following the increased popularity of deep neural networks.

(2) The training of neural networks as well as linear architectures.

Research on both of these issues has been extensive and is continuing. An important direction, not discussed here, is how to take advantage of distributed computation, particularly in conjunction with partitioned architectures (see [BeT96], Section 3.1.3, [BeY10], [BeY12], [Ber19c]).

Methods for selection of basis functions have received much attention, particularly in the context of neural network research and deep reinforcement learning (see e.g., the book by Goodfellow, Bengio, and Courville [GBC16]). For discussions that are focused outside the neural network area, see Bertsekas and Tsitsiklis [BeT96], Keller, Mannor, and Precup [KMP06], Jung and Polani [JuP07], Bertsekas and Yu [BeY09], and Bhatnagar, Borkar, and Prashanth [BBP13]. Moreover, there has been considerable research on the optimal feature selection within given parametric classes (see Menache, Mannor, and Shimkin [MMS05], Yu and Bertsekas [YuB09b], Busoniu et al. [BBD10], and Di Castro and Mannor [DiM10]).

Incremental algorithms are supported by substantial theoretical analysis, which addresses issues of convergence, rate of convergence, stepsize selection, and component order selection. Moreover, incremental algorithms have been extended to constrained optimization settings, where the constraints are also treated incrementally, first by Nedić [Ned11], and then by several other authors: Bertsekas [Ber11c], Wang and Bertsekas [WaB14], [WabB16], Bianchi [Bia16], Iusem, Jofre, and Thompson [IJT18]. It is beyond our scope to cover this analysis. The author's surveys [Ber10] and

[Ber15b], and convex optimization and nonlinear programming textbooks [Ber15a], [Ber16a], collectively contain an extensive account of incremental methods, including the Kaczmarz, and incremental gradient, subgradient, aggregated gradient, Newton, Gauss-Newton, and extended Kalman filtering methods, and give many references. We also refer to the book [BeT96] and paper [BeT00] by Bertsekas and Tsitsiklis, and the survey by Bottou, Curtis, and Nocedal [BCN18] for a theoretically oriented treatments.

Section 3.2: The publicly and commercially available neural network training programs incorporate heuristics for scaling and preprocessing data, stepsize selection, initialization, etc, which can be very effective in specialized problem domains. We refer to books on neural networks such as Bishop [Bis95], Goodfellow, Bengio, and Courville [GBC16], and Haykin [Hay08].

Deep neural networks have created a lot of excitement in the machine learning field, in view of some high profile successes in image and speech recognition, and in RL with the AlphaGo and AlphaZero programs. One question is whether and for what classes of target functions we can enhance approximation power by increasing the number of layers while keeping the number of weights constant. For analysis and speculation around this question, see Bengio [Ben09], Liang and Srikant [LiS16], Yarotsky [Yar17], Daubechies et al. [DDF19], and the references quoted there. Another research question relates to the role of overparametrization in the success of deep neural networks. With more weights than training data, the training problem has infinitely many solutions. The question then is how to select a solution that works well on test data (i.e., data outside the training set); see [ZBH16], [BMM18], [BRT18], [SJL18], [BLL19], [HMR19], [MVS19].

Section 3.3: Fitted value iteration has a long history; it has been mentioned by Bellman among others. It has interesting properties, and at times exhibits pathological/unstable behavior due to accumulation of errors over a long horizon. We will discuss this behavior in Section 5.2.

Section 3.4: Advantage updating was proposed by Baird [Bai93], [Bai94], and is discussed in Section 6.6 of the book [BeT96].

Section 3.5: Classification (sometimes called "pattern classification" or "pattern recognition") is a major subject in machine learning, for which there are many approaches and an extensive literature; see e.g. the textbooks by Bishop [Bis95], [Bis06], and Duda, Hart, and Stork [DHS12]. Approximation in policy space was formulated as a classification problem in the context of DP by Lagoudakis and Parr [LaP03], and was followed up by several other authors; see our subsequent discussion of rollout-based policy iteration for infinite horizon problems (cf. Section 5.7.3). While we have focused on a classification approach that makes use of least squares regression and a parametric architecture, other classification methods may also be used. For example the paper [LaP03] discusses the use of nearest neighbor schemes, support vector machines, as well as neural networks.

4

Infinite Horizon Dynamic Programming

Contents

In this chapter, we provide an introduction to the theory of infinite horizon problems. We focus on exact solution by DP methods, leaving the discussion of approximations to subsequent chapters. Infinite horizon problems differ from their finite horizon counterparts in two main respects:

(a) The number of stages is infinite.

(b) The system is stationary, i.e., the system equation, the cost per stage, and the disturbance probability distributions do not change from one stage to the next.

The assumption of an infinite number of stages is hardly ever satisfied in practice, but is a reasonable approximation for problems involving a finite but very large number of stages. The assumption of stationarity is often satisfied in practice, and in other cases it approximates well a situation where the system parameters vary relatively slowly with time. Moreover, nonstationary (including periodic) problems can be converted to stationary ones using a conceptually simple reformulation; see [Ber12], Section 4.6.

Infinite horizon problems give rise to elegant and insightful analysis, and their optimal policies are often simpler than their finite horizon counterparts. For example, optimal policies are typically stationary, i.e., the optimal rule for choosing controls does not change over time.

On the other hand, infinite horizon problems generally require a more sophisticated mathematical treatment. Our discussion will be limited to relatively simple finite-state problems (the theory of problems with an infinite state space is considerably more intricate; see the books [BeS78], [Ber12], [Ber18a]). Still some theoretical results are needed in this chapter. They will be explained intuitively to the extent possible, and their mathematical proofs will be provided in the end-of-chapter appendix.

4.1 AN OVERVIEW OF INFINITE HORIZON PROBLEMS

We will focus on two types of infinite horizon problems, where we aim to minimize the total cost over an infinite number of stages, given by

$$J_\pi(x_0) = \lim_{N\to\infty} \underset{\substack{w_k\\ k=0,1,\ldots}}{E} \left\{ \sum_{k=0}^{N-1} \alpha^k g\big(x_k, \mu_k(x_k), w_k\big) \right\};$$

see Fig. 4.1.1. Here, $J_\pi(x_0)$ denotes the cost associated with an initial state x_0 and a policy $\pi = \{\mu_0, \mu_1, \ldots\}$, and α is a scalar in the interval $[0,1]$. When α is strictly less that 1, it has the meaning of a *discount factor*, and its effect is that future costs matter to us less than the same costs incurred at the present time.

Thus the infinite horizon cost of a policy is the limit of its finite horizon cost as the horizon tends to infinity. (We assume that the limit exists and is finite for the moment, and address the issue later.) The two types of problems, considered in Sections 4.2 and 4.3, respectively, are:

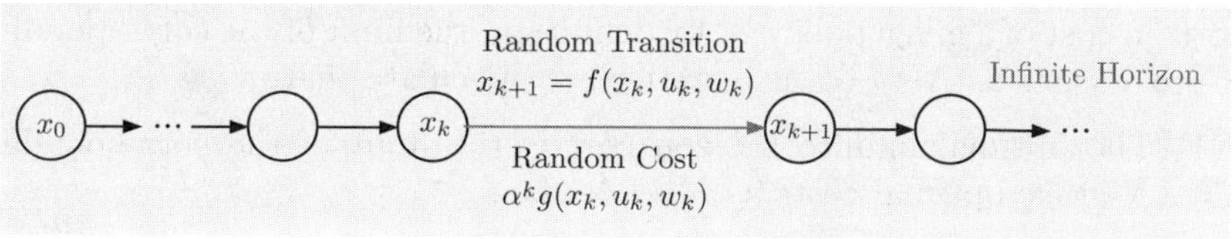

Figure 4.1.1 Illustration of an infinite horizon problem. The system and cost per stage are stationary, except for the use of a discount factor α. If $\alpha = 1$, there is typically a special cost-free termination state that we aim to reach.

(a) *Stochastic shortest path problems* (SSP for short). Here, $\alpha = 1$ but there is a special cost-free termination state; once the system reaches that state it remains there at no further cost. We will assume a problem structure such that termination is inevitable. Thus the horizon is in effect finite, but its length is random and may be affected by the policy being used.

(b) *Discounted problems*. Here, $\alpha < 1$ and there need not be a termination state. However, we will see that a discounted problem can be readily converted to an SSP problem. This can be done by introducing an artificial termination state to which the system moves with probability $1 - \alpha$ at every stage, thus making termination inevitable. As a result, our algorithms and analysis for SSP problems can be easily adapted to discounted problems.

A Preview of Infinite Horizon Theory

There are several analytical and computational issues regarding our infinite horizon problems. Many of them revolve around the relation between the optimal cost function J^* of the infinite horizon problem and the optimal cost functions of the corresponding N-stage problems.

In particular, consider the undicounted case and let $J_N(x)$ denote the optimal cost of the problem involving N stages, initial state x, cost per stage $g(x, u, w)$, and zero terminal cost. This cost is generated after N iterations of the DP algorithm

$$J_{k+1}(x) = \min_{u \in U(x)} \mathop{E}_{w} \Big\{ g(x, u, w) + J_k\big(f(x, u, w)\big) \Big\}, \qquad k = 0, 1, \ldots, \quad (4.1)$$

starting from the initial condition $J_0(x) = 0$ for all x.† The algorithm (4.1) is known as the *value iteration* algorithm (VI for short). Since the infinite

† This is just the finite horizon DP algorithm of Chapter 1. However, we have reversed the time indexing to suit our purposes. Thus the index of the cost functions produced by the algorithm is incremented with each iteration, and not decremented as in the case of finite horizon.

horizon cost of a given policy is, by definition, the limit of the corresponding N-stage costs as $N \to \infty$, it is natural to speculate that:

(1) The optimal infinite horizon cost is the limit of the corresponding N-stage optimal costs as $N \to \infty$; i.e.,

$$J^*(x) = \lim_{N\to\infty} J_N(x) \tag{4.2}$$

for all states x.

(2) The following equation should hold for all states x,

$$J^*(x) = \min_{u\in U(x)} \mathop{E}_{w} \Big\{ g(x,u,w) + J^*\big(f(x,u,w)\big)\Big\}. \tag{4.3}$$

This is obtained by taking the limit as $N \to \infty$ in the VI algorithm (4.1) using Eq. (4.2). Equation (4.3) is really a system of equations (one equation per state x), which has as solution the optimal costs-to-go of all the states. It can also be viewed as a *functional equation* for the optimal cost function J^*, and it is called *Bellman's equation.*

(3) If $\mu(x)$ attains the minimum in the right-hand side of the Bellman equation (4.3) for each x, then the policy $\{\mu, \mu, \ldots\}$ should be optimal. This type of policy is called *stationary*. Intuitively, optimal policies can be found within this class of policies, since the future optimization problem when starting at a given state looks the same regardless of the time when we start.

All three of the preceding results hold for SSP problems under our assumptions, as we will state later in Section 4.2 and prove in the appendix to this chapter. They also hold for discounted problems in suitably modified form that incorporates the discount factor. In fact the algorithms and analysis of this chapter are quite similar for discounted problems and SSP problems (under our assumptions), to the point where we may discuss a particular method for one of the two problems with the understanding that its application to the other problem can be straightforwardly adapted.

Transition Probability Notation for Finite-State Infinite Horizon Problems

Throughout this chapter we assume a finite-state discrete-time dynamic system, and we will use a special transition probability notation that is suitable for such a system. We generally denote states by the symbol i and successor states by the symbol j. We will assume that there are n states (in addition to the termination state for SSP problems). These states are denoted $1, \ldots, n$, and the termination state is denoted t. The control u is constrained to take values in a given finite constraint set $U(i)$, which may depend on the current state i. The use of a control u at state i specifies the transition probability $p_{ij}(u)$ to the next state j, at a cost $g(i,u,j)$.†

Given an admissible policy $\pi = \{\mu_0, \mu_1, \ldots\}$ [one with $\mu_k(i) \in U(i)$ for all i and k] and an initial state i_0, the system becomes a Markov chain whose generated trajectory under π, denoted $\{i_0, i_1, \ldots\}$, has a well-defined probability distribution. The total expected cost associated with an initial state i is

$$J_\pi(i) = \lim_{N\to\infty} E\left\{\sum_{k=0}^{N-1} \alpha^k g\big(i_k, \mu^k(i_k), i_{k+1}\big) \;\Big|\; i_0 = i,\ \pi\right\},$$

where α is either 1 (for SSP problems) or less than 1 for discounted problems. The expected value is taken with respect to the joint distribution of the states $i_1, i_2, \ldots$, conditioned on $i_0 = i$ and the use of π. The optimal cost from state i, i.e., the minimum of $J_\pi(i)$ over all policies π, is denoted by $J^*(i)$.‡

The cost function of a stationary policy $\pi = \{\mu, \mu, \ldots\}$ is denoted by $J_\mu(i)$. For brevity, we refer to π as the stationary policy μ. We say that μ is optimal if

$$J_\mu(i) = J^*(i) = \min_\pi J_\pi(i), \qquad \text{for all states } i.$$

As noted earlier, under our assumptions, we will show that there will always exist an optimal policy, which is stationary.

4.2 STOCHASTIC SHORTEST PATH PROBLEMS

In this section, we consider the SSP problem, where we assume that there is no discounting ($\alpha = 1$), and that *there is a special cost-free termination state* t. Once the system reaches that state, it remains there at no further cost, i.e.,

$$p_{tt}(u) = 1, \quad g(t,u,t) = 0, \qquad \text{for all } u \in U(t).$$

We denote by $1, \ldots, n$ the states other than the termination state t; see Fig. 4.2.1.

† To convert from the transition probability format to the system equation format used in the preceding chapters, we can simply use the system equation $x_{k+1} = w_k$, where w_k is the disturbance that takes values according to the transition probabilities $p_{x_k w_k}(u_k)$.

‡ Because of the underlying Markov chain, stochastic DP problems are called *Markovian decision problems* by many authors, including Bellman. We will use instead the term "DP problem" to refer to both deterministic and stochastic problems.

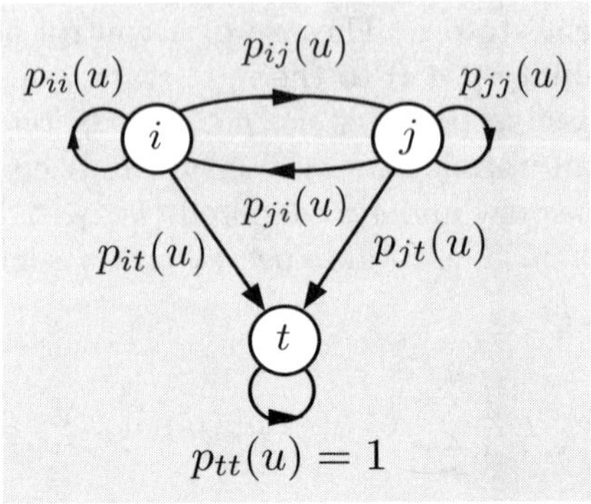

Figure 4.2.1 The transition graph of an SSP problem. There are n states, plus the termination state t, with transition probabilities $p_{ij}(u)$. The termination state is cost-free and absorbing.

With this notation, the Bellman equation (4.3) and the VI algorithm (4.1) take the following form.

Bellman Equation and Value Iteration for SSP Problems:

For all $i = 1, \ldots, n$, we have

$$J^*(i) = \min_{u \in U(i)} \left[p_{it}(u)g(i,u,t) + \sum_{j=1}^{n} p_{ij}(u)\big(g(i,u,j) + J^*(j)\big) \right]. \tag{4.4}$$

For all $i = 1, \ldots, n$, and any initial conditions $J_0(1), \ldots, J_0(n)$, the VI algorithm generates the sequence $\{J_k\}$ according to

$$J_{k+1}(i) = \min_{u \in U(i)} \left[p_{it}(u)g(i,u,t) + \sum_{j=1}^{n} p_{ij}(u)\big(g(i,u,j) + J_k(j)\big) \right].$$

The right-hand bracketed expression in the Bellman equation (4.4) represents an expected value, which is similar to the one we have seen in earlier DP expressions. It is the sum of three terms:

(a) The contribution

$$p_{it}(u)g(i,u,t)$$

to the expected cost of the current stage of the terminating i-to-t transition.

(b) The contribution

$$\sum_{j=1}^{n} p_{ij}(u)g(i,u,j)$$

to the expected cost of the current stage of the nonterminating i-to-j transitions.

(c) The optimal expected cost-to-go

$$\sum_{j=1}^{n} p_{ij}(u)J^*(j)$$

starting from the next state j [the optimal cost of t is $J^*(t) = 0$, so it does not appear in the preceding sum].

Note that the deterministic shortest path problem of Section 1.3.1 is obtained as the special case of the SSP problem where for each state-control pair (i, u), the transition probability $p_{ij}(u)$ is equal to 1 for a unique state j that depends on (i, u). Moreover, any deterministic or stochastic finite-state, finite horizon problem with a termination state (cf. Section 1.3.3) can be converted to an SSP problem. In particular, the reader may verify that the finite-state N-step horizon stochastic DP problem of Section 1.2 can be obtained as a special case of an SSP problem by viewing as state the pair (x_k, k) and introducing an artificial zero-cost transition from each pair (x_N, N) to an artificial termination state t.

We are interested in problems where reaching the termination state t is inevitable. Thus, the essence of the problem is to reach t with minimum expected cost. Throughout this chapter, when discussing SSP problems, we will make the following assumption, which will be shown to guarantee eventual termination under *all* policies.†

---

† The main analytical and algorithmic results for SSP problems are valid under more general conditions, which involve the notion of a proper policy (see the end-of-chapter references). In particular, a stationary policy is called *proper* if starting from every state, it is guaranteed to eventually reach the destination. The policy is called *improper* if it is not proper.

The subsequent four propositions can also be shown under the weaker assumption that there is at least one proper policy, and furthermore, every improper policy is "bad" in the sense that it results in infinite expected cost from at least one initial state (see [BeT89], [BeT91], or [Ber12], Chapter 3). These assumptions, when specialized to deterministic shortest path problems, are similar to the assumptions of Section 1.3.1. They imply that there is at least one path to the destination from every starting state and that all cycles have positive cost. In the absence of these assumptions, serious complications may occur. In particular, Bellman's equation may have no solution or an infinite number of solutions. Moreover, it may have a unique solution, which is different than J^*; see [BeY16],

Assumption 4.2.1: (Termination is Inevitable Under All Policies) There exists an integer m such that regardless of the policy used and the initial state, there is positive probability that the termination state will be reached after no more than m stages; i.e., for all admissible policies π we have

$$\rho_\pi = \max_{i=1,\ldots,n} P\{x_m \neq t \mid x_0 = i, \pi\} < 1.$$

Let ρ be the maximum probability of not reaching t within m stages, over all starting states and policies:

$$\rho = \max_\pi \rho_\pi.$$

Note that ρ_π depends only on the first m components of the policy π. Furthermore, since the number of controls available at each state is finite, the number of distinct m-stage policies is also finite. Thus there can be only a finite number of distinct values of ρ_π, so that $\rho < 1$. This implies that *the probability of not reaching t over a finite horizon diminishes to 0 as the horizon becomes longer*, regardless of the starting state and policy.

To see this, note that for any π and any initial state i

$$\begin{aligned} P\{x_{2m} \neq t \mid x_0 = i, \pi\} &= P\{x_{2m} \neq t \mid x_m \neq t,\, x_0 = i, \pi\} \\ &\quad \cdot P\{x_m \neq t \mid x_0 = i, \pi\} \\ &\leq \rho^2. \end{aligned}$$

More generally, for each π, the probability of not reaching the termination state after km stages diminishes like ρ^k regardless of the initial state, i.e.,

$$P\{x_{km} \neq t \mid x_0 = i, \pi\} \leq \rho^k, \qquad i = 1, \ldots, n. \tag{4.5}$$

This implies that the limit defining the associated total cost vector J_π exists and is finite, and is central in the proof of the following results.

We now describe the main theoretical results for SSP problems; the proofs are given in the appendix to this chapter. Our first result is that the infinite horizon version of the DP algorithm, which is VI [cf. Eq. (4.1)], converges to the optimal cost function J^*. The optimal cost $J^*(t)$ starting from

---

and [Ber18a], Section 3.1.2. The book [Ber18a], Section 3.1.1, discusses a simple example, which in addition to t, involves a single state 1 at which we can either stay at cost a or move to t at cost b; anomalies occur when $a = 0$ and when $a < 0$. For analysis and extensions for different or more general SSP problems, see [DeR79], [Pat01], [Pat07], [Ber16b], [Ber18a], [Ber18e], [GuS18], [Ber19b].

t is of course 0, so it is just neglected where appropriate in the subsequent discussion. Generally, J^* is obtained in the limit, after an infinite number of iterations. However, there are important cases where convergence is obtained in finitely many iterations (see [Ber12], Chapter 3; a simple example is finite horizon problems that are formulated as SSP problems by adding an artificial termination state at the end of the horizon).

Proposition 4.2.1: (Convergence of VI) The optimal cost $J^*(i)$ is finite for each i. Moreover, given any initial conditions $J_0(1), \ldots, J_0(n)$, the sequence $\{J_k(i)\}$ generated by the VI algorithm

$$J_{k+1}(i) = \min_{u \in U(i)} \left[p_{it}(u) g(i,u,t) + \sum_{j=1}^{n} p_{ij}(u) \big(g(i,u,j) + J_k(j) \big) \right],$$

converges to $J^*(i)$ for each $i = 1, \ldots, n$.

Our next result is that the limiting form of the DP equation, Bellman's equation, has J^* as its unique solution.

Proposition 4.2.2: (Bellman's Equation) The optimal cost function

$$J^* = \big(J^*(1), \ldots, J^*(n) \big)$$

satisfies for all $i = 1, \ldots, n$, the equation

$$J^*(i) = \min_{u \in U(i)} \left[p_{it}(u) g(i,u,t) + \sum_{j=1}^{n} p_{ij}(u) \big(g(i,u,j) + J^*(j) \big) \right], \tag{4.6}$$

and is the unique solution of this equation.

The next proposition is a special case of the preceding two. It expresses that by restricting attention to a single policy μ, we obtain a VI algorithm and a Bellman equation that are specific to μ.

Proposition 4.2.3: (VI and Bellman's Equation for Policies) For any stationary policy μ, the corresponding cost function $J_\mu = \big(J_\mu(1), \ldots, J_\mu(n) \big)$ satisfies for all $i = 1, \ldots, n$, the equation

$$J_\mu(i) = p_{it}\big(\mu(i)\big)g\big(i, \mu(i), t\big) + \sum_{j=1}^{n} p_{ij}\big(\mu(i)\big)\Big(g\big(i, \mu(i), j\big) + J_\mu(j)\Big),$$

and is the unique solution of this equation. Furthermore, given any initial conditions $J_0(1), \ldots, J_0(n)$, the sequence $\{J_k(i)\}$ generated by the VI algorithm that is specific to μ,

$$J_{k+1}(i) = p_{it}\big(\mu(i)\big)g\big(i, \mu(i), t\big) + \sum_{j=1}^{n} p_{ij}\big(\mu(i)\big)\Big(g\big(i, \mu(i), j\big) + J_k(j)\Big),$$

converges to the cost $J_\mu(i)$ for each i.

Our final result provides a necessary and sufficient condition for optimality of a stationary policy.

Proposition 4.2.4: (Optimality Condition) A stationary policy μ is optimal if and only if for every state i, $\mu(i)$ attains the minimum in the Bellman equation (4.6).

We provide an example illustrating Bellman's equation.

Example 4.2.1 (Maximum Expected Time to Termination)

The case where

$$g(i, u, j) = -1, \qquad \text{for all } i,\ u \in U(i), \text{ and } j,$$

corresponds to a problem where the objective is to terminate as late as possible on the average, while the opposite of the optimal cost, $-J^*(i)$, is the maximum expected time to termination starting from state i. Under our assumptions, the optimal costs $J^*(i)$ uniquely solve Bellman's equation, which has the form

$$J^*(i) = \min_{u \in U(i)} \left[-1 + \sum_{j=1}^{n} p_{ij}(u) J^*(j)\right], \qquad i = 1, \ldots, n.$$

In the special case of a single policy μ, where there is only one control at each state, $-J_\mu(i)$ represents the expected time to reach t starting from i. This is known as the mean first passage time from i to t, and is given as the unique solution of the corresponding Bellman equation

$$J_\mu(i) = -1 + \sum_{j=1}^{n} p_{ij}\big(\mu(i)\big) J_\mu(j), \qquad i = 1, \ldots, n.$$

Bellman Operators and Contraction Properties

We will now provide an insightful mathematical result about SSP problems, which is proved in the appendix with the aid of the preceding example. To this end let us introduce for any vector $J = \big(J(1),\ldots,J(n)\big)$, the notation

$$(TJ)(i) = \min_{u\in U(i)}\left[p_{it}(u)g(i,u,t) + \sum_{j=1}^{n} p_{ij}(u)\big(g(i,u,j)+J(j)\big)\right], \qquad (4.7)$$

for all i, and

$$(T_\mu J)(i) = p_{it}\big(\mu(i)\big)g\big(i,\mu(i),t\big) + \sum_{j=1}^{n} p_{ij}\big(\mu(i)\big)\Big(g\big(i,\mu(i),j\big)+J(j)\Big), \quad (4.8)$$

for all μ and i. Here T and T_μ are the DP operators (or *Bellman operators* as we will call them), which map the vector J into the vectors

$$TJ = \big((TJ)(1),\ldots,(TJ)(n)\big), \qquad T_\mu J = \big((T_\mu J)(1),\ldots,(T_\mu J)(n)\big),$$

respectively. This is a very helpful shorthand notation that greatly simplifies the statement and proofs of analytical and algorithmic results. For example, Bellman's equations can be written in terms of these operators as the fixed point equations $J^* = TJ^*$ and $J_\mu = T_\mu J_\mu$.

The next proposition states that T and T_μ are contraction mappings, so the unique fixed point property of this mapping follows from general mathematical results about contraction mappings (see e.g., [Ber12], [Ber18a]). Moreover the contraction property provides a convergence rate estimate for VI, and is the basis for further analysis of exact and approximate methods for SSP problems (see the author's monograph [Ber18a] for a theoretical development of DP, which is based on fixed point theory and an abstract Bellman operator viewpoint).

Proposition 4.2.5: (Contraction Property of the Bellman Operator) The Bellman operator T of Eq. (4.7) is a contraction mapping with respect to some weighted norm

$$\|J\| = \max_{i=1,\ldots,n} \frac{\big|J(i)\big|}{v(i)},$$

defined by some vector $v = \big(v(1),\ldots,v(n)\big)$ with positive components. In other words, there exists a positive scalar $\rho < 1$ such that for any two n-dimensional vectors J and J', we have

$$\|TJ - TJ'\| \leq \rho\,\|J - J'\|.$$

The same is true for T_μ, the Bellman operator (4.8) of any policy μ.

The proof of the proposition, given in the appendix, shows that the weight $v(i)$ is equal to the maximum expected number of steps to reach t from i (cf. Example 4.2.1). Moreover the modulus of contraction is

$$\rho = \max_{i=1,\ldots,n} \frac{v(i) - 1}{v(i)}.$$

Thus, in problems where to reach t from some states requires a very large number of steps, the contraction modulus ρ is very close to 1. Note that the weight vector v and the corresponding weighted norm may be different for T_μ and for T, since the maximum expected number of steps to reach t from i may be different under μ and under an optimal policy.

Among others, the preceding contraction property provides a convergence rate estimate for VI, namely that the generated sequence $\{J_k\}$ satisfies

$$\|J_k - J^*\| \leq \rho^k \|J_0 - J^*\|.$$

This follows from the fact that J_k and J^* can be viewed as the results of the k-fold application of T to the vectors J_0 and J^*, respectively.

Bellman Equation and Value Iteration for Q-Factors

The results just given have counterparts involving Q-factors. The optimal Q-factors are defined for all $i = 1, \ldots, n$, and $u \in U(i)$ by

$$Q^*(i, u) = p_{it}(u) g(i, u, t) + \sum_{j=1}^{n} p_{ij}(u)\big(g(i, u, j) + J^*(j)\big).$$

As in the finite horizon case, $Q^*(i, u)$ can be interpreted as the cost of starting at i, using u for the first stage, and using an optimal policy afterwards. Once Q^* is computed by some method, an optimal policy μ^* can be obtained from the minimization

$$\mu^*(i) \in \arg\min_{u \in U(i)} Q^*(i, u), \qquad i = 1, \ldots, n.$$

Similarly, if approximately optimal Q-factors $\tilde{Q}(i, u)$ are obtained by some method (model-based or model-free), a suboptimal policy $\tilde{\mu}$ can be obtained from the minimization

$$\tilde{\mu}(i) \in \arg\min_{u \in U(i)} \tilde{Q}(i, u), \qquad i = 1, \ldots, n.$$

Our basic results relating Bellman's equation and the VI algorithm are stated as follows.

Bellman Equation and Value Iteration for Q-Factors and SSP Problems:

For all $i = 1, \ldots, n$, and $u \in U(i)$ we have

$$Q^*(i,u) = p_{it}(u)g(i,u,t) + \sum_{j=1}^{n} p_{ij}(u)\left(g(i,u,j) + \min_{v\in U(j)} Q^*(j,v)\right).$$

For all $i = 1, \ldots, n$, and $u \in U(i)$, and any initial conditions $Q_0(i,u)$, the VI algorithm generates the sequence $\{Q_k\}$ according to

$$Q_{k+1}(i,u) = p_{it}(u)g(i,u,t) + \sum_{j=1}^{n} p_{ij}(u)\left(g(i,u,j) + \min_{v\in U(j)} Q_k(j,v)\right).$$

Actually, the optimal Q-factors $Q^*(i,u)$ can be viewed as optimal state costs associated with a modified SSP problem, which involves a new state for each pair (i,u) with transition probabilities $p_{ij}(u)$ to the states $j = 1, \ldots, n, t$; see Fig. 4.2.2. Then the preceding Bellman equation for the optimal Q-factors, together with the Bellman equation (4.6) for the optimal costs $J^*(j)$, can be viewed as the Bellman equation for the modified SSP problem.

Temporal Differences and Cost Shaping

Bellman's equation can be written in an alternative form, which involves the differential

$$\hat{J} = J^* - V,$$

where $V = \big(V(1), \ldots, V(n)\big)$ is *any* n-dimensional vector and $V(t) = 0$. In particular, by subtracting $V(i)$ from both sides of the Bellman equation (4.6), and adding and subtracting $V(j)$ within the right-hand side summation, we obtain

$$\hat{J}(i) = \min_{u\in U(i)} \left[p_{it}(u)\hat{g}(i,u,t) + \sum_{j=1}^{n} p_{ij}(u)\big(\hat{g}(i,u,j) + \hat{J}(j)\big) \right], \qquad (4.9)$$

for all $i = 1, \ldots, n$, where

$$\hat{g}(i,u,j) = \begin{cases} g(i,u,j) + V(j) - V(i) & \text{if } i,j = 1,\ldots,n, \\ g(i,u,t) - V(i) & \text{if } i = 1,\ldots,n,\ j = t. \end{cases} \qquad (4.10)$$

We refer to Eq. (4.9) as the *variational form of Bellman's equation*, and to the modified cost per stage $\hat{g}$ as the *temporal difference* corresponding to

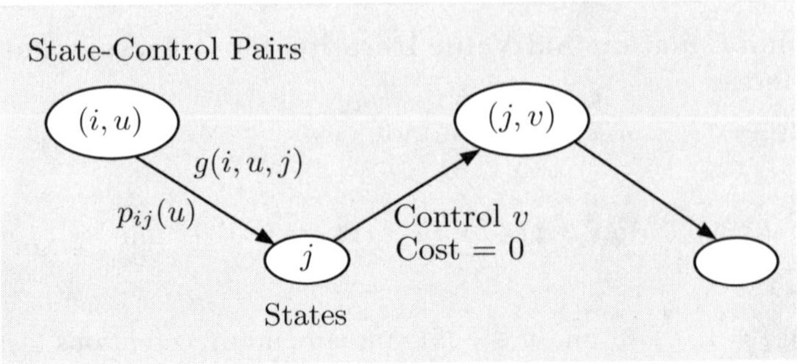

Figure 4.2.2 States, transition probabilities, and stage costs corresponding to a modified SSP problem, which yields the optimal Q-factors as well as the optimal costs. The states of this problem are the pairs (i, u), $u \in U(i)$, the original problem states $i = 1, \ldots, n$, and the termination state t. A control $v \in U(j)$ is available only at the original system states j, leading to the pair (j, v) at cost 0. The transition from a pair (i, u) leads to j with probability $p_{ij}(u)$ and cost 0. The Bellman equation for this modified problem is

$$Q^*(i,u) = p_{it}(u)g(i,u,t) + \sum_{j=1}^{n} p_{ij}(u)\left(g(i,u,j) + \min_{v \in U(j)} Q^*(j,v)\right),$$

for the states (i, u), $u \in U(i)$, and

$$J^*(j) = \min_{v \in U(j)} Q^*(j,v),$$

for the states $j = 1, \ldots, n$. Note that a policy μ for this problem leads from a state j to the state $\big(j, \mu(j)\big)$, so in any system trajectory, only pairs of the form $\big(j, \mu(j)\big)$ are visited after the first transition.

V. Temporal differences play a significant role in several algorithmic RL contexts; see Section 5.5, and the approximate DP/RL books noted earlier.

Note that Eq. (4.9) is the Bellman equation for a cost-modified problem, where the cost per stage g has been replaced by the temporal difference $\hat{g}$. Thus by applying Prop. 4.2.2 we have that $\hat{J} = J^* - V$ is the unique solution of this equation, so that J^* can be obtained by solving either the original or the cost-modified version of the problem. Moreover, a policy μ, has cost function $\hat{J}_\mu = J_\mu - V$ in the cost-modified problem. It follows that the original and the cost-modified SSP problems are essentially equivalent, and the choice of V does not matter when exact DP methods are used to solve them. However, when approximations are used, different results may be obtained, which can be more favorable with an appropriate choice of V.

In particular, we have the option to choose V and an approximation architecture methodology that matches the differential $\hat{J} = J^* - V$ better than it matches J^*. For example, we may obtain V with some problem approximation scheme as a rough estimate of J^*, and then use a different approximation in value space scheme, based on different principles, for the corresponding cost-modified problem. We refer to this as *cost shaping* (the name "reward shaping" in used in the RL literature, for problems involving reward maximization). While cost shaping does not change the optimal policies of the original DP problem, it may change significantly the suboptimal policies produced by approximate DP methods, such as the ones we have discussed and will discuss further in the next two chapters; see the material on baselines and advantage updating in Section 3.4, policy gradient methods in Section 5.7.1, and biased aggregation in Section 6.5. A key fact is that with a choice of V close to J^*, the range of values of $\hat{J}$ is reduced, and this tends to enhance the ability of approximation methods to capture small scale variations in J^*.

4.3 DISCOUNTED PROBLEMS

In this section, we consider the discounted problem, where there is a discount factor $\alpha < 1$. Using our transition probability notation, the Bellman equation and the VI algorithm take the following form.

Bellman Equation and Value Iteration for Discounted Problems:

For all $i = 1, \ldots, n$, we have

$$J^*(i) = \min_{u \in U(i)} \sum_{j=1}^{n} p_{ij}(u)\big(g(i,u,j) + \alpha J^*(j)\big).$$

For all $i = 1, \ldots, n$, and any initial conditions $J_0(1), \ldots, J_0(n)$, the VI algorithm generates the sequence $\{J_k\}$ according to

$$J_{k+1}(i) = \min_{u \in U(i)} \sum_{j=1}^{n} p_{ij}(u)\big(g(i,u,j) + \alpha J_k(j)\big).$$

We will show that the discounted problem can be converted to an SSP problem for which the analysis of the preceding section applies. To see this, let $i = 1, \ldots, n$ be the states, and consider an associated SSP problem involving the states $1, \ldots, n$ plus an *artificial termination state* t, with state transitions and costs obtained as follows: From a state $i \neq t$, when control u is applied, the next state is j with probability $\alpha p_{ij}(u)$ at a cost $g(i, u, j)$, and t with probability $1 - \alpha$ at zero cost; see Fig. 4.3.1. Note that Assumption 4.2.1 of the preceding section is satisfied for this

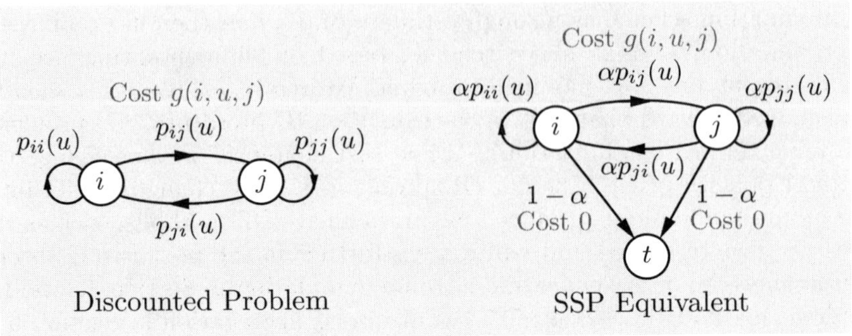

Figure 4.3.1 Transition probabilities for an α-discounted problem and its associated SSP problem. In the latter problem, the probability that the state is not t after k stages is α^k. The transition costs at the kth stage are $g(i, u, j)$ for both problems, but they must be multiplied by α^k because of discounting (in the discounted case) or because they are incurred with probability α^k when termination has not yet been reached (in the SSP case).

SSP problem, since t is reached with probability $1 - \alpha > 0$ from every state in a single step.

Suppose now that we use the same policy in the discounted problem and in the associated SSP problem. Then, as long as termination has not occurred, *the state evolution in the two problems is governed by the same transition probabilities*. Furthermore, the expected cost of the kth stage of the associated shortest path problem is the expected value of $g\big(i_k, \mu^k(i_k), i_{k+1}\big)$ multiplied by the probability that state t has not yet been reached, which is α^k. This is also the expected cost of the kth stage for the discounted problem. Thus starting from a given state, *the cost of any given policy (as well as the optimal cost) is the same for the original discounted problem and for the associated SSP problem.*

It follows that we can apply Props. 4.2.1-4.2.5 of the preceding section to the associated SSP problem and obtain corresponding results for the discounted problem, which properly incorporate the discount factor in accordance with the discounted-to-SSP equivalence just established.

Proposition 4.3.1: (Convergence of VI) The optimal cost $J^*(i)$ is finite for each i. Moreover, given any initial conditions $J_0(1), \ldots, J_0(n)$, the sequence $\big\{J_k(i)\big\}$ generated by the VI algorithm

$$J_{k+1}(i) = \min_{u \in U(i)} \sum_{j=1}^{n} p_{ij}(u)\big(g(i, u, j) + \alpha J_k(j)\big), \qquad i = 1, \ldots, n,$$

converges to $J^*(i)$ for each i.

Proposition 4.3.2: (Bellman's Equation) The optimal cost function

$$J^* = \big(J^*(1), \ldots, J^*(n)\big)$$

satisfies for all $i = 1, \ldots, n$, the equation

$$J^*(i) = \min_{u \in U(i)} \sum_{j=1}^{n} p_{ij}(u)\big(g(i,u,j) + \alpha J^*(j)\big), \tag{4.11}$$

and is the unique solution of this equation.

Proposition 4.3.3: (VI and Bellman's Equation for Policies) For any stationary policy μ, the corresponding cost function $J_\mu = \big(J_\mu(1), \ldots, J_\mu(n)\big)$ is the unique solution of the equation

$$J_\mu(i) = \sum_{j=1}^{n} p_{ij}\big(\mu(i)\big)\Big(g\big(i,\mu(i),j\big) + \alpha J_\mu(j)\Big), \qquad i = 1, \ldots, n.$$

Furthermore, given any initial conditions $J_0(1), \ldots, J_0(n)$, the sequence $\big\{J_k(i)\big\}$ generated by the VI algorithm

$$J_{k+1}(i) = \sum_{j=1}^{n} p_{ij}\big(\mu(i)\big)\Big(g\big(i,\mu(i),j\big) + \alpha J_k(j)\Big), \qquad i = 1, \ldots, n,$$

converges to the cost $J_\mu(i)$ for each i.

Proposition 4.3.4: (Optimality Condition) A stationary policy μ is optimal if and only if for every state i, $\mu(i)$ attains the minimum in the Bellman equation (4.11).

Bellman's equation (4.11) has a familiar DP interpretation. At state i, the optimal cost $J^*(i)$ is the minimum over all controls of the sum of the expected current stage cost and the expected optimal cost of all future stages. The former cost is $g(i,u,j)$. The latter cost is $J^*(j)$, but since this cost starts accumulating after one stage, it is discounted by multiplication with α.

Similar to Prop. 4.2.5, there is a contraction mapping result and convergence rate estimate for VI. To this end we introduce the Bellman operators

$$(TJ)(i) = \min_{u\in U(i)} \sum_{j=1}^{n} p_{ij}(u)\big(g(i,u,j) + \alpha J(j)\big), \qquad i = 1,\dots,n, \quad (4.12)$$

and

$$(T_\mu J)(i) = \sum_{j=1}^{n} p_{ij}\big(\mu(i)\big)\Big(g\big(i,\mu(i),j\big) + \alpha J(j)\Big), \qquad i = 1,\dots,n, \quad (4.13)$$

in analogy with their SSP counterparts of Eqs. (4.7) and (4.8). Similar to the SSP case, Bellman's equations can be written in terms of these operators as the fixed point equations $J^* = TJ^*$ $J_\mu = T_\mu J_\mu$. The following contraction result, shown in the appendix, is useful for the analysis of exact and approximate methods for discounted problems.

Proposition 4.3.5: (Contraction Property of the Bellman Operator) The Bellman operators T and T_μ of Eqs. (4.12) and (4.13) are contraction mappings of modulus α with respect to the maximum norm

$$\|J\| = \max_{i=1,\dots,n} \big|J(i)\big|.$$

In particular, for any two n-dimensional vectors J and J', we have

$$\|TJ - TJ'\| \le \alpha\|J - J'\|, \qquad \|T_\mu J - T_\mu J'\| \le \alpha\|J - J'\|.$$

Example 4.3.1 (Asset Selling)

Consider a problem of selling an asset over an infinite number of periods. At each period an offer becomes available. We assume that offers at different periods are independent and that they can take n values $v_1,\dots,v_n$ with corresponding probabilities $p(1),\dots,p(n)$. Here, if accepted, the amount i_k offered in period k, will be invested at a rate of interest r. By depreciating the sale amount to period 0 dollars, we view $(1+r)^{-k} i_k$ as the reward for selling the asset in period k at a price i_k, where $r > 0$ is the rate of interest. Then we have a discounted reward problem with discount factor $\alpha = 1/(1+r)$. The analysis of the present section is applicable, and the optimal value function J^* is the unique solution of Bellman's equation

$$J^*(i) = \max\left[v_i, \frac{1}{1+r}\sum_{j=1}^{n} p_j J^*(j)\right].$$

Thus the optimal reward function is characterized by the critical number

$$c = \frac{1}{1+r} \sum_{j=1}^{n} p_j J^*(j).$$

An optimal policy is obtained by minimizing over the two controls. It is to sell if and only if the current offer i is greater than c. The critical number c can be obtained by a simple form of VI (see [Ber17], Section 3.4).

A far more difficult version of the problem is one where the offers are correlated, so the offer at each stage may be viewed as an observation that provides information about future offers. A related difficult version of the problem is one where the probability distribution $p = \big(p(1), \ldots, p(n)\big)$ of the offers is unknown, and is estimated as new offers are revealed. In both cases the problem can be formulated as a partial state information problem involving a belief state: the estimate of the distribution p given the past offers (suitable conditions are of course needed to ensure that the estimate of p can be in principle computed exactly or can be approximated as a practical matter). Some of the finite horizon approximation methods of Chapters 2 and 3 can be adapted to solve such a problem. However, an exact solution is practically impossible, since this would involve DP calculations over an infinite dimensional space of belief states.

Cost Shaping for Discounted Problems

Let us also mention that the cost shaping idea discussed for SSP problems, extends readily to discounted problems. In particular, the variational form of Bellman's equation takes the form

$$\hat{J}(i) = \min_{u \in U(i)} \sum_{j=1}^{n} p_{ij}(u)\big(\hat{g}(i,u,j) + \alpha \hat{J}(j)\big), \qquad i = 1, \ldots, n,$$

for any given vector V, where

$$\hat{g}(i,u,j) = g(i,u,j) + \alpha V(j) - V(i), \qquad i = 1, \ldots, n,$$

is the temporal difference corresponding to V; cf. Eqs. (4.9) and (4.10).

Bellman Equation and Value Iteration for Q-Factors

As in the SSP case, the results just given have counterparts involving the optimal Q-factors, defined by

$$Q^*(i,u) = \sum_{j=1}^{n} p_{ij}(u)\big(g(i,u,j) + \alpha J^*(j)\big), \qquad i = 1, \ldots, n, \ u \in U(i).$$

They can be obtained from the corresponding SSP results, by viewing the discounted problem as a special case of the SSP problem. Once Q^* or an approximation $\tilde{Q}$ is computed by some method (model-based or model-free), an optimal policy μ^* or approximately optimal policy $\tilde{\mu}$ can be obtained from the minimization

$$\mu^*(i) \in \arg \min_{u \in U(i)} Q^*(i,u), \qquad i = 1, \ldots, n,$$

or the approximate version

$$\tilde{\mu}(i) \in \arg \min_{u \in U(i)} \tilde{Q}(i,u), \qquad i = 1, \ldots, n.$$

Our basic results relating Bellman's equation and the VI algorithm are stated as follows.

Bellman Equation and Value Iteration for Q-Factors and Discounted Problems:

For all $i = 1, \ldots, n$, and $u \in U(i)$ we have

$$Q^*(i,u) = \sum_{j=1}^{n} p_{ij}(u) \left(g(i,u,j) + \alpha \min_{v \in U(j)} Q^*(j,v) \right). \tag{4.14}$$

For all $i = 1, \ldots, n$, and $u \in U(i)$, and for any initial conditions $Q_0(i,u)$, the VI algorithm generates the sequence $\{Q_k\}$ according to

$$Q_{k+1}(i,u) = \sum_{j=1}^{n} p_{ij}(u) \left(g(i,u,j) + \alpha \min_{v \in U(j)} Q_k(j,v) \right). \tag{4.15}$$

The VI algorithm (4.15) forms the basis for various Q-learning methods to be discussed in Section 5.4.

4.4 SEMI-MARKOV DISCOUNTED PROBLEMS

In this section we will briefly discuss a variant of the discounted problem, called *semi-Markov* problem, where the transition between states may take a random amount of time, thereby affecting the cost per stage and the factor that the cost-to-go from the next state must be discounted. We will provide a fairly straightforward extension of our earlier discounted problem analysis for such problems. In particular, we will see that all the analytical results for discounted problems (cf. Props. 4.3.1-4.3.5) can be extended to semi-Markov problems, simply by replacing the discounted transition probabilities $\alpha p_{ij}(u)$ in the DP/Bellman operator with positive scalars $m_{ij}(u) < 1$ that depend on the control u and the transition (i, j). Moreover, the VI and PI algorithms of the next two sections, and their approximations of the next chapter admit straightforward extensions.†

We continue to assume that there are n states, denoted by $1, \ldots, n$, and that state transitions and control selections take place at discrete times. However, we allow the length of the time interval from one transition to the next to be random. The state and control at any time t are denoted by $x(t)$ and $u(t)$, respectively, and stay constant between transitions. We use the following notation:

t_k: The time of occurrence of the kth transition. By convention, we denote $t_0 = 0$.

$x_k = x(t_k)$: We have $x(t) = x_k$ for $t_k \leq t < t_{k+1}$.

$u_k = u(t_k)$: We have $u(t) = u_k$ for $t_k \leq t < t_{k+1}$.

In place of transition probabilities, we have *transition distributions* $\xi_{ij}(\tau, u)$, which for a given pair (i, u), specify the joint distribution of the transition interval and the next state:

$$\xi_{ij}(\tau, u) = P\{t_{k+1} - t_k \leq \tau,\ x_{k+1} = j \mid x_k = i,\ u_k = u\}.$$

Note that the transition distributions specify the ordinary transition probabilities via

$$p_{ij}(u) = P\{x_{k+1} = j \mid x_k = i,\ u_k = u\} = \lim_{\tau \to \infty} \xi_{ij}(\tau, u).$$

Note also that the conditional cumulative distribution function (CDF) of τ given i, j, u is

$$P\{t_{k+1} - t_k \leq \tau \mid x_k = i,\ x_{k+1} = j,\ u_k = u\} = \frac{\xi_{ij}(\tau, u)}{p_{ij}(u)} \tag{4.16}$$

[assuming that $p_{ij}(u) > 0$]. Thus, $\xi_{ij}(\tau, u)$ can be viewed as a "scaled CDF", i.e., a CDF multiplied by $p_{ij}(u)$ (see Fig. 4.4.1).

An advantage of working with the transition distributions $\xi_{ij}(\tau, u)$ is that they can be used to model discrete, continuous, and mixed distributions for the transition time τ. Generally, expected values of functions of

† Semi-Markov discounted problems occur commonly in practice, and in this section we aim to establish that they can be treated just like the ordinary discounted problems of Section 4.3. They also arise in the context of approximations, when a sequence of controls, to be selected over several time periods, are lumped together into a single control (see [PSS98], [CiS15]). Otherwise, semi-Markov problems are peripheral to our development, so the reader may wish to skip the present section at first reading, and return to it as needed.

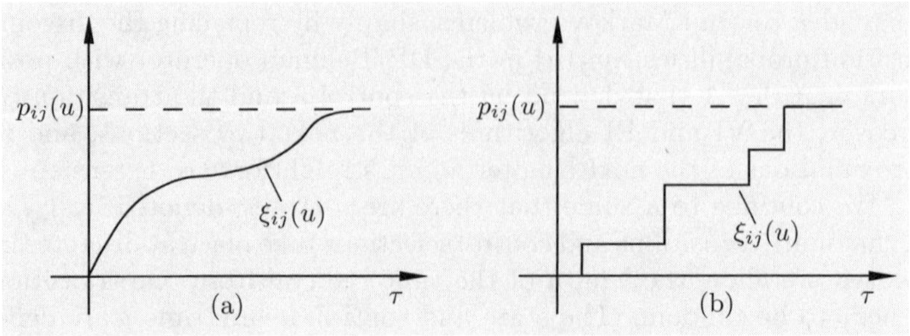

Figure 4.4.1 Illustration of the transition distributions $\xi_{ij}(\tau, u)$ and the conditional CDF of τ. Figures (a) and (b) correspond to the cases where τ is a continuous and a discrete random variable, respectively.

τ can be written as integrals involving the differential of ξ_{ij} with respect to τ, denoted $d\xi_{ij}(\tau, u)$. For example, the conditional expected value of τ given i, j, and u is written using the conditional CDF (4.16) as

$$E\{\tau \mid i, j, u\} = \int_0^\infty \tau \frac{d\xi_{ij}(\tau, u)}{p_{ij}(u)}. \tag{4.17}$$

If $\xi_{ij}(\tau, u)$ is continuous and piecewise differentiable with respect to τ, its partial derivative

$$\zeta_{ij}(\tau, u) = \frac{d\xi_{ij}(\tau, u)}{d\tau}$$

can be viewed as a "scaled" probability density function for τ. Then $d\xi_{ij}(\tau, u)$ may be replaced by $\zeta_{ij}(\tau, u)d\tau$, and expected values of functions of τ can be written in terms of $\zeta_{ij}(\tau, u)$. For example, Eq. (4.17) is written as

$$E\{\tau \mid i, j, u\} = \int_0^\infty \tau \frac{\zeta_{ij}(\tau, u)}{p_{ij}(u)} d\tau.$$

If $\xi_{ij}(\tau, u)$ is discontinuous and "staircase-like," then τ is a discrete random variable, and expected values of functions of τ can be written as summations.

Using Eq. (4.17), for each state i and control $u \in U(i)$, the expected transition time, denoted $\overline{\tau}_i(u)$, is given by

$$\overline{\tau}_i(u) = \sum_{j=1}^n p_{ij}(u) E\{\tau \mid i, j, u\} = \sum_{j=1}^n \int_0^\infty \tau \, d\xi_{ij}(\tau, u).$$

A technical assumption needed here is that $\overline{\tau}_i(u)$ is nonzero and finite:

$$0 < \overline{\tau}_i(u) < \infty. \tag{4.18}$$

We assume that for given state i and control $u \in U(i)$, the cost that is incurred in a small time interval dt is $g(i,u)dt$. Thus, we may view $g(i,u)$ as *cost per unit time* (for simplicity, we assume that this cost does not depend on j). The cost function has the form

$$\lim_{T\to\infty} E\left\{\int_0^T e^{-\beta t} g\big(x(t),u(t)\big)\,dt\right\},$$

where β is a given positive discount parameter. Since the cost per unit time, $g\big(x(t),u(t)\big)$, remains constant between transitions, the expected cost of a single transition from state i under control u is given by

$$\begin{aligned}
G(i,u) &= E\left\{\int_0^\tau e^{-\beta t} g(i,u)dt\right\}\\
&= g(i,u)\,E\left\{\int_0^\tau e^{-\beta t}dt\right\}\\
&= g(i,u)\,E_j\left\{E_\tau\left\{\int_0^\tau e^{-\beta t}dt \;\Big|\; j\right\}\right\}\\
&= g(i,u)\sum_{j=1}^n p_{ij}(u)\int_0^\infty\left(\int_0^\tau e^{-\beta t}dt\right)\frac{d\xi_{ij}(\tau,u)}{p_{ij}(u)}
\end{aligned}$$

or equivalently, since $\int_0^\tau e^{-\beta t}dt = (1-e^{-\beta\tau})/\beta$,

$$G(i,u) = g(i,u)\sum_{j=1}^n\int_0^\infty \frac{1-e^{-\beta\tau}}{\beta}\,d\xi_{ij}(\tau,u). \tag{4.19}$$

The cost of a policy $\pi = \{\mu_0,\mu_1,\ldots\}$ starting from state i is given by

$$J_\pi(i) = \lim_{N\to\infty}\sum_{k=0}^{N-1} E\left\{\int_{t_k}^{t_{k+1}} e^{-\beta t} g\big(x_k,\mu_k(x_k)\big)dt \;\Big|\; x_0 = i\right\}.$$

This cost can be broken down into the sum of the expected cost of the first transition, which is $G\big(i,\mu_0(i)\big)$, plus the expected cost-to-go starting from the next state, discounted by the factor $e^{-\beta\tau}$, where τ is the (random) time when the first transition occurs:

$$J_\pi(i) = G\big(i,\mu_0(i)\big) + E\big\{e^{-\beta\tau}J_{\pi_1}(j) \mid x_0 = i, u_0 = \mu_0(i)\big\}. \tag{4.20}$$

The last term in the above equation can be calculated as

$$\begin{aligned}
&E\big\{e^{-\beta\tau}J_{\pi_1}(j) \mid x_0 = i, u_0 = \mu_0(i)\big\}\\
&\qquad = E\big\{E\{e^{-\beta\tau} \mid j\}\,J_{\pi_1}(j) \;\big|\; x_0 = i, u_0 = \mu_0(i)\big\}\\
&\qquad = \sum_{j=1}^n p_{ij}\big(\mu_0(i)\big)\left(\int_0^\infty e^{-\beta\tau}\frac{d\xi_{ij}\big(\tau,\mu_0(i)\big)}{p_{ij}\big(\mu_0(i)\big)}\right)J_{\pi_1}(j)\\
&\qquad = \sum_{j=1}^n m_{ij}\big(\mu_0(i)\big)J_{\pi_1}(j),
\end{aligned}$$

where for any $u \in U(i)$, $m_{ij}(u)$ is given by

$$m_{ij}(u) = \int_0^\infty e^{-\beta\tau} d\xi_{ij}(\tau, u). \tag{4.21}$$

Thus, combining Eqs. (4.19)-(4.21), we see that $J_\pi(i)$ can be written as

$$J_\pi(i) = G\big(i, \mu_0(i)\big) + \sum_{j=1}^n m_{ij}\big(\mu_0(i)\big) J_{\pi_1}(j), \tag{4.22}$$

which is similar to the corresponding equation for discounted discrete-time problems $[m_{ij}\big(\mu_0(i)\big)$ replaces $\alpha p_{ij}\big(\mu_0(i)\big)]$.

In analogy with the discrete-time case, we may associate Eq. (4.22) with an SSP problem involving an artificial termination state t. Under control u, from state i the system moves to state j with probability $m_{ij}(u)$ and to the termination state t with probability

$$1 - \sum_{j=1}^n m_{ij}(u).$$

The assumption of a positive expected transition time [cf. Eq. (4.18)] implies that

$$\sum_{j=1}^n m_{ij}(u) < 1, \qquad \text{for all } i, u \in U(i),$$

so that the Assumption 4.2.1, which is required for the validity of our SSP analysis of Section 4.2, is satisfied. By using an essentially identical approach to the one of Section 4.3, we can derive analogs of all the discounted cost results. In particular, the optimal cost function J^* is the unique solution of Bellman's equation

$$J^*(i) = \min_{u \in U(i)} \left[G(i, u) + \sum_{j=1}^n m_{ij}(u) J^*(j) \right], \qquad i = 1, \ldots, n.$$

Moreover the VI algorithm takes the form

$$J_{k+1}(i) = \min_{u \in U(i)} \left[G(i, u) + \sum_{j=1}^n m_{ij}(u) J_k(j) \right], \qquad i = 1, \ldots, n.$$

What is happening here is that essentially we have the equivalent of a discrete-time discounted problem where the (effective) discount factor is $1 - \sum_{j=1}^n m_{ij}(u)$ and depends on i and u.

We finally note that in some problems, in addition to the cost per stage $G(i,u)$, there is an extra (instantaneous) one-stage cost $H(i,u)$ that is incurred at the time the control u is chosen at state i, and is independent of the length of the transition interval. In this case, Bellman's equation takes the form

$$J^*(i) = \min_{u \in U(i)} \left[H(i,u) + G(i,u) + \sum_{j=1}^n m_{ij}(u) J^*(j) \right].$$

Another problem variation arises when the cost per stage function g depends on the next state j. Here, once the system goes into state i, a control $u \in U(i)$ is selected, the next state is determined to be j with probability $p_{ij}(u)$, and the cost incurred is $g(i,u,j)$. In this case, $G(i,u)$ should be defined by

$$G(i,u) = \sum_{j=1}^n \int_0^\infty g(i,u,j) \frac{1-e^{-\beta\tau}}{\beta}\, d\xi_{ij}(\tau,u),$$

[cf. Eq. (4.19)] and the preceding development goes through without modification.

Other problem formulations that combine various aspects of the preceding ideas are possible. For a simple illustration, consider a version of the α-discounted problem of Section 4.3, with the only difference that when we are at state i and select control u, the next state will be j with probability $p_{ij}(u)$ and cost $g(i,u,j)$, but the transition will occur after a (nonrandom) integer number of time periods $d(i,u,j)$. Then it can be seen that the Bellman equation takes the intuitive form

$$J^*(i) = \min_{u \in U(i)} \sum_{j=1}^n p_{ij}(u)\big(g(i,u,j) + \alpha^{d(i,u,j)} J^*(j)\big), \qquad i = 1,\ldots,n.$$

In this case, the problem is essentially a discounted problem, but with state and control-dependent discount factor $\alpha^{d(i,u,j)}$.

4.5 ASYNCHRONOUS DISTRIBUTED VALUE ITERATION

We have already encountered the VI algorithm for SSP problems,

$$J_{k+1}(i) = \min_{u \in U(i)} \left[p_{it}(u) g(i,u,t) + \sum_{j=1}^n p_{ij}(u)\big(g(i,u,j) + J_k(j)\big) \right], \quad (4.23)$$

and its discounted version

$$J_{k+1}(i) = \min_{u \in U(i)} \sum_{j=1}^n p_{ij}(u)\big(g(i,u,j) + \alpha J_k(j)\big). \quad (4.24)$$

It is one of the principal methods for calculating the optimal cost function J^*.

Each VI, as described above, operates at all states simultaneously in every iteration. In a more general form of VI, at any one iteration, $J(i)$ may be updated only for a subset of states. An example is the *Gauss-Seidel method*, where at each iteration, $J(i)$ is updated only for a single selected state $\bar{i}$ and it is left unchanged for all other states $i \neq \bar{i}$. In the standard version of this method the states are taken up for iteration in a cyclic order, but more general iteration orders are possible, deterministic as well as randomized, as we discuss next.

In more complex versions of VI, one may use iterates that were generated several iterations earlier, i.e., iterates that are outdated. Methods of this type are called *asynchronous VI algorithms* and may be motivated by several considerations such as:

(a) *Faster convergence.* Generally, computational experience with DP as well as analysis have shown that convergence is accelerated by incorporating the results of VI updates for some states as early as possible into subsequent VI updates for other states. This is known as the *Gauss-Seidel effect*, which is discussed in detail in the distributed algorithms book by Bertsekas and Tsitsiklis [BeT89].

(b) *Parallel and distributed asynchronous computation.* In this context, we have several processors, each applying VI for a subset of states, and communicating the results to other processors (perhaps with some delay). One objective here may be faster computation by taking advantage of parallelism. Another objective may be computational convenience in problems where useful information is generated and processed locally at geographically dispersed points. An example is data or sensor network computations, where nodes, gateways, sensors, and data collection centers collaborate to route and control the flow of data, using DP or shortest path-type computations.

(c) *Simulation-based implementations.* Here we assume that iterations at various states are performed in the order that the states are generated by some form of simulation. Generally, within this context, it is unlikely that a VI will be performed simultaneously for all states i as the methods (4.23) and (4.24) require.

With these contexts in mind, models of asynchronous distributed VI methods have been developed. To describe one of these models, let us consider a computation framework involving a partition of the set of states $\{1, \ldots, n\}$ into disjoint nonempty subsets $I_1, \ldots, I_m$, and a network of m processors, each updating corresponding components of J. In particular, let J be partitioned as $J = (J_1, \ldots, J_m)$, where J_ℓ is the restriction of J on the set I_ℓ. Then in a (synchronous) distributed VI algorithm for the discounted problem, each processor $\ell = 1, \ldots, m$ updates J_ℓ at iteration t

according to

$$J_\ell^{t+1}(i) = \min_{u\in U(i)} \sum_{s=1}^{m} \sum_{j\in I_s} p_{ij}(u)\big(g(i,u,j) + \alpha J_s^t(j)\big), \qquad \text{for all } i \in I_\ell.$$

Here to accommodate the distributed algorithmic framework and its overloaded notation, we will use superscript t to denote iterations/times where some (but not all) processors update their corresponding components, reserving the subscript ℓ to denote component/processor index.

In an asynchronous VI algorithm, processor ℓ updates J_ℓ only for t in a selected subset $\mathcal{R}_\ell$ of iterations, and with components J_j, $j \neq \ell$, supplied by other processors with communication "delays" $t - \tau_{\ell j}(t)$ for all $i \in I_\ell$,

$$J_\ell^{t+1}(i) = \begin{cases} \min_{u\in U(i)} \sum_{s=1}^{m} \sum_{j\in I_s} p_{ij}(u)\big(g(i,u,j) + \alpha J_s^{\tau_{\ell j}}(j)\big) & \text{if } t \in \mathcal{R}_\ell, \\ J_\ell^t(i) & \text{if } t \notin \mathcal{R}_\ell. \end{cases} \tag{4.25}$$

[Here $\tau_{\ell j}(t) = t$ corresponds to "no delay" in information transfer from processor j to processor ℓ.] Communication delays arise naturally in the context of asynchronous distributed computing systems of the type described in many sources (an extensive reference is the book [BeT89]). Such systems are interesting for solution of large DP problems, particularly for methods that are based on simulation, which is naturally well-suited for distributed computation. On the other hand, if the entire algorithm is centralized at a single physical processor, the algorithm (4.25) ordinarily will not involve communication delays, i.e., $\tau_{\ell j}(t) = t$ for all ℓ, j, and t.

The simpler case where I_ℓ consists of a single element ℓ arises often, particularly in the context of simulation.† In this case we may simplify the notation of iteration (4.25) by writing J_ℓ^t in place of the scalar component $J_\ell^t(\ell)$, as we do in the following example.

Example 4.5.1 (One-State-at-a-Time Iterations)

Let us consider a generalization of the Gauss-Seidel method for the discounted problem. We view each state as a processor by itself, so that $I_\ell = \{\ell\}$, $\ell = 1,\ldots,n$, and execute VI one-state-at-a-time, according to some state sequence $\{i^0, i^1, \ldots\}$, which is generated in some way, possibly by simulation. Thus, starting from some initial vector J^0, we generate a sequence $\{J^t\}$, with $J^t = (J_1^t,\ldots,J_n^t)$, as follows:

$$J_\ell^{t+1} = \begin{cases} T(J_1^t,\ldots,J_n^t)(\ell) & \text{if } \ell = i^t, \\ J_\ell^t & \text{if } \ell \neq i^t, \end{cases}$$

† Note that we can assume without loss of generality that each state is assigned to a separate processor. The reason is that a physical processor that updates a group of states may be replaced by a group of fictitious processors, each assigned to a single state, and updating their corresponding components of J simultaneously.

where for simplicity we write J_ℓ^t instead of $J_\ell^t(\ell)$. This algorithm is a special case of iteration (4.25) where the set of times at which J_ℓ is updated is $\mathcal{R}_\ell = \{t \mid i^t = \ell\}$, and there are no communication delays (as in the case where the entire algorithm is centralized at a single physical processor).

The asynchronous algorithm (4.25) has some remarkable convergence properties. In particular let us assume the following:

(1) The set of times $\mathcal{R}_\ell$ at which processor ℓ updates J_ℓ is infinite, for each $\ell = 1, \ldots, m$.

(2) $\lim_{t\to\infty} \tau_{\ell j}(t) = \infty$ for all $\ell, j = 1, \ldots, m$.

These assumptions are natural and essential for proving any kind of convergence result about the algorithm. In particular, the condition $\tau_{\ell j}(t) \to \infty$ guarantees that outdated information about the processor updates will eventually be purged from the computation. It is also natural to assume that $\tau_{\ell j}(t)$ is monotonically increasing with t, but this assumption is not necessary for convergence.

Under the preceding assumptions it can be shown that the iterates generated by the asynchronous algorithm (4.25) converge to J^* ($J_\ell^t \to J_\ell^*$ for all ℓ) starting from any initial conditions J_ℓ^0. The proof is based on a general convergence theorem for totally asynchronous iterations from the author's paper [Ber83], which has served as the basis for the treatment of totally asynchronous iterations in the book [BeT89] (Chapter 6), and their application to DP (i.e., VI and PI), and asynchronous gradient-based optimization. A detailed treatment is beyond the scope of this text, and is given in several of the author's textbooks [BeT89], [Ber12], [Ber18a].

4.6 POLICY ITERATION

The major alternative to VI is *policy iteration* (PI for short). This algorithm starts with a stationary policy μ^0, and generates iteratively a sequence of new policies $\mu^1, \mu^2, \ldots$. The algorithm has solid convergence guarantees when implemented in its exact form, as we will show shortly.

The closest analog to PI that we have encountered so far is the rollout algorithm of Section 2.4. There we have started with some policy and produced an improved policy though a process of cost function evaluation and one-step or multistep lookahead minimization. This idea is extended in the context of PI, which consists of multiple successive policy evaluations and policy improvements, i.e., a form of *perpetual rollout*.

4.6.1 Exact Policy Iteration

Consider first the SSP problem. Here, each PI consists of two phases: *policy evaluation* and *policy improvement*; see Fig. 4.6.1. The algorithm takes the following form.

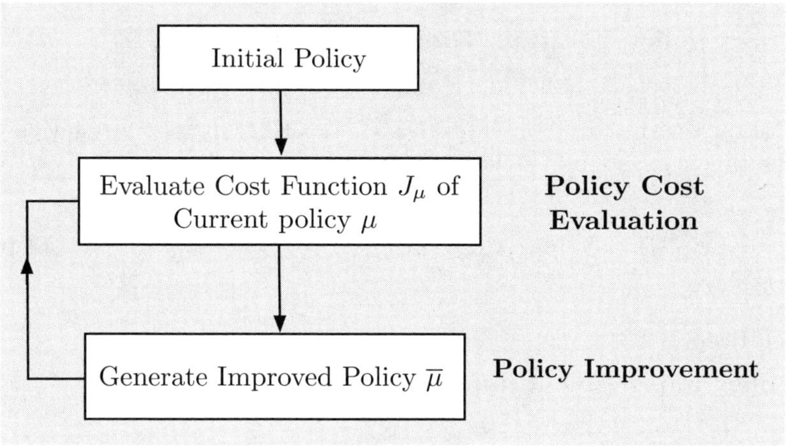

Figure 4.6.1 Illustration of exact PI. Each iteration consists of a policy evaluation using the current policy μ, followed by generation of an improved policy $\overline{\mu}$.

Exact Policy Iteration: SSP Problems

Given the typical policy μ^k:

Policy evaluation computes $J_{\mu^k}(i)$, $i = 1, \ldots, n$, as the solution of the (linear) system of Bellman equations

$$J_{\mu^k}(i) = p_{it}\big(\mu^k(i)\big)g\big(i, \mu^k(i), t\big) + \sum_{j=1}^{n} p_{ij}\big(\mu^k(i)\big)\Big(g\big(i, \mu^k(i), j\big) + J_{\mu^k}(j)\Big),$$

(cf. Prop. 4.2.3).

Policy improvement then computes a new policy μ^{k+1} as

$$\mu^{k+1}(i) \in \arg\min_{u \in U(i)} \left[p_{it}(u)g(i, u, t) + \sum_{j=1}^{n} p_{ij}(u)\big(g(i, u, j) + J_{\mu^k}(j)\big) \right],$$

for $i = 1, \ldots, n$.

The process is repeated with μ^{k+1} used in place of μ^k, unless we have

$$J_{\mu^{k+1}}(i) = J_{\mu^k}(i)$$

for all i, in which case the algorithm terminates with the policy μ^k.

The counterpart for discounted problems is as follows.

Exact Policy Iteration: Discounted Problems

Given the typical policy μ^k:

Policy evaluation computes $J_{\mu^k}(i)$, $i = 1, \ldots, n$, as the solution of the (linear) system of Bellman equations

$$J_{\mu^k}(i) = \sum_{j=1}^{n} p_{ij}\big(\mu^k(i)\big)\Big(g\big(i, \mu^k(i), j\big) + \alpha J_{\mu^k}(j)\Big), \tag{4.26}$$

(cf. Prop. 4.3.3).

Policy improvement then computes a new policy μ^{k+1} as

$$\mu^{k+1}(i) \in \arg\min_{u \in U(i)} \sum_{j=1}^{n} p_{ij}(u)\big(g(i, u, j) + \alpha J_{\mu^k}(j)\big), \quad i = 1, \ldots, n, \tag{4.27}$$

for $i = 1, \ldots, n$.

The process is repeated with μ^{k+1} used in place of μ^k, unless we have $J_{\mu^{k+1}}(i) = J_{\mu^k}(i)$ for all i, in which case the algorithm terminates with the policy μ^k.

The following proposition, shown in the appendix, establishes the validity of PI, including finite termination with an optimal policy.

Proposition 4.6.1: (Convergence of Exact PI) For both the SSP and the discounted problems, the exact PI algorithm generates an improving sequence of policies, i.e.,

$$J_{\mu^{k+1}}(i) \leq J_{\mu^k}(i), \qquad \text{for all } i \text{ and } k, \tag{4.28}$$

and terminates with an optimal policy.

The proof of the policy improvement property (4.28) is quite intuitive and is worth summarizing for the discounted problem. Let μ be a policy and $\overline{\mu}$ be the policy obtained from μ via a PI. We want to show that $J_{\overline{\mu}} \leq J_\mu$. To this end, let us *denote by J_N the cost function of a policy that applies $\overline{\mu}$ for the first N stages and applies μ at every subsequent stage.* We have the Bellman equation

$$J_\mu(i) = \sum_{j=1}^{n} p_{ij}\big(\mu(i)\big)\Big(g\big(i, \mu(i), j\big) + \alpha J_\mu(j)\Big),$$

which together with the policy improvement equation (4.27) implies that

$$J_1(i) = \sum_{j=1}^{n} p_{ij}\big(\overline{\mu}(i)\big)\Big(g\big(i, \overline{\mu}(i), j\big) + \alpha J_\mu(j)\Big) \leq J_\mu(i). \tag{4.29}$$

From the definition of J_2 and J_1 we have

$$J_2(i) = \sum_{j=1}^{n} p_{ij}\big(\overline{\mu}(i)\big)\Big(g\big(i, \overline{\mu}(i), j\big) + \alpha J_1(j)\Big), \tag{4.30}$$

so the preceding two relations imply that

$$J_2(i) \leq J_1(i) \leq J_\mu(i), \qquad \text{for all } i. \tag{4.31}$$

Continuing similarly, we obtain

$$J_{N+1}(i) \leq J_N(i) \leq J_\mu(i), \qquad \text{for all } i \text{ and } N. \tag{4.32}$$

Since $J_N \to J_{\overline{\mu}}$ (cf. Prop. 4.3.2), it follows that $J_{\overline{\mu}} \leq J_\mu$.

In practice, a lot of cost improvement is often obtained with the first few policies that are generated by the PI algorithm. This may happen even if the number of iterations needed for termination is quite large. The following is an example where termination occurs after just two iterations.

Example 4.6.1 (Treasure Hunting)

A treasure hunter has obtained a lease to search a site that contains n treasures, and wants to find a searching policy that maximizes his expected gain over an infinite number of days. At each day, knowing the current number of treasures not yet found, he may decide to continue searching for more treasures at a cost c per day, or to permanently stop searching. If he searches on a day when there are i treasures on the site, he finds $m \in [0, i]$ treasures with given probability $p(m \mid i)$, where we assume that $p(0 \mid i) < 1$ for all $i \geq 1$, and that the expected number of treasures found,

$$r(i) = \sum_{m=0}^{i} m\, p(m \mid i),$$

is monotonically increasing with i. Each found treasure is worth 1 unit.

We formulate the problem as an SSP problem, with state equal to the number of treasures not yet found. The termination state is state 0 (all treasures found), where the hunter stops searching. When the hunter decides to search at a state $i \geq 1$, the state moves to $i - m$ with probability $p(m \mid i)$. Here the inevitable termination Assumption 4.2.1 is satisfied, in view of our condition $p(0 \mid i) < 1$ for all i. Bellman's equation is

$$J^*(i) = \max\left[0, r(i) - c + \sum_{m=0}^{i} p(m \mid i) J^*(i - m)\right], \qquad i = 1, \ldots, n,$$

with $J^*(0) = 0$.

Let us apply PI starting with the policy μ^0 that never searches. This policy has value function

$$J_{\mu^0}(i) = 0, \qquad \text{for all } i.$$

The policy μ^1 subsequently produced by PI is the one that searches at a state i if and only if $r(i) > c$, and has value function satisfying the Bellman equation

$$J_{\mu^1}(i) = \begin{cases} 0 & \text{if } r(i) \leq c, \\ r(i) - c + \sum_{m=0}^{i} p(m \mid i) J_{\mu^1}(i-m) & \text{if } r(i) > c. \end{cases} \tag{4.33}$$

Note that the values $J_{\mu^1}(i)$ are nonnegative for all i, since by Prop. 4.6.1, we have

$$J_{\mu^1}(i) \geq J_{\mu^0}(i) = 0.$$

The next policy generated by PI is obtained from the maximization

$$\mu^2(i) \in \arg\max \left[0, r(i) - c + \sum_{m=0}^{i} p(m \mid i) J_{\mu^1}(i-m)\right], \qquad i = 1, \ldots, n.$$

For i such that $r(i) \leq c$, we have $r(j) \leq c$ for all $j < i$ because $r(i)$ is monotonically nondecreasing in i. Moreover, using Eq. (4.33), we have $J_{\mu^1}(i-m) = 0$ for all $m \geq 0$. It follows that for i such that $r(i) \leq c$,

$$0 \geq r(i) - c + \sum_{m=0}^{i} p(m \mid i) J_{\mu^1}(i-m),$$

and $\mu^2(i) =$ *stop searching*.

For i such that $r(i) > c$, we have $\mu^2(i) =$ *search*, since $J_{\mu^1}(i) \geq 0$ for all i, so that

$$0 < r(i) - c + \sum_{m=0}^{i} p(m \mid i) J_{\mu^1}(i-m).$$

Thus, μ^2 is the same as μ^1 and the PI algorithm terminates. By Prop. 4.6.1, it follows that μ^2 [the policy that searches with i treasures present if and only if the expected reward $r(i)$ exceeds the searching cost c] is optimal.

In the preceding example the convergence of PI is very fast. Let us also give an example where the convergence of PI is slow both near convergence and far from convergence. In this example, each PI optimizes the control of only one state, and termination of the PI algorithm requires n iterations.

Example 4.6.2 (An Example of Slow PI Convergence)

Consider a deterministic shortest path problem with states $1, \ldots, n$ and destination 0. At each state $i = 1, \ldots, n-1$ we can move to either state $i-1$ or state $i+1$ with cost 1. At state n we must move to the destination with cost n. The optimal policy is to move from $i = 1, \ldots, n-1$ to the state $i-1$, and the optimal cost function is $J^*(i) = i$ for all $i = 1, \ldots, n$, and $J^*(0) = 0$.

Assume that we start the PI algorithm with the policy μ^0 that moves from state $i = 1, \ldots, n-1$ to state $i+1$. The corresponding cost function is

$$J_{\mu^0}(i) = 2n - i, \qquad \text{for all } i = 1, \ldots, n,$$

and $J_{\mu^0}(0) = 0$. To compute the next policy μ^1 by policy improvement for states $i = 1, \ldots, n-1$, we must compare $1 + J_{\mu^0}(i-1)$ and $1 + J_{\mu^0}(i+1)$, so that we prefer state $i+1$ for all $i \neq 1$. Thus according to μ^1, we move to 0 from states 1 and n, and to $i+1$ for all states $i = 2, \ldots, n$. Proceeding similarly, we see that the policy μ^k generated by the kth iteration moves to 0 from state n, moves to $i-1$ from states $i = 1, \ldots, k$, and moves to state $i+1$ from states $i = k+1, \ldots, n-1$. Thus the optimal policy will be obtained after exactly $n-1$ iterations, and it will take one more iteration to verify its optimality, for a total of n.

The preceding example shows that despite the fact that PI often makes fast initial progress in practice, there is no guarantee that it will do so for a given problem. Let us also note that one may prove polynomial estimates of the number of iterations needed for the PI algorithm to terminate. However, for large n, these estimates are not very helpful, and they are usually very conservative predictors of practical behavior.

4.6.2 Optimistic and Multistep Lookahead Policy Iteration

The PI algorithm that we have discussed so far uses exact policy evaluation of the current policy μ^k and one-step lookahead policy improvement, i.e., it computes exactly J_{μ^k}, and it obtains the next policy μ^{k+1} by a one-step lookahead minimization using J_{μ^k} as an approximation to J^*. We will now describe a more flexible algorithm, called *optimistic PI*, whereby J_{μ^k} is approximated by any number of VIs corresponding to μ^k (cf. Prop. 4.3.3) and the policy improvement is done using multistep lookahead (alternative names used in the literature are"modified PI" and "generalized PI").

Optimistic PI aspires (somewhat optimistically) to obtain a good successor policy using an inaccurate evaluation of the current policy. It starts with a function J_0, an initial guess of J^*. It generates a sequence $\{J_k\}$ and an associated sequence of policies $\{\mu^k\}$. The kth iteration starts with a function J_k and first generates μ^k. It then generates J_{k+1} using m_k iterations of the VI algorithm that corresponds to μ^k, starting with J_k. We will prove that $\{J_k\}$ and $\{\mu^k\}$ asymptotically converge to J^* and an optimal policy, respectively.

Optimistic Policy Iteration: Discounted Problems

Given the typical function J_k:

Policy improvement computes a policy μ^k such that

$$\mu^k(i) \in \arg \min_{u \in U(i)} \sum_{j=1}^{n} p_{ij}(u)\big(g(i,u,j)+\alpha J_k(j)\big), \quad i = 1,\ldots,n. \tag{4.34}$$

Optimistic policy evaluation starts with $\hat{J}_{k,0} = J_k$, and uses m_k VI iterations for policy μ^k to compute $\hat{J}_{k,1},\ldots,\hat{J}_{k,m_k}$ according to

$$\hat{J}_{k,m+1}(i) = \sum_{j=1}^{n} p_{ij}(\mu^k(i))\Big(g\big(i,\mu^k(i),j\big) + \alpha \hat{J}_{k,m}(j)\Big), \tag{4.35}$$

for all $i = 1,\ldots,n$, $m = 0,\ldots,m_k - 1$, and finally sets $J_{k+1} = \hat{J}_{k,m_k}$.

From Eq. (4.35), it can be seen that one way to interpret optimistic PI is that *we approximate J_{μ^k} by using μ^k for m_k stages, and adding a terminal cost function equal to the current cost estimate J_k* instead of using μ^k for an additional infinite number of stages. Accordingly, *simulation-based approximations of optimistic PI, evaluate the cost function J_{μ^k} by using m_k-stage trajectories, with the cost of future stages accounted using J_k, which plays the role of a terminal cost function approximation.*

The convergence properties of optimistic PI are solid, although it may require an infinite number of iterations to converge to J^*. To see why this is so, suppose that we evaluate each policy with a *single* VI (i.e., $m_k \equiv 1$). Then Eqs. (4.34) and (4.35) yield $J_{k+1} = TJ_k$, and the method is essentially identical to the VI method, which requires an infinite number of iterations to converge.

The following proposition, shown in the appendix, establishes the validity of optimistic PI. There is a corresponding convergence property for SSP problems, but its currently available proof is fairly complicated. It is given in Section 3.5.1 of the book [Ber12]. Asynchronous versions of optimistic PI also involve theoretical convergence difficulties, which are discussed in Section 2.6.2 of [Ber12] and Section 2.6.3 of [Ber18a].

Proposition 4.6.2: (Convergence of Optimistic PI) For the discounted problem, the sequence $\{J_k\}$ generated by the optimistic PI algorithm satisfies $J_k \to J^*$, while the generated policies μ^k are optimal for all k sufficiently large.

The proof of the proposition is based on the policy improvement line of proof we gave earlier, following Prop. 4.6.1. In particular, if J_0 satisfies

$T_{\mu_0} J_0 \le J_0$, the argument of Eqs. (4.29)-(4.32) can be used to show that $J^* \le J_{k+1} \le J_k$ for all k. Moreover, the proof of the appendix argues that we may assume that $T_{\mu_0} J_0 \le J_0$ holds without loss of generality, since we may add a constant to J_0 without affecting the sequence $\{\mu^k\}$ generated by the algorithm. The proof of the appendix also shows that the generated policies μ^k are optimal after some k, but this fact cannot ordinarily be exploited in practice because the verification that μ^k is optimal requires additional computations that complicate the method.

Multistep Policy Improvement

The motivation for multistep policy improvement is that it may yield a better policy μ^{k+1} than the one obtained with one-step policy improvement. In fact this makes even more sense when the evaluation of μ^k is approximate, since then the longer lookahead may compensate for errors in the policy evaluation. We give below the method in its exact nonoptimistic form (in a different version it may be combined with optimistic PI, i.e., with policy evaluation performed approximately with a finite number of value iterations starting from some cost function approximation).

Multistep Lookahead Exact Policy Iteration: Discounted Problems

Given the typical policy μ^k:

Policy evaluation computes $J_{\mu^k}(i)$, $i = 1, \ldots, n$, as the solution of the (linear) system of Bellman equations

$$J_{\mu^k}(i) = \sum_{j=1}^{n} p_{ij}\big(\mu^k(i)\big)\Big(g\big(i, \mu^k(i), j\big) + \alpha J_{\mu^k}(j)\Big), \qquad i = 1, \ldots, n,$$

(cf. Prop. 4.2.3).

Policy improvement with ℓ-step lookahead then solves the ℓ-stage problem with terminal cost function J_{μ^k}. If $\{\hat\mu_0, \ldots, \hat\mu_{\ell-1}\}$ is the optimal policy of this problem, then the new policy μ^{k+1} is $\hat\mu_0$.

The process is repeated with μ^{k+1} used in place of μ^k, unless we have $J_{\mu^{k+1}}(i) = J_{\mu^k}(i)$ for all i, in which case the algorithm terminates with the policy μ^k.

Exact PI with multistep lookahead has the same solid convergence properties as its one-step lookahead counterpart: it terminates with an optimal policy, and the generated sequence of policies is monotonically improving. The proof is similar to the ones for other forms of PI and will not be given.

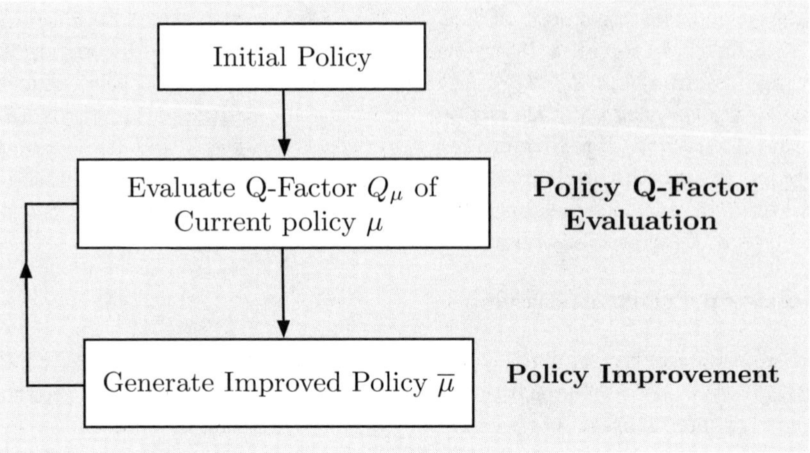

Figure 4.6.2 Block diagram of exact PI for Q-factors. Each iteration consists of a policy evaluation using the current policy μ, followed by generation of an improved policy $\overline{\mu}$.

4.6.3 Policy Iteration for Q-factors

Similar to VI, we may also equivalently implement PI through the use of Q-factors. To see this, note that the policy improvement step may be implemented by minimizing over $u \in U(i)$ the expression

$$Q_\mu(i,u) = \sum_{j=1}^{n} p_{ij}(u)\big(g(i,u,j) + \alpha J_\mu(j)\big), \qquad i = 1,\ldots,n,\ u \in U(i),$$

which we view as the *Q-factor of the pair* (i,u) *corresponding to* μ. Moreover we have

$$J_\mu(j) = Q_\mu\big(j,\mu(j)\big),$$

(cf. Prop. 4.3.3).

The following algorithm is thus obtained; see Fig. 4.6.2.

Exact Policy Iteration for Q-Factors: Discounted Problems

Given the typical policy μ^k:

Policy evaluation computes $Q_{\mu^k}(i,u)$, for all $i = 1,\ldots,n$, and $u \in U(i)$, as the solution of the (linear) system of equations

$$Q_{\mu^k}(i,u) = \sum_{j=1}^{n} p_{ij}(u)\Big(g(i,u,j) + \alpha Q_{\mu^k}\big(j,\mu^k(j)\big)\Big). \tag{4.36}$$

Policy improvement then computes a new policy μ^{k+1} as

$$\mu^{k+1}(i) \in \arg\min_{u \in U(i)} Q_{\mu^k}(i,u), \qquad i = 1,\ldots,n. \tag{4.37}$$

The process is repeated with μ^{k+1} used in place of μ^k, unless we have $Q_{\mu^k}\big(i,\mu^{k+1}(i)\big) = Q_{\mu^k}\big(i,\mu^k(i)\big)$ for all i, in which case the algorithm terminates with the policy μ^k.

Note that the system (4.36) has a unique solution, since from the uniqueness of solution of Bellman's equation, any solution must satisfy

$$Q_{\mu^k}\big(j,\mu^k(j)\big) = J_{\mu^k}(j).$$

Hence the Q-factors $Q_{\mu^k}\big(j,\mu^k(j)\big)$ are uniquely determined, and then the remaining Q-factors $Q_{\mu^k}(i,u)$ are also uniquely determined from Eq. (4.36).

The PI algorithm for Q-factors is mathematically equivalent to PI for costs, as given in the preceding subsection. The only difference is that we calculate *all* the Q-factors $Q_{\mu^k}(i,u)$, rather than just the costs $J_{\mu^k}(j) = Q_{\mu^k}\big(j,\mu^k(j)\big)$, i.e., just the Q-factors corresponding to the controls chosen by the current policy. However, the remaining Q-factors $Q_{\mu^k}(i,u)$ are needed for the policy improvement step (4.37), so no extra computation is required. It can be verified also that the PI algorithm (4.36)-(4.37) can be viewed as the PI algorithm for the discounted version of the modified problem of Fig. 4.2.2.

We finally note that asynchronous and optimistic PI algorithms for Q-factors involve substantial theoretical convergence complications (even without Q-factor approximations), as first shown by Williams and Baird [WiB93]. These difficulties have been resolved in papers by Bertsekas and Yu [BeY12], [YuB13a], by introducing special modifications to the policy evaluation Eq. (4.36). The modifications involve the solution of a stopping-type DP problem and may also be incorporated in Q-learning algorithms with Q-factor approximations.

4.7 NOTES AND SOURCES

Sections 4.1, 4.2, 4.3, 4.4: In this chapter we have provided an introduction to infinite horizon DP with a view towards approximate solution methods that are suitable for large-scale problems. We have restricted ourselves to discounted and SSP finite-state problems with perfect state information. These problems and their semi-Markov versions are discussed in several textbooks; see e.g., [Ros70], [Put94], [Ber12], [Ber17].

Finite-state problems involving minimization of an average cost per stage are more challenging than discounted problems. The book by Puterman [Put94] and the author's [Ber12] contain a detailed account of average cost finite-state Markovian decision problems. The corresponding approximate DP methodology is discussed in Section 7.1 of the book by Bertsekas and Tsitsiklis [BeT96], and the more recent papers by Tsitsiklis and Van Roy [TsV99a], and by Yu and Bertsekas [YuB09b]. Infinite-state versions of these problems exhibit very complex behaviors (see [Ber12], Section 5.6, for counterexamples).

Generally, infinite-state DP problems exhibit more complex behaviors and present many largely unresolved challenges for approximate DP methods. The theory of these problems is developed in a few books, including the author's [Ber12] and [Ber18a]. The latter book contains much recent research, including deterministic and stochastic infinite-state shortest path problems, and highlights pathological behavior that may hamper the application of RL methods. An advanced treatment, which also addresses the complex probabilistic questions introduced by infinite disturbance spaces is given by Bertsekas and Shreve [BeS78]; see also Yu and Bertsekas [YuB15], and Yu [Yu15] for recent supplementary work.

Partial state information problems are often formulated in a belief space of conditional state distributions, and accordingly are viewed as infinite-state problems with special structure. Their analysis and algorithmic solution is challenging, particularly when the original state space is itself infinite. They are not covered in much detail in this book.

The variational form of Bellman's equation has been used in various contexts, involving error bounds for VI, since the early days of DP theory, see e.g., [Ber12], Section 2.1.1. Within RL, the variational equation has been used in algorithmic settings under the name *reward shaping* or *potential-based shaping* (we have used the term "cost shaping" here as we are focusing on cost minimization); see the papers by Ng, Harada, and Russell [NHR99], Wiewiora [Wie03], Asmuth, Littman, and Zinkov [ALZ08], Devlin and Kudenko [DeK11], Grzes [Grz17] for some representative works. While reward shaping does not change the optimal policies of the original DP problem, it may change significantly the suboptimal policies produced by approximation in value space methods. Basically, with reward shaping and a linear feature-based approximation architecture, *V is used as an extra feature*. This is closely related to the idea of using approximate cost functions of policies as basis functions in approximation architectures, as suggested in the book [BeT96], Section 3.1.4.

Section 4.5, 4.6: The methods of VI and PI, and their optimistic variants are the cornerstones of infinite horizon computational DP, and they serve as the principal points of departure for approximation in value space. Both of these methods were introduced by Bellman in his 1957 book [Bel57], and in earlier reports on which the book was based.†

Asynchronous distributed VI methods were introduced by the author [Ber82], with further analysis given in [Ber83], and textbook descriptions in [BeT89], [Ber12], [Ber18a]. The idea of optimistic PI seems to have occurred simultaneously to several authors. It was first published and formally analyzed by van Nunen [Van76]. The papers by Puterman and Shin [PuS78], [PuS82] have also been influential. For recent complexity analysis, see Feinberg, Huang, and Scherrer [FHS14].

We finally note that in addition to VI, optimistic PI can also be implemented via distributed asynchronous computation; see the author's papers on DP and fixed point computations [Ber82], [Ber83], the paper by Williams and Baird on asynchronous PI [WiB93], and the series of papers by Bertsekas and Yu [BeY10], [BeY12], [YuB13a] on asynchronous optimistic PI and Q-learning. Generally, asynchronous and distributed algorithms are natural in computational contexts involving simulation, which by its nature is well suited to both multiprocessing and asynchronous implementation.

4.8 APPENDIX: MATHEMATICAL ANALYSIS

In this appendix we provide proofs of the mathematical results stated in this chapter. We also prove some supplementary results that are described in the chapter without formal statement.

We will make extensive use of the Bellman operators T and T_μ, given below for the α-discounted problem:

$$(TJ)(i) = \min_{u \in U(i)} \sum_{j=1}^{n} p_{ij}(u)\big(g(i,u,j) + \alpha J(j)\big), \qquad i = 1,\ldots,n,$$

$$(T_\mu J)(i) = \sum_{i=1}^{n} p_{ij}\big(\mu(i)\big)\Big(g\big(i,\mu(i),j\big) + \alpha J(j)\Big), \qquad i = 1,\ldots,n.$$

We note the key monotonicity property of these operators:

$$TJ \geq TJ', \quad T_\mu J \geq T_\mu J', \qquad \text{for all } J \text{ and } J' \text{ with } J \geq J',$$

---

† Bellman used the term "successive approximation" for VI, and "approximation in policy space" for (nonoptimistic) PI in his book [Bel57] (in earlier reports, written at the RAND corporation, he also used the term "approximation in strategy space" for PI). Of course Bellman introduced just the basic synchronous nonoptimistic methods, and proved their properties and their convergence in a limited number of special contexts. Extensions and variations of VI and PI have been discussed and analyzed in many subsequent works. The names "value iteration" and "policy iteration" are universally used at present.

where, here and later, the inequalities are meant to be pointwise, i.e., for each state. For the α-discounted problem, we also note the "constant shift" property, which states that if the function J is increased uniformly by a constant c, then the functions TJ and $T_\mu J$ are also increased uniformly by the constant αc.

4.8.1 Proofs for Stochastic Shortest Path Problems

We provide the proofs of Props. 4.2.1-4.2.5 from Section 4.2. A key insight for the analysis is that the expected cost incurred within an m-stage block vanishes exponentially as the start of the block moves forward (here m is the integer of Assumption 4.2.1, i.e., the termination state can be reached within m steps with positive probability from every starting state).

In particular, the cost in the m stages between km and $(k+1)m-1$ is bounded in absolute value by $\rho^k C$, where

$$C = m \max_{\substack{i=1,\ldots,n \\ j=1,\ldots,n,t \\ u\in U(i)}} \big|g(i,u,j)\big|.$$

Thus, we have

$$\big|J_\pi(i)\big| \le \sum_{k=0}^{\infty} \rho^k C = \frac{1}{1-\rho} C. \tag{4.38}$$

This shows that the "tail" of the cost series,

$$\sum_{k=mK}^{\infty} E\Big\{g\big(x_k, \mu_k(x_k), w_k\big)\Big\},$$

vanishes as K increases to ∞, since the probability that $x_{mK} \neq t$ decreases like ρ^K [cf. Eq. (4.5)].

Intuitively, since the "tail" of the cost series can be neglected as $K \to \infty$, it is valid to take the limit in the finite horizon DP algorithm (as the length of the horizon tends to ∞), and obtain the infinite horizon Bellman equation and VI convergence. This is the mathematical essence of the following proofs.

Proposition 4.2.1: (Convergence of VI) The optimal cost $J^*(i)$ is finite for each i. Moreover, given any initial conditions $J_0(1), \ldots, J_0(n)$, the sequence $\big\{J_k(i)\big\}$ generated by the VI algorithm

$$J_{k+1}(i) = \min_{u\in U(i)} \left[p_{it}(u) g(i,u,t) + \sum_{j=1}^{n} p_{ij}(u)\big(g(i,u,j) + J_k(j)\big) \right], \tag{4.39}$$

converges to $J^*(i)$ for each $i = 1, \ldots, n$.

Proof: From Eq. (4.38), we see that $J^*(i)$ is finite. For every $K > 0$, x_0, and $\pi = \{\mu_0, \mu_1, \ldots\}$, we break down the cost $J_\pi(x_0)$ into the portions incurred over the first mK stages and over the remaining stages:

$$
\begin{aligned}
J_\pi(x_0) &= \lim_{N\to\infty} E\left\{\sum_{k=0}^{N-1} g\big(x_k, \mu_k(x_k), w_k\big)\right\} \\
&= E\left\{\sum_{k=0}^{mK-1} g\big(x_k, \mu_k(x_k), w_k\big)\right\} + \lim_{N\to\infty} E\left\{\sum_{k=mK}^{N-1} g\big(x_k, \mu_k(x_k), w_k\big)\right\}.
\end{aligned}
$$

The expected cost during the Kth m-stage cycle [stages Km to $(K+1)m-1$] is upper bounded by $C\rho^K$ [cf. Eqs. (4.5) and (4.38)], so that

$$
\left|\lim_{N\to\infty} E\left\{\sum_{k=mK}^{N-1} g\big(x_k, \mu_k(x_k), w_k\big)\right\}\right| \leq C \sum_{k=K}^{\infty} \rho^k = \frac{\rho^K C}{1-\rho}.
$$

Also, consider J_0, the initial iterate of the VI algorithm, and bound its expected value under π after mK stages. We have

$$
\begin{aligned}
\big|E\{J_0(x_{mK})\}\big| &= \left|\sum_{i=1}^{n} P(x_{mK} = i \mid x_0, \pi) J_0(i)\right| \\
&\leq \left(\sum_{i=1}^{n} P(x_{mK} = i \mid x_0, \pi)\right) \max_{i=1,\ldots,n} \big|J_0(i)\big| \\
&\leq \rho^K \max_{i=1,\ldots,n} \big|J_0(i)\big|,
\end{aligned}
$$

since the probability that $x_{mK} \neq t$ is less or equal to ρ^K for any policy.

Combining the preceding relations, we obtain

$$
\begin{aligned}
-\rho^K &\max_{i=1,\ldots,n} \big|J_0(i)\big| + J_\pi(x_0) - \frac{\rho^K C}{1-\rho} \\
&\leq E\left\{J_0(x_{mK}) + \sum_{k=0}^{mK-1} g\big(x_k, \mu_k(x_k), w_k\big)\right\} \qquad (4.40) \\
&\leq \rho^K \max_{i=1,\ldots,n} \big|J_0(i)\big| + J_\pi(x_0) + \frac{\rho^K C}{1-\rho}.
\end{aligned}
$$

Note that the expected value in the middle term of the above inequalities is the mK-stage cost of policy π starting from state x_0, with a terminal cost $J_0(x_{mK})$; the minimum of this cost over all π is equal to the value $J_{mK}(x_0)$, which is generated by the VI algorithm (4.39) after mK iterations. Thus, by taking the minimum over π in Eq. (4.40), we obtain for all x_0 and K,

$$
\begin{aligned}
-\rho^K &\max_{i=1,\ldots,n} \big|J_0(i)\big| + J^*(x_0) - \frac{\rho^K C}{1-\rho} \\
&\leq J_{mK}(x_0) \\
&\leq \rho^K \max_{i=1,\ldots,n} \big|J_0(i)\big| + J^*(x_0) + \frac{\rho^K C}{1-\rho},
\end{aligned}
$$

and by taking the limit as $K \to \infty$, it follows that

$$\lim_{K \to \infty} J_{mK}(x_0) = J^*(x_0)$$

for all x_0. It can be seen that for all $\ell = 1, \ldots, m$, we have

$$\big|J_{mK+\ell}(x_0) - J_{mK}(x_0)\big| \leq \rho^K \left(C + \max_{i=1,\ldots,n} J_0(i) - \min_{i=1,\ldots,n} J_0(i)\right),$$

implying that $\lim_{K\to\infty} J_{mK+\ell}(x_0)$ is the same for all $\ell = 1, \ldots, m$, so that $\lim_{k\to\infty} J_k(x_0) = J^*(x_0)$. **Q.E.D.**

Proposition 4.2.2: (Bellman's Equation) The optimal cost function $J^* = \big(J^*(1), \ldots, J^*(n)\big)$ satisfies for all $i = 1, \ldots, n$, the equation

$$J^*(i) = \min_{u \in U(i)} \left[p_{it}(u) g(i, u, t) + \sum_{j=1}^{n} p_{ij}(u)\big(g(i, u, j) + J^*(j)\big) \right],$$

and in fact it is the unique solution of this equation.

Proof: By taking the limit as $k \to \infty$ in the VI algorithm (4.39) and using its convergence property shown in Prop. 4.2.1, we see that $J^*(1), \ldots, J^*(n)$ satisfy Bellman's equation (we are using here the fact that the limit and minimization operations commute when the minimization is over a finite number of alternatives).

To show uniqueness, observe that if $J(1), \ldots, J(n)$ satisfy Bellman's equation, then the VI algorithm (4.39) starting from $J(1), \ldots, J(n)$ just replicates $J(1), \ldots, J(n)$. It follows from the convergence result of Prop. 4.2.1 that $J(i) = J^*(i)$ for all i. **Q.E.D.**

Proposition 4.2.3: (VI and Bellman's Equation for Policies) For any stationary policy μ, the corresponding cost function $J_\mu = \big(J_\mu(1), \ldots, J_\mu(n)\big)$ satisfies for all $i = 1, \ldots, n$, the equation

$$J_\mu(i) = p_{it}\big(\mu(i)\big) g\big(i, \mu(i), t\big) + \sum_{j=1}^{n} p_{ij}\big(\mu(i)\big)\Big(g\big(i, \mu(i), j\big) + J_\mu(j)\Big),$$

and is in fact the unique solution of this equation. Furthermore, given any initial conditions $J_0(1), \ldots, J_0(n)$, the sequence $\{J_k(i)\}$ generated by the VI algorithm that is specific to μ,

$$J_{k+1}(i) = p_{it}\big(\mu(i)\big)g\big(i, \mu(i), t\big) + \sum_{j=1}^{n} p_{ij}\big(\mu(i)\big)\Big(g\big(i, \mu(i), j\big) + J_k(j)\Big),$$

converges to the cost $J_\mu(i)$ for each i.

Proof: Given μ, consider a modified stochastic shortest path problem, which is the same as the original except that the control constraint set contains only one element for each state i, the control $\mu(i)$; i.e., the control constraint set is

$$\tilde{U}(i) = \{\mu(i)\}$$

instead of $U(i)$. From Prop. 4.2.2, J_μ solves uniquely Bellman's equation for this modified problem, i.e., for all i,

$$J_\mu(i) = p_{it}\big(\mu(i)\big)g\big(i, \mu(i), t\big) + \sum_{j=1}^{n} p_{ij}\big(\mu(i)\big)\Big(g\big(i, \mu(i), j\big) + J_\mu(j)\Big).$$

From Prop. 4.2.1, the VI algorithm converges to $J_\mu(i)$. **Q.E.D.**

Proposition 4.2.4: (Optimality Condition) A stationary policy μ is optimal if and only if for every state i, $\mu(i)$ attains the minimum in Bellman's equation (4.6).

Proof: We have that $\mu(i)$ attains the minimum in Eq. (4.6) if and only if for all $i = 1, \ldots, n$, we have

$$\begin{aligned} J^*(i) &= \min_{u \in U(i)} \left[p_{it}(u)g(i, u, t) + \sum_{j=1}^{n} p_{ij}(u)\big(g(i, u, j) + J^*(j)\big) \right] \\ &= p_{it}\big(\mu(i)\big)g\big(i, \mu(i), t\big) + \sum_{j=1}^{n} p_{ij}\big(\mu(i)\big)\Big(g\big(i, \mu(i), j\big) + J^*(j)\Big). \end{aligned}$$

Proposition 4.2.3 and this equation imply that $J_\mu(i) = J^*(i)$ for all i. Conversely, if $J_\mu(i) = J^*(i)$ for all i, Props. 4.2.2 and 4.2.3 imply this equation. **Q.E.D.**

Proposition 4.2.5: (Contraction Property of the Bellman Operator) The Bellman operator T of Eq. (4.7) is a contraction mapping with respect to some weighted norm

$$\|J\| = \max_{i=1,\ldots,n} \frac{\big|J(i)\big|}{v(i)},$$

defined by some vector $v = \big(v(1), \ldots, v(n)\big)$ with positive components. In other words, there exists a positive scalar $\rho < 1$ such that for any two n-dimensional vectors J and J', we have

$$\|TJ - TJ'\| \leq \rho\, \|J - J'\|.$$

The same is true for T_μ, the Bellman operator (4.8) of any policy μ.

Proof: We first define the vector v by using the problem of Example 4.2.1. In particular, we let $v(i)$ be the maximal expected number of steps to termination starting from state i. From the Bellman equation in Example 4.2.1, we have for all $i = 1, \ldots, n$, and stationary policies μ,

$$v(i) = 1 + \max_{u \in U(i)} \sum_{j=1}^{n} p_{ij}(u) v(j) \geq 1 + \sum_{j=1}^{n} p_{ij}\big(\mu(i)\big) v(j), \qquad i = 1, \ldots, n.$$

Thus we obtain for all μ,

$$\sum_{j=1}^{n} p_{ij}\big(\mu(i)\big) v(j) \leq v(i) - 1 \leq \rho\, v(i), \qquad i = 1, \ldots, n, \tag{4.41}$$

where ρ is defined by

$$\rho = \max_{i=1,\ldots,n} \frac{v(i) - 1}{v(i)}.$$

Since $v(i) \geq 1$ for all i, we have $\rho < 1$.

We will now show that Eq. (4.41) implies the desired contraction property. Indeed, consider the operator T_μ, which when applied to a vector $J = \big(J(1), \ldots, J(n)\big)$ produces the vector $T_\mu J = \big((T_\mu J)(1), \ldots, (T_\mu J)(n)\big)$ defined by

$$(T_\mu J)(i) = p_{it}\big(\mu(i)\big) g\big(i, \mu(i), t\big) + \sum_{j=1}^{n} p_{ij}\big(\mu(i)\big) \Big(g\big(i, \mu(i), j\big) + J(j) \Big),$$

for all $i = 1, \ldots, n$. We have for all J, J', and i

$$\begin{aligned}
(T_\mu J)(i) &= (T_\mu J')(i) + \sum_{j=1}^{n} p_{ij}\big(\mu(i)\big)\big(J(j) - J'(j)\big) \\
&= (T_\mu J')(i) + \sum_{j=1}^{n} p_{ij}\big(\mu(i)\big)v(j)\frac{\big(J(j) - J'(j)\big)}{v(j)} \\
&\leq (T_\mu J')(i) + \sum_{j=1}^{n} p_{ij}\big(\mu(i)\big)v(j)\,\|J - J'\| \\
&\leq (T_\mu J')(i) + \rho\, v(i)\,\|J - J'\|,
\end{aligned}$$

where the last inequality follows from Eq. (4.41). By minimizing both sides over all $\mu(i) \in U(i)$, we obtain

$$(TJ)(i) \leq (TJ')(i) + \rho\, v(i)\,\|J - J'\|, \qquad i = 1, \ldots, n.$$

Thus we have

$$\frac{(TJ)(i) - (TJ')(i)}{v(i)} \leq \rho\,\|J - J'\|, \qquad i = 1, \ldots, n.$$

Similarly, by reversing the roles of J and J', it follows that

$$\frac{(TJ')(i) - (TJ)(i)}{v(i)} \leq \rho\,\|J - J'\|, \qquad i = 1, \ldots, n.$$

By combining the preceding two inequalities, we have

$$\frac{\big|(TJ)(i) - (TJ')(i)\big|}{v(i)} \leq \rho\,\|J - J'\|, \qquad i = 1, \ldots, n,$$

and by maximizing the left-hand side over i, we obtain the contraction property $\|TJ - TJ'\| \leq \rho\|J - J'\|$. **Q.E.D.**

4.8.2 Proofs for Discounted Problems

Since we have shown that the discounted problem can be converted to the equivalent SSP problem of Fig. 4.3.1, we can apply Props. 4.2.1-4.2.4. Then Props. 4.3.1-4.3.4 are obtained from the construction of Fig. 4.3.1. The contraction property of Prop. 4.3.5 using the proof of Prop. 4.2.5: in the SSP problem of Fig. 4.3.1, the expected number of steps to terminate starting from a state $i \neq t$ can be obtained as the mean of a geometrically distributed random variable with parameter $1 - \alpha$:

$$v(i) = 1 \cdot (1 - \alpha) + 2 \cdot \alpha(1 - \alpha) + 3 \cdot \alpha^2(1 - \alpha) + \cdots = \frac{1}{1 - \alpha}, \quad i = 1, \ldots, n,$$

so that the modulus of contraction is (cf. the proof of Prop. 4.2.5)

$$\rho = \frac{v(i) - 1}{v(i)} = \alpha.$$

Thus by applying Prop. 4.2.5, we obtain Prop. 4.3.5. Note that there is similar contraction property for T_μ. The proof of the contraction property for the discounted case can actually be obtained more simply, without going through the SSP counterpart of Prop. 4.2.5 (see e.g., [Ber12], Section 1.2).

4.8.3 Convergence of Exact and Optimistic Policy Iteration

We provide a proof of the convergence of exact PI for the case of a discounted problem. The proof for the SSP problem is similar. The intuition behind the proof has been summarized earlier [cf. Eq. (4.29)].

> **Proposition 4.6.1: (Convergence of Exact PI)** For both the SSP and the discounted problems, the exact PI algorithm generates an improving sequence of policies [i.e., $J_{\mu^{k+1}}(i) \leq J_{\mu^k}(i)$ for all i and k] and terminates with an optimal policy.

Proof: For any k, consider the sequence $\{J_N\}$ generated by the VI algorithm for policy μ^{k+1}:

$$J_{N+1}(i) = \sum_{j=1}^{n} p_{ij}\big(\mu^{k+1}(i)\big)\Big(g\big(i, \mu^{k+1}(i), j\big) + \alpha J_N(j)\Big), \qquad i = 1, \ldots, n,$$

where $N = 0, 1, \ldots$, and

$$J_0(i) = J_{\mu^k}(i), \qquad i = 1, \ldots, n.$$

Using the policy evaluation and improvement relations (4.26) and (4.27),

$$\begin{aligned} J_0(i) &= \sum_{j=1}^{n} p_{ij}\big(\mu^{k}(i)\big)\Big(g\big(i, \mu^{k}(i), j\big) + \alpha J_0(j)\Big) \\ &\geq \sum_{j=1}^{n} p_{ij}\big(\mu^{k+1}(i)\big)\Big(g\big(i, \mu^{k+1}(i), j\big) + \alpha J_0(j)\Big) \qquad i = 1, \ldots, n. \\ &= J_1(i), \end{aligned}$$

By using the above inequality we obtain

$$\begin{aligned} J_1(i) &= \sum_{j=1}^{n} p_{ij}\big(\mu^{k+1}(i)\big)\Big(g\big(i, \mu^{k+1}(i), j\big) + \alpha J_0(j)\Big) \\ &\geq \sum_{j=1}^{n} p_{ij}\big(\mu^{k+1}(i)\big)\Big(g\big(i, \mu^{k+1}(i), j\big) + \alpha J_1(j)\Big) \qquad i = 1, \ldots, n, \\ &= J_2(i), \end{aligned}$$

and by continuing similarly we have

$$J_0(i) \geq J_1(i) \geq \cdots \geq J_N(i) \geq J_{N+1}(i) \geq \cdots, \qquad i = 1, \ldots, n. \tag{4.42}$$

Since by Prop. 4.3.3, $J_N(i) \to J_{\mu^{k+1}}(i)$, we obtain $J_0(i) \geq J_{\mu^{k+1}}(i)$ or

$$J_{\mu^k}(i) \geq J_{\mu^{k+1}}(i), \qquad i = 1, \ldots, n, \quad k = 0, 1, \ldots.$$

Thus the sequence of generated policies is improving, and since the number of stationary policies is finite, we must after a finite number of iterations, say $k+1$, obtain $J_{\mu^k}(i) = J_{\mu^{k+1}}(i)$ for all i. Then we will have equality throughout in Eq. (4.42), which means that

$$J_{\mu^k}(i) = \min_{u \in U(i)} \sum_{j=1}^{n} p_{ij}(u)\big(g(i,u,j) + \alpha J_{\mu^k}(j)\big), \qquad i = 1, \ldots, n.$$

Thus the costs $J_{\mu^k}(1), \ldots, J_{\mu^k}(n)$ solve Bellman's equation, and by the uniqueness part of Prop. 4.3.2, it follows that $J_{\mu^k} = J^*$ and that μ^k is optimal. **Q.E.D.**

We provide a proof of convergence of optimistic PI for discounted problems. The proof for SSP problems is somewhat more complicated and will not be given (see [Ber12], Section 2.5.5).

Proposition 4.6.2: (Convergence of Optimistic PI) For the discounted problem, the sequence $\{J_k\}$ generated by the optimistic PI algorithm satisfies $J_k \to J^*$, while the generated policies μ^k are optimal for all k sufficiently large.

Proof: First we choose a scalar r such that the vector $\bar{J}_0$ defined by

$$\bar{J}_0 = J_0 + r\,e,$$

satisfies $T\bar{J}_0 \leq \bar{J}_0$, where e is the unit vector,

$$e = (1, 1, \ldots, 1)'.$$

This can be done since if r is large enough so that

$$TJ_0 - J_0 \leq (1-\alpha) r\,e,$$

we have

$$T\bar{J}_0 = TJ_0 + \alpha r\,e \leq J_0 + r\,e = \bar{J}_0.$$

With $\bar{J}_0$ so chosen, define for all k, $\bar{J}_{k+1} = T_{\mu^k}^{m_k} \bar{J}_k$. Then, since

$$T(J + re) = TJ + \alpha r\, e, \qquad T_\mu(J + re) = T_\mu J + \alpha r\, e,$$

for any J, r, and μ, it can be seen by induction that for all k and $m = 0, 1, \ldots, m_k$, the vectors $J_{k+1} = T_{\mu^k}^m J_k$ and $\bar{J}_{k+1} = T_{\mu^k}^m \bar{J}_k$ differ by a multiple of the unit vector, namely

$$r\alpha^{m_0+\cdots+m_{k-1}+m} e. \tag{4.43}$$

It follows that if J_0 is replaced by $\bar{J}_0$ as the starting vector in the algorithm, the same sequence of policies $\{\mu^k\}$ will be obtained, i.e.,

$$T_{\mu^k} \bar{J}_k = T\bar{J}_k, \qquad \text{for all } k.$$

Moreover, the difference $\bar{J}_{k+1} - J_{k+1}$, i.e., the vector (4.43), converges to 0. Thus we have

$$\lim_{k\to\infty} \bar{J}_k = \lim_{k\to\infty} J_k.$$

Next we will show that for all k,

$$J^* \leq \bar{J}_k \leq T^k \bar{J}_0. \tag{4.44}$$

Since $\lim_{k\to\infty} T^k \bar{J}_0 = J^*$ (by the convergence property of VI), the preceding two relations will imply that $J^* = \lim_{k\to\infty} \bar{J}_k = \lim_{k\to\infty} J_k$.

Indeed, we have $T_{\mu^0} \bar{J}_0 = T\bar{J}_0 \leq \bar{J}_0$, from which by repeated application of T_{μ^0}, we obtain

$$T_{\mu^0}^m \bar{J}_0 \leq T_{\mu^0}^{m-1} \bar{J}_0, \qquad m = 1, 2, \ldots,$$

so that we can write

$$T_{\mu^1} \bar{J}_1 = T\bar{J}_1 \leq T_{\mu^0} \bar{J}_1 = T_{\mu^0}^{m_0+1} \bar{J}_0 \leq T_{\mu^0}^{m_0} \bar{J}_0 = \bar{J}_1 \leq T_{\mu^0} \bar{J}_0 = T\bar{J}_0.$$

As part of this relation, we have $\bar{J}_1 \leq T\bar{J}_0$, and similarly we obtain $\bar{J}_k \leq T\bar{J}_{k-1}$ for all k, so that

$$\bar{J}_k \leq T^k \bar{J}_0, \qquad k = 0, 1, \ldots.$$

On the other hand, since $T\bar{J}_0 \leq \bar{J}_0$, the sequence $T^k \bar{J}_0$ is monotonically nonincreasing, converges to J^*, and is bounded from below by J^*. It follows that $\bar{J}_k$, being the result of successive applications of operators of the form T_μ to $\bar{J}_0$, is also bounded from below by J^*. Thus, Eq. (4.44) follows, implying that $J^* = \lim_{k\to\infty} J_k$.

Finally, from the finiteness of the state and control spaces, it follows that there exists $\epsilon > 0$ such that if $\|J - J^*\| \leq \epsilon$ and μ attains the minimum of the Bellman equation expression for J, i.e., $T_\mu J = TJ$, then μ also attains the minimum when J is replaced by J^*, i.e., $T_\mu J^* = TJ^*$, so that μ is optimal. Since $J_k \to J^*$ and $T_{\mu^k} J_k = TJ_k$, this shows that for all sufficiently large k, we have $T_{\mu^k} J^* = TJ^*$, so that by Prop. 4.3.4, μ^k is optimal. **Q.E.D.**

5

Infinite Horizon Reinforcement Learning

Contents

In this chapter, we consider the use of approximate DP/RL methods for suboptimal solution of the infinite horizon SSP and discounted problems of the preceding chapter. In particular, we will consider approximate versions of the fundamental value iteration (VI) and policy iteration (PI) algorithms. In the process, we will make frequent references to the DP operators T and T_μ (or Bellman operators), which map an n-dimensional vector J into the n-dimensional vectors TJ and $T_\mu J$, and provide shorthand notation for algorithms and analysis. For the purpose of easy reference, we state these operators here:

For SSP problems:

$$(TJ)(i) = \min_{u\in U(i)} \left[p_{it}(u)g(i,u,t) + \sum_{j=1}^n p_{ij}(u)\big(g(i,u,j)+J(j)\big) \right], \tag{5.1}$$

for all $i = 1,\ldots,n$, and

$$(T_\mu J)(i) = p_{it}\big(\mu(i)\big)g\big(i,\mu(i),t\big) + \sum_{j=1}^n p_{ij}\big(\mu(i)\big)\Big(g\big(i,\mu(i),j\big)+J(j)\Big), \tag{5.2}$$

for all policies μ and states $i = 1,\ldots,n$.

For discounted problems:

$$(TJ)(i) = \min_{u\in U(i)} \sum_{j=1}^n p_{ij}(u)\big(g(i,u,j)+\alpha J(j)\big), \tag{5.3}$$

and

$$(T_\mu J)(i) = \sum_{j=1}^n p_{ij}\big(\mu(i)\big)\Big(g\big(i,\mu(i),j\big)+\alpha J(j)\Big), \tag{5.4}$$

for all policies μ and states $i = 1,\ldots,n$.

5.1 APPROXIMATION IN VALUE SPACE - PERFORMANCE BOUNDS

In this section we will discuss the general framework for approximation in value space for infinite horizon DP, beginning with discounted problems. Consistently with the corresponding finite horizon schemes of Chapter 2, the general idea is to compute some approximation $\tilde{J}$ of the optimal cost function J^*, and then use one-step or multistep lookahead to implement a suboptimal policy $\tilde{\mu}$. Thus, a *one-step lookahead policy* applies at state i the control $\tilde{\mu}(i)$ that attains the minimum in the expression

$$\min_{u\in U(i)} \sum_{j=1}^n p_{ij}(u)\big(g(i,u,j)+\alpha \tilde{J}(j)\big), \tag{5.5}$$

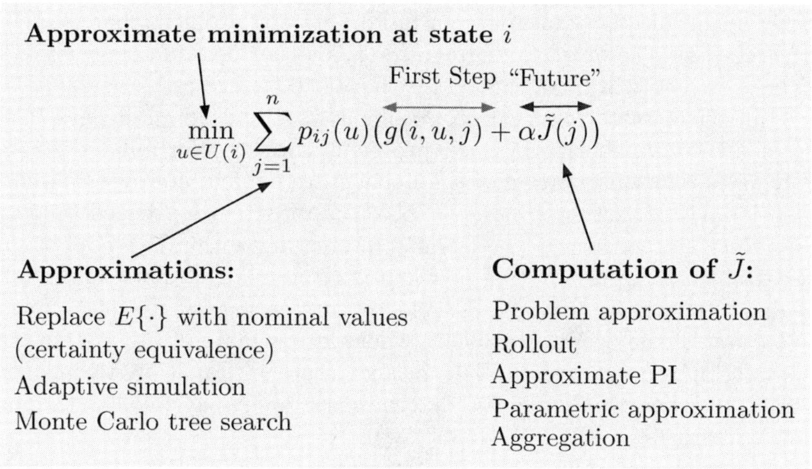

Figure 5.1.1 Schematic illustration of various options for approximation in value space with one-step lookahead in infinite horizon problems. The lookahead function values $\tilde{J}(j)$ approximate the optimal cost-to-go values $J^*(j)$, and can be computed by a variety of methods. There may be additional approximations in the minimization over u_k and the computation of the expected value.

see Fig. 5.1.1.

Similarly, at state i, a *two-step lookahead policy* applies the control $\tilde{\mu}(i)$ attaining the minimum in the preceding equation, where now $\tilde{J}$ is obtained itself on the basis of a one-step lookahead approximation. In other words, for all states j that can be reached from i, we have

$$\tilde{J}(j) = \min_{u \in U(j)} \sum_{m=1}^{n} p_{jm}(u)\big(g(j,u,m) + \alpha \hat{J}(m)\big),$$

where $\hat{J}$ is some approximation of J^*. Thus $\tilde{J}$ is the result of a single value iteration starting from $\hat{J}$. Policies with lookahead of more than two stages are similarly defined. *The "effective one-step" cost approximation $\tilde{J}$ in ℓ-step lookahead is the result of $\ell - 1$ successive value iterations starting from some initial approximation $\hat{J}$.* Otherwise expressed, *ℓ-step lookahead with $\hat{J}$ at the end is the same as one-step lookahead with $T^{\ell-1}\hat{J}$ at the end,* where T is the Bellman operator [cf. Eq. (5.3)].

Types of Approximation in Value Space Schemes

In Chapter 2 we described several types of limited lookahead schemes, where $\tilde{J}$ is obtained in different ways, such as problem approximation, rollout, and others. Some of these schemes can be fruitfully adapted to infinite horizon problems; see Fig. 5.1.1. For example, the problem approximation approaches of Section 2.3 admit straightforward extensions to infinite horizon settings, whereby the function $\tilde{J}(j)$ in Eq. (5.5) is obtained by solving exactly an infinite horizon (or even finite horizon) problem that is related to the original in some way. Aggregation is another possible approximation approach, which will be discussed in Chapter 6.

Imperfect state observation problems pose special challenges in the case of an infinite horizon. Such problems can be reformulated to ones involving perfect observation of the belief state, which, however, is infinite dimensional (cf. Section 1.3.6).† Problem approximation based on forms of certainty equivalence is particularly interesting within this context, because a natural approximating problem is often evident. For example, the lookahead function $\tilde{J}$ may be derived by solving a perfect state information variant of the original problem, where an estimate of the system state is used as if it were exact. The perfect state information variant may be tractable, because it may be deterministic or it may involve a modest number of states.

In this chapter, the approximation in value space schemes that we will focus on in Sections 5.1-5.5 are primarily based on approximate (optimistic) PI and operate as follows, starting with an initial policy μ^0:

(a) We generate (usually off-line) several policies $\mu^0, \mu^1, \ldots, \mu^m$.

(b) We evaluate each policy μ^k approximately, with a cost function $\tilde{J}_{\mu^k}$, often by using a parametric approximation/neural network approach, and possibly including the use of truncated rollout.

(c) We generate the next policy μ^{k+1} using one-step or multistep policy improvement based on $\tilde{J}_{\mu^k}$.

(d) We use (on-line) the approximate evaluation $\tilde{J}_{\mu^m}$ of the last policy in the sequence as the lookahead approximation $\tilde{J}$ in the one-step lookahead minimization (5.5), or its multistep counterpart.

We will view rollout as a simple form of approximate PI, which involves a single policy iteration, executed with the aid of simulation. The rollout may be truncated and supplemented with a (potentially sophisticated) terminal cost function approximation (cf. Section 2.4). We will give some performance bounds for limited lookahead schemes and for rollout in Sections 5.1.1 and 5.1.2, respectively. We will discuss performance bounds for the general approximate PI scheme in Section 5.1.3.

5.1.1 Limited Lookahead

We will now consider performance bounds for ℓ-step lookahead. In particular, for a given state i_0, let $\hat{\mu}_0, \ldots, \hat{\mu}_{\ell-1}$ attain the minimum in the ℓ-step

---

† Because of the infinite dimensionality of the belief space, the theory of Chapter 4 has to be extended before it can be applied, since it was developed for finite state space. Usually this is fairly straightforward for discounted problems, but less so for SSP problems. We will not provide further discussion in this book.

lookahead minimization

$$\min_{\mu_0,\ldots,\mu_{\ell-1}} E\left\{\sum_{k=0}^{\ell-1} \alpha^k g\big(i_k, \mu_k(i_k), j_k\big) + \alpha^\ell \tilde{J}(i_\ell)\right\}.$$

We focus on the suboptimal policy that applies control $\tilde{\mu}(i_0) = \hat{\mu}_0(i_0)$, and we refer to $\tilde{\mu}$ as the *ℓ-step lookahead policy corresponding to $\tilde{J}$*. Equivalently, in the shorthand notation of the Bellman operators T and $T_{\tilde{\mu}}$ of Eqs. (5.3) and (5.4), the ℓ-step lookahead policy $\tilde{\mu}$ is defined by

$$T_{\tilde{\mu}}(T^{\ell-1}\tilde{J}) = T^\ell \tilde{J}.$$

We will derive a bound for the performance of $\tilde{\mu}$ in part (a) of the following proposition, proved in the appendix.

We will also derive a bound for the case of a useful generalized one-step lookahead scheme [part (b) of the following proposition]. This scheme aims to reduce the computation to obtain $\tilde{\mu}(i)$, by performing the lookahead minimization over a subset $\overline{U}(i) \subset U(i)$. Thus, the control $\tilde{\mu}(i)$ used in this scheme is one that attains the minimum in the expression

$$\min_{u\in\overline{U}(i)} \sum_{j=1}^{n} p_{ij}(u)\big(g(i,u,j) + \alpha\tilde{J}(j)\big).$$

This is attractive when by using some heuristic, we can identify a subset $\overline{U}(i)$ of promising controls, and to save computation, we restrict attention to this subset in the one-step lookahead minimization.

Proposition 5.1.1: (Limited Lookahead Performance Bounds)

(a) Let $\tilde{\mu}$ be the ℓ-step lookahead policy corresponding to $\tilde{J}$. Then

$$\|J_{\tilde{\mu}} - J^*\| \leq \frac{2\alpha^\ell}{1-\alpha}\|\tilde{J} - J^*\|, \tag{5.6}$$

where $\|\cdot\|$ denotes the maximum norm $\|J\| = \max_{i=1,\ldots,n}\big|J(i)\big|$.

(b) Define

$$\hat{J}(i) = \min_{u\in\overline{U}(i)} \sum_{j=1}^{n} p_{ij}(u)\big(g(i,u,j) + \alpha\tilde{J}(j)\big), \qquad i = 1,\ldots,n, \tag{5.7}$$

where $\overline{U}(i) \subset U(i)$ for all $i = 1,\ldots,n$, and let $\tilde{\mu}$ be the one-step lookahead policy obtained by minimization in the right side of this equation. Then we have

$$J_{\tilde{\mu}}(i) \leq \tilde{J}(i) + \frac{c}{1-\alpha}, \qquad i = 1, \ldots, n, \tag{5.8}$$

where

$$c = \max_{i=1,\ldots,n} \big(\hat{J}(i) - \tilde{J}(i)\big).$$

An important point regarding the bound (5.6) is that $\tilde{\mu}$ is unaffected by a constant shift in $\tilde{J}$ [an addition of a constant β to all values $\tilde{J}(i)$]. Thus $\|\tilde{J} - J^*\|$ in Eq. (5.6) can be replaced by the potentially smaller number

$$\min_{\beta \in \Re} \max_{i=1,\ldots,n} \big|\tilde{J}(i) + \beta - J^*(i)\big|. \tag{5.9}$$

Another interesting point is that it is sufficient to take maximum in the preceding expression over just the states i that can occur following the ℓ steps of lookahead, thus potentially improving the bound of Eq. (5.6) even further. This can be used to obtain sharper versions of Prop. 5.1.1(a), as well as the related subsequent Prop. 5.1.3(a), although we will not pursue this analysis further.

The bound (5.6) suggests that performance is improved when the length ℓ of the lookahead is increased, and also when the lookahead cost approximation $\tilde{J}$ is closer to the optimal cost J^* (when modified with an optimal constant shift β). Both of these conclusions are intuitive and also consistent with practical experience. Note that we are not guaranteeing that multistep lookahead will lead to better performance than one-step lookahead; we know that this is not necessarily true (cf. Example 2.2.1). It is the performance *bound* that is improved with multistep lookahead.

Regarding the bound (5.8), we note that it guarantees that when $c \leq 0$, the cost $J_{\tilde{\mu}}$ of the one-step lookahead policy is no larger than $\tilde{J}$. The case $c = 0$ is equivalent to $\hat{J} \leq \tilde{J}$, which resembles the consistent improvement condition for deterministic rollout methods (cf. Section 2.4.1). If $\tilde{J} = J_\mu$ for some policy μ with $\mu(i) \in \overline{U}(i)$ for all i (as in the case of the pure form of rollout to be discussed in Section 5.1.2), then $c = 0$, and from Eq. (5.8) it follows that we have cost improvement, i.e., $J_{\tilde{\mu}} \leq J_\mu$.

Unfortunately, the bound (5.6) is not very reassuring when α is close to 1. Nonetheless, the following example shows that the bound can be tight even in very simple problems with just two states. What is happening here is that an $O(\epsilon)$ difference in single stage cost between two controls can generate an $O\big(\epsilon/(1-\alpha)\big)$ difference in policy costs (through discounted accumulation over an infinite number of steps), yet it can be "nullified" in Bellman's equation by an $O(\epsilon)$ difference between J^* and $\tilde{J}$.

Example 5.1.1

Consider the two-state discounted problem shown in Fig. 5.1.2, where ϵ is a positive scalar and $\alpha \in [0, 1)$ is the discount factor. Here there are two control

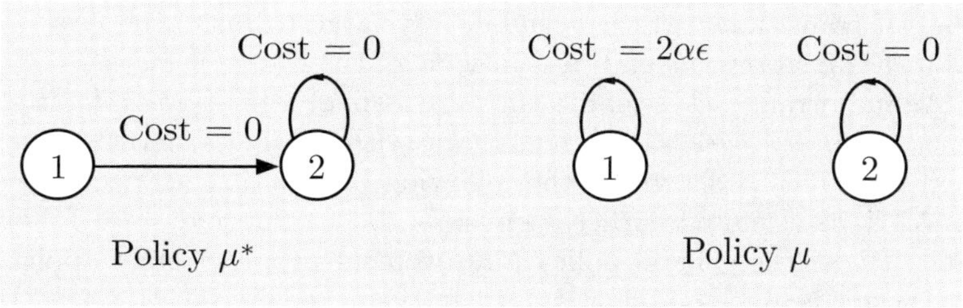

Figure 5.1.2 A two-state problem for proving the tightness of the performance bound of Prop. 5.1.1(b) (cf. Example 5.1.1). All transitions are deterministic as shown. At state 1 there are two possible decisions: move to state 2 at cost 0 (policy μ^*) or stay at state 1 at cost $2\alpha\epsilon$ (policy μ).

choices at state 1: move to state 2 at cost 0 (policy μ^*) or stay at state 1 at cost $2\alpha\epsilon$ (policy μ). The optimal policy is μ^*, and the optimal cost-to-go function is $J^*(1) = J^*(2) = 0$. Consider the cost function approximation $\tilde{J}$

$$\tilde{J}(1) = -\epsilon, \qquad \tilde{J}(2) = \epsilon,$$

so that

$$\|\tilde{J} - J^*\| = \epsilon.$$

The policy μ that decides to stay at state 1 is a one-step lookahead policy based on $\tilde{J}$, because

$$2\alpha\epsilon + \alpha\tilde{J}(1) = \alpha\epsilon = 0 + \alpha\tilde{J}(2).$$

Moreover, we have

$$J_\mu(1) = \frac{2\alpha\epsilon}{1-\alpha} = \frac{2\alpha}{1-\alpha}\|\tilde{J} - J^*\|,$$

so the bound of Eq. (5.6) holds with equality when $\ell = 1$.

5.1.2 Rollout and Approximate Policy Improvement

Let us first consider rollout in its pure form, where $\tilde{J}$ in Eq. (5.5) is the cost function of some stationary policy μ (also called the *base policy* or *base heuristic*), i.e., $\tilde{J} = J_\mu$. Then, *the rollout policy is the result of a single policy iteration starting from* μ. The policy evaluation that yields the costs $J_\mu(j)$ needed for policy improvement may be done in any suitable way. Monte-Carlo simulation (averaging the costs of many trajectories starting from j) is one major possibility. Of course if the problem is deterministic, a single simulation trajectory starting from j is sufficient, in which case the rollout policy is much less computationally demanding. Note also that in discounted problems the simulated trajectories can be truncated after a number of transitions, which is sufficiently large to make the cost of the remaining transitions insignificant in view of the discount factor.

An important fact is that in the pure form of rollout, *the rollout policy improves over the base policy*, consistent with the finite horizon case; cf. Section 2.4. This is shown by the following proposition [a special case of Prop. 5.1.1(b) as noted earlier], and it is to be expected since rollout is one-step PI, so the general policy improvement property of PI applies. A related result is given as Lemma 5.9.1 in the appendix (Section 5.9.3).

Proposition 5.1.2: (Cost Improvement by Rollout) Let $\tilde{\mu}$ be the rollout policy obtained by the one-step lookahead minimization

$$\min_{u\in \overline{U}(i)} \sum_{j=1}^{n} p_{ij}(u)\big(g(i,u,j)+\alpha J_\mu(j)\big),$$

where μ is a base policy [cf. Eq. (5.7) with $\tilde{J}=J_\mu$] and we assume that $\mu(i)\in \overline{U}(i)\subset U(i)$ for all $i=1,\ldots,n$. Then $J_{\tilde{\mu}}\le J_\mu$.

Let us also mention the variation of rollout that uses multiple base heuristics, and simultaneously improves on all of them. This variant, also called *parallel rollout* because of its evident parallelization potential, is similar to its finite horizon counterpart; cf. Section 2.4.1.

Example 5.1.2 (Rollout with Multiple Heuristics)

Let $\mu_1,\ldots,\mu_M$ be stationary policies, let

$$\tilde{J}(i)=\min\big\{J_{\mu_1}(i),\ldots,J_{\mu_M}(i)\big\},\qquad i=1,\ldots,n,$$

let $\overline{U}(i)\subset U(i)$, and assume that

$$\mu_1(i),\ldots,\mu_M(i)\in \overline{U}(i),\qquad i=1,\ldots,n.$$

Then, for all i and $m=1,\ldots,M$, we have

$$\begin{aligned}
\hat{J}(i) &= \min_{u\in \overline{U}(i)} \sum_{j=1}^{n} p_{ij}(u)\big(g(i,u,j)+\alpha \tilde{J}(j)\big)\\
&\le \min_{u\in \overline{U}(i)} \sum_{j=1}^{n} p_{ij}(u)\big(g(i,u,j)+\alpha J_{\mu_m}(j)\big)\\
&\le \sum_{j=1}^{n} p_{ij}\big(\mu_m(i)\big)\Big(g\big(i,\mu_m(i),j\big)+\alpha J_{\mu_m}(j)\Big)\\
&= J_{\mu_m}(i),
\end{aligned}$$

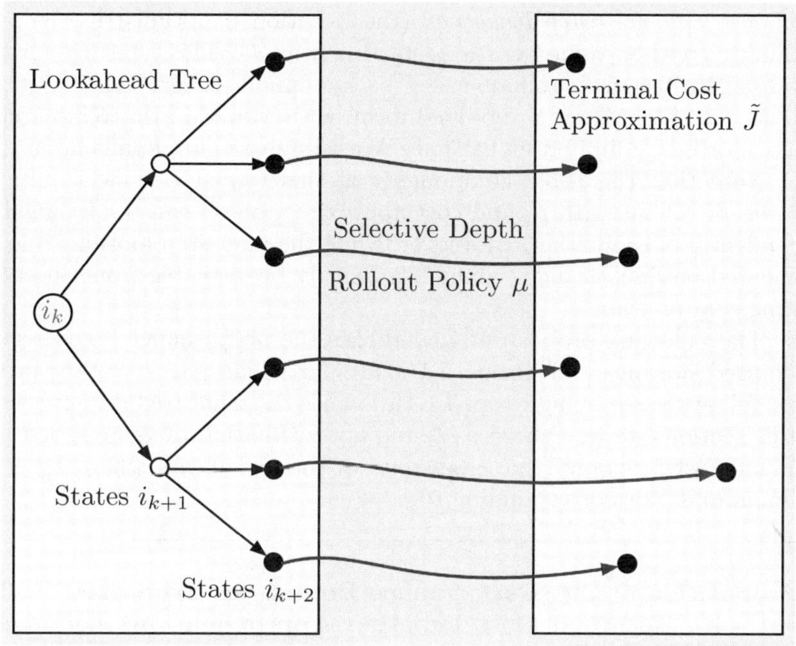

Figure 5.1.3 Illustration of two-step lookahead, rollout with a policy μ for a limited and state-dependent number of steps, and a terminal cost function approximation $\tilde{J}$. A Monte Carlo tree search scheme may also be used for multistep lookahead; cf. Section 2.4.2.

which, by taking minimum of the right-hand side over m, yields

$$\hat{J}(i) \leq \tilde{J}(i), \qquad i = 1, \ldots, n.$$

Using Prop. 5.1.1(b), we see that the rollout policy $\tilde{\mu}$, obtained by using $\tilde{J}$ as one-step lookahead approximation satisfies

$$J_{\tilde{\mu}}(i) \leq \tilde{J}(i) = \min\big\{J_{\mu_1}(i), \ldots, J_{\mu_M}(i)\big\}, \qquad i = 1, \ldots, n,$$

i.e., it improves over each of the policies $\mu_1, \ldots, \mu_M$.

Truncated Rollout with Multistep Lookahead and Terminal Cost Function Approximation

Let us next discuss a truncated variant of the rollout approach, whereby we use ℓ-step lookahead, we then apply rollout with policy μ for a limited number of steps, and finally we approximate the cost of the remaining steps using some terminal cost approximation $\tilde{J}$; see Fig. 5.1.3 for a case where $\ell = 2$. We can view this form of rollout as *a single optimistic policy iteration combined with multistep lookahead* (the iteration is optimistic because it evaluates μ with m value iterations starting from $\tilde{J}$). This type of algorithm was used in Tesauro's rollout-based backgammon player [TeG96] (it was also used in AlphaGo in a modified form, with Monte Carlo tree search in place of ordinary limited lookahead). We will give more details later.

Note that the three components of this rollout scheme (multistep lookahead, rollout with μ, and cost approximation $\tilde{J}$) can be designed independently of each other. Moreover, while the multistep lookahead is implemented on-line, μ and $\tilde{J}$ should ordinarily be available from an earlier off-line computation.

The following proposition generalizes the performance bounds given for limited lookahead (cf. Props. 5.1.1). In particular, part (a) of the proposition follows by applying Prop. 5.1.1(a), since the truncated rollout scheme of this section can be viewed as ℓ-step approximation in value space with terminal cost function $T_\mu^m \tilde{J}$ at the end of the lookahead, where T_μ is the Bellman operator corresponding to μ.

Proposition 5.1.3: (Performance Bounds for Truncated Rollout with Terminal Cost Function Approximation) Let ℓ and m be positive integers, let μ be a policy, and let $\tilde{J}$ be a function of the state. Consider a truncated rollout scheme consisting of ℓ-step lookahead, followed by rollout with a policy μ for m steps, and a terminal cost function approximation $\tilde{J}$ at the end of the m steps. Let $\tilde{\mu}$ be the policy generated by this scheme.

(a) We have

$$\|J_{\tilde{\mu}} - J^*\| \le \frac{2\alpha^\ell}{1-\alpha}\|T_\mu^m \tilde{J} - J^*\|, \tag{5.10}$$

where T_μ is the Bellman operator of Eq. (5.4), and $\|\cdot\|$ denotes the maximum norm $\|J\| = \max_{i=1,\ldots,n} \big|J(i)\big|$.

(b) We have

$$J_{\tilde{\mu}}(i) \le \tilde{J}(i) + \frac{c}{1-\alpha}, \qquad i = 1,\ldots,n,$$

where

$$c = \max_{i=1,\ldots,n} \big((T_\mu \tilde{J})(i) - \tilde{J}(i)\big).$$

(c) We have

$$J_{\tilde{\mu}}(i) \le J_\mu(i) + \frac{2}{1-\alpha}\|\tilde{J} - J_\mu\|, \qquad i = 1,\ldots,n.$$

Some of the insights provided by the preceding proposition are:

(a) Part (a) of the proposition implies that as the size of lookahead ℓ increases, the bound on the performance of the rollout policy also improves. Moreover, assuming that μ is close to optimal (so that $T_\mu^m \tilde{J}$ is close to J^* as $m \to \infty$), the bound on the performance of the rollout policy $\tilde{\mu}$ improves as m increases (the improvement is enhanced if $\tilde{J}$ is also close to J_μ).†

(b) Part (c) suggests that if $\tilde{J}$ is close to J_μ, the performance of the rollout policy $\tilde{\mu}$ is nearly improved relative to the performance of the base policy μ. This is consistent with the cost improvement property of rollout, cf. Prop. 5.1.2. Part (b) admits a similar interpretation.

In summary, the guidelines for truncated rollout that these results suggest are to *choose as large lookahead as practical, and choose $\tilde{J}$ as close to J_μ or J^* as possible.* It is not clear to what extent increasing m, the length of rollout of μ, improves performance, and some examples suggest that m should not be chosen to be very large. A small value of m is also beneficial in another way: it limits the amount of computation needed for rollout, and reduces the variance of the cost estimates. In practice, for an infinite horizon problem, m is ordinarily chosen in some empirical fashion.

Regarding the terminal cost approximation $\tilde{J}$ in truncated rollout schemes, it may be heuristic, based on problem approximation, or based on a more systematic simulation methodology. For example, the values $J_\mu(i)$ may be computed by simulation for all i in a subset of representative states, and $\tilde{J}$ may be selected from a parametric class of vectors by a least squares regression of the computed values. This approximation may be performed off-line, outside the time-sensitive restrictions of a real-time implementation, and the result $\tilde{J}$ may be used on-line in place of J_μ as a terminal cost function approximation. Note that a good choice of terminal cost approximation is crucial for some types of SSP problems where most or all of the cost is incurred upon reaching the termination state (for example winning or losing a game). Note also that once cost function approximation is introduced at the end of the rollout, the cost improvement property of the rollout policy over the base policy may be lost [cf. Prop. 5.1.3(c)].

The truncated rollout scheme of Fig. 5.1.3 has been adopted in the rollout backgammon algorithm of Tesauro and Galperin [TeG96]. The policy μ and the terminal cost function approximation $\tilde{J}$ were provided by the TD-Gammon algorithm of Tesauro [Tes94], which was based on a neural network, trained using a form of optimistic policy iteration and TD(λ). A similar algorithm (with Monte Carlo tree search instead of ℓ-step lookahead) was used in the AlphaGo program (Silver et al. [SHM16]), with μ and $\tilde{J}$ obtained using an approximate PI scheme and a deep neural network.

---

† Note that $\tilde{\mu}$ is unaffected by a constant shift in $\tilde{J}$, so that an optimized constant β may be added to $\tilde{J}$ in the bound (5.10) [cf. Eq. (5.9)].

5.1.3 Approximate Policy Iteration

When the number of states is very large, the policy evaluation step and/or the policy improvement step of the PI method may be implementable only through approximations. In an approximate PI scheme, each policy μ^k is evaluated approximately, with a cost function $\tilde J_{\mu^k}$, often with the use of a feature-based architecture or a neural network, and the next policy μ^{k+1} is generated by (perhaps approximate) policy improvement based on $\tilde J_{\mu^k}$.

To formalize this type of procedure, we assume a policy evaluation error satisfying

$$\max_{i=1,\ldots,n} \big|\tilde J_{\mu^k}(i) - J_{\mu^k}(i)\big| \le \delta, \tag{5.11}$$

and a policy improvement error satisfying

$$\max_{i=1,\ldots,n} \Bigg|\sum_{j=1}^{n} p_{ij}\big(\mu^{k+1}(i)\big)\big(g(i,\mu^{k+1}(i),j)+\alpha \tilde J_{\mu^k}(j)\big) - \min_{u\in U(i)} \sum_{j=1}^{n} p_{ij}(u)\big(g(i,u,j)+\alpha\tilde J_{\mu^k}(j)\big)\Bigg| \le \epsilon, \tag{5.12}$$

where δ and ϵ are some nonnegative scalars. Here ϵ includes the simulation errors, plus any additional errors due to function approximation. Also δ is a measure of the accuracy of the lookahead minimization in the policy improvement operation (in many cases $\delta = 0$).

The following proposition, proved in the appendix (and also in the original source [BeT96], Section 6.2.2), provides a performance bound for discounted problems (a similar result is available for SSP problems; see [BeT96], Section 6.2.2).

Proposition 5.1.4: (Performance Bound for Approximate PI) Consider the discounted problem, and let $\{\mu^k\}$ be the sequence generated by the approximate PI algorithm defined by the approximate policy evaluation (5.11) and the approximate policy improvement (5.12). Then the policy error

$$\max_{i=1,\ldots,n} \big|J_{\mu^k}(i) - J^*(i)\big|,$$

becomes less or equal to

$$\frac{\epsilon + 2\alpha\delta}{(1-\alpha)^2},$$

asymptotically as $k \to \infty$.

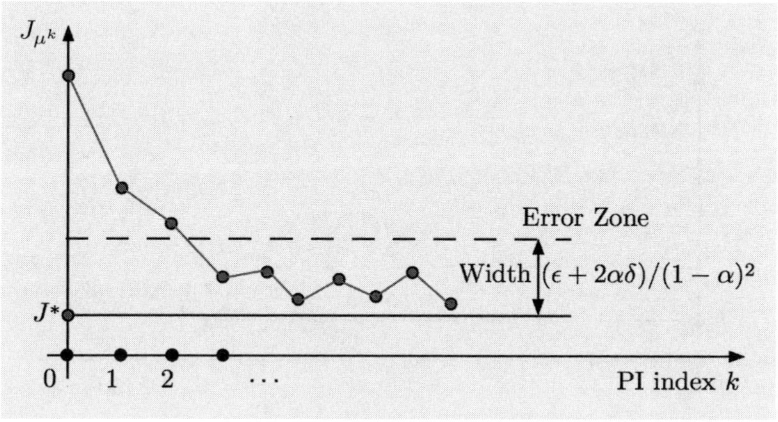

Figure 5.1.4 Illustration of typical behavior of approximate PI. In the early iterations, the method tends to make rapid and fairly monotonic progress, until J_{μ^k} gets within an error zone of size less than

$$\frac{\epsilon + 2\alpha\delta}{(1-\alpha)^2}.$$

After that J_{μ^k} oscillates randomly within that zone. The figure is oversimplified since it shows the difference $J_{\mu^k} - J^*$ at a single state. For different states the nature of the error oscillation may be different.

The preceding performance bound is important, because it is in qualitative agreement with the empirical behavior of approximate PI. Typically, in the beginning, the method tends to make rapid and fairly monotonic progress, but eventually it gets into an oscillatory pattern. This happens after J_{μ^k} gets within an error zone of size

$$\frac{\epsilon + 2\alpha\delta}{(1-\alpha)^2}$$

or smaller, and then J_{μ^k} oscillates fairly randomly within that zone; see Fig. 5.1.4. In practice, the error bound of Prop. 5.1.4 tends to be pessimistic, so the zone of oscillation is usually much narrower than what is suggested by the bound. However, the bound itself can be proved to be tight, in worst case. This is shown with an example in the book [BeT96], Section 6.2.3. Note that the bound of Prop. 5.1.4 also holds in the case of infinite state and control spaces discounted problems, when there are infinitely many policies (see [Ber18a], Prop. 2.4.3).

Performance Bound for the Case Where Policies Converge

Generally, the policy sequence $\{\mu^k\}$ generated by approximate PI may oscillate between several policies, as noted earlier. However, under some

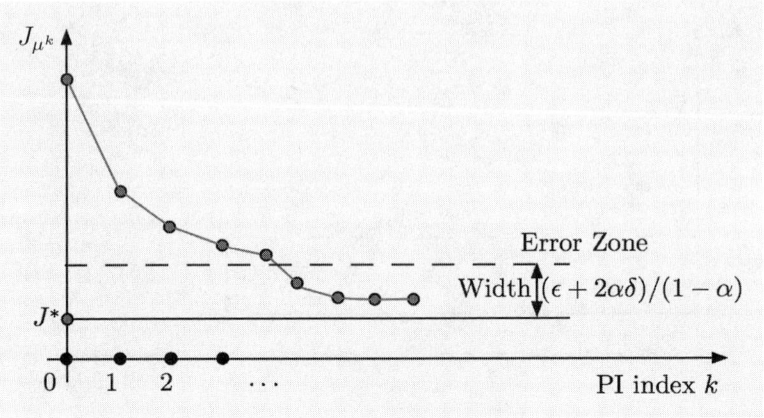

Figure 5.1.5 Illustration of typical behavior of approximate PI when policies converge. The method tends to make monotonic progress, and J_{μ^k} converges within an error zone of size less than $(\epsilon + 2\alpha\delta)/(1 - \alpha)$.

circumstances the sequence will converge to some policy $\tilde{\mu}$, i.e.,

$$\mu^{\overline{k}+1} = \mu^{\overline{k}} = \tilde{\mu} \qquad \text{for some } \overline{k}. \tag{5.13}$$

An important case where this happens is aggregation methods, which will be discussed in Chapter 6. In this case the behavior of the method is more regular, and we can show the following bound, which is more favorable than the one of Prop. 5.1.4 by a factor $1/(1-\alpha)$, as illustrated in Fig. 5.1.5. For the proof, see the appendix (or the original source [BeT96], Section 6.2.2).

Proposition 5.1.5: (Performance Bound for Approximate PI when Policies Converge) Let $\tilde{\mu}$ be a policy generated by the approximate PI algorithm under conditions (5.11), (5.12), and (5.13). Then we have

$$\max_{i=1,\ldots,n} \big| J_{\tilde{\mu}}(i) - J^*(i) \big| \leq \frac{\epsilon + 2\alpha\delta}{1 - \alpha}.$$

We finally note that related performance bounds hold for optimistic PI methods, where the policy evaluation is performed with just a few approximate value iterations (cf. Section 4.6.2). These bounds are similar to the ones of the nonoptimistic method, and do not suggest superiority of one type of PI method over the other. Their derivation is quite complicated; see [Ber12], Chapter 2, or [Ber18a], Section 2.5.2, and the end-of-chapter references. Moreover, there are versions of the preceding bounds, which apply to more general abstract DP problems, such as the semi-Markov problems of Section 4.4 and others; see [Ber18a], Section 2.4.1.

5.2 FITTED VALUE ITERATION

In Chapter 4 we discussed the VI algorithm for SSP problems,

$$J_{k+1}(i) = \min_{u \in U(i)} \left[p_{it}(u) g(i,u,t) + \sum_{j=1}^{n} p_{ij}(u)\big(g(i,u,j) + J_k(j)\big) \right], \tag{5.14}$$

and its discounted version

$$J_{k+1}(i) = \min_{u \in U(i)} \sum_{j=1}^{n} p_{ij}(u)\big(g(i,u,j) + \alpha J_k(j)\big). \tag{5.15}$$

It is one of the principal methods for calculating the optimal cost function J^*.

Unfortunately, when the number of states is large, the iterations (5.14) and (5.15) may be prohibitively time consuming. This motivates an approximate version of VI, which is patterned after the least squares regression/fitted VI scheme of Section 3.3 for finite horizon problems.

We start with some initial approximation to J^*, call it $\tilde J_0$. Then we generate a sequence $\{\tilde J_k\}$, where $\tilde J_{k+1}$ is equal to the exact value iterate $T\tilde J_k$ plus some error [we are using here the shorthand notation for the Bellman operator T given in Eqs. (5.1) and (5.3)]. Assuming that values $(T\tilde J_k)(i)$ may be generated for sample states i, we may obtain $\tilde J_{k+1}$ by some form of least squares regression. We will now discuss how the error $(\tilde J_k - J^*)$ is affected by this type of approximation process.

Error Bounds and Pathologies of Fitted Value Iteration

We will focus on fitted VI for discounted problems. The analysis for SSP problems is qualitatively similar. We first consider estimates of the *cost function error*

$$\max_{i=1,\ldots,n} \big|\tilde J_k(i) - J^*(i)\big|, \tag{5.16}$$

and the *policy error*

$$\max_{i=1,\ldots,n} \big|J_{\tilde\mu^k}(i) - J^*(i)\big|, \tag{5.17}$$

where the policy $\tilde\mu^k$ is obtained from the one-step lookahead minimization

$$\tilde\mu^k(i) \in \arg\min_{u \in U(i)} \sum_{j=1}^{n} p_{ij}(u)\big(g(i,u,j) + \alpha \tilde J_k(j)\big).$$

It turns out that such estimates are possible, but under assumptions whose validity may be hard to guarantee. In particular, it is natural to assume that the error in generating the value iterates $(T\tilde{J}_k)(i)$ is within some $\delta > 0$ for every state i and iteration k, i.e., that

$$\max_{i=1,\ldots,n} \left| \tilde{J}_{k+1}(i) - \min_{u\in U(i)} \sum_{j=1}^{n} p_{ij}(u)\big(g(i,u,j) + \alpha \tilde{J}_k(j)\big) \right| \leq \delta. \tag{5.18}$$

It is then possible to show that asymptotically, as $k \to \infty$, the cost error (5.16) becomes less or equal to $\delta/(1-\alpha)$, while the policy error (5.17) becomes less or equal to $2\delta/(1-\alpha)^2$.

Such error bounds are given in Section 6.5.3 of the book [BeT96] (see also Prop. 2.5.3 of [Ber12]), but it is important to note that the condition (5.18) may not be satisfied by the natural least squares regression/fitted VI scheme of Section 3.3. This is illustrated by the following example due to [TsV96] (see also [BeT96], Section 6.5.3), which shows that the errors from successive approximate value iterations can accumulate to the point where the condition (5.18) cannot be maintained (with a scalar δ that is independent of k), and the approximate value iterates $\tilde{J}_k$ can grow unbounded.

Example 5.2.1 (Error Amplification in Approximate Value Iteration)

Consider a two-state discounted problem with states 1 and 2, and a single policy. The transitions are deterministic: from state 1 to state 2, and from state 2 to state 2. The transitions are also cost-free; see Fig. 5.2.1. Thus the Bellman equation is

$$J(1) = \alpha J(2), \qquad J(2) = \alpha J(2),$$

and its unique solution is $J^*(1) = J^*(2) = 0$. Moreover, exact VI has the form

$$J_{k+1}(1) = \alpha J_k(2), \qquad J_{k+1}(2) = \alpha J_k(2).$$

We consider a VI approach that approximates cost functions within the one-dimensional subspace of linear functions $S = \big\{(r, 2r) \mid r : \text{scalar}\big\}$; this is a favorable choice since the optimal cost function $J^* = (0,0)$ belongs to S. We use a weighted least squares regression scheme. In particular, given $\tilde{J}_k = (r_k, 2r_k)$, we find $\tilde{J}_{k+1} = (r_{k+1}, 2r_{k+1})$ as follows; see Fig. 5.2.2:

(a) We compute the exact VI iterate from $\tilde{J}_k$:

$$T\tilde{J}_k = \big(\alpha \tilde{J}_k(2), \alpha \tilde{J}_k(2)\big) = (2\alpha r_k, 2\alpha r_k).$$

(b) For some weights $\xi_1, \xi_2 > 0$, we obtain the scalar r_{k+1} as

$$r_{k+1} \in \arg\min_r \left[\xi_1\big(r - (T\tilde{J}_k)(1)\big)^2 + \xi_2\big(2r - (T\tilde{J}_k)(2)\big)^2\right],$$

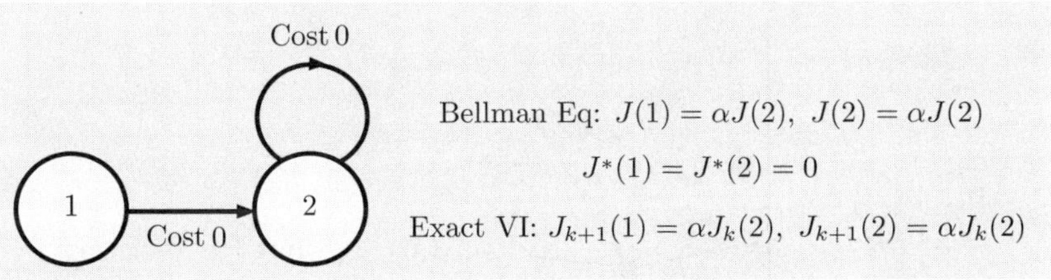

Figure 5.2.1 Illustration of the discounted problem of Example 5.2.1. There are two states, 1 and 2, and a single policy. The transitions are deterministic: from state 1 to state 2, and from state 2 to state 2. These transitions are also cost-free.

or equivalently

$$r_{k+1} \in \arg\min_r \left[\xi_1 (r - 2\alpha r_k)^2 + \xi_2 (2r - 2\alpha r_k)^2\right].$$

To perform the preceding minimization, we set to zero the derivative with respect to r of the quadratic cost function, and obtain after some calculation

$$r_{k+1} = \alpha\zeta r_k \quad \text{where} \quad \zeta = \frac{2(\xi_1 + 2\xi_2)}{\xi_1 + 4\xi_2} > 1. \tag{5.19}$$

Thus if ξ_1 and ξ_2 are chosen so that $\alpha > 1/\zeta$, the sequence $\{r_k\}$ diverges and so does $\{\tilde J_k\}$. In particular, for the natural choice $\xi_1 = \xi_2 = 1$, we have $\zeta = 6/5$, so the approximate VI scheme diverges for α in the range $(5/6, 1)$; see Fig. 5.2.2.

The difficulty here is that the approximate VI operator that generates $\tilde J_{k+1}$ by a weighted least squares-based approximation of $T\tilde J_k$ is not a contraction (even though T itself is a contraction). At the same time there is no δ such that the condition (5.18) is satisfied for all k, because of error amplification in each approximate VI.

The preceding example indicates that the choice of the least squares weights is important in determining the success of least squares-based approximate VI schemes. Generally, in regression-based parametric architecture training schemes of the type discussed in Section 3.1.2, the weights are related to the way samples are collected: the weight ξ_i for state i is the proportion of the number of samples in the least squares summation that correspond to state i. Thus $\xi_1 = \xi_2 = 1$ in the preceding example means that we use an equal number of samples for each of the two states 1 and 2.

Now let us consider an approximation architecture $\tilde J(i, \cdot)$ and a sampling process for approximating the value iterates. In particular, let

$$\tilde J_k(i) = \tilde J(i, r_k), \qquad i = 1, \ldots, n,$$

where r_k is the parameter vector corresponding to iteration k. Then the parameter r_{k+1} used to represent the next value iterate as

$$\tilde J_{k+1}(i) = \tilde J(i, r_{k+1}), \qquad i = 1, \ldots, n,$$

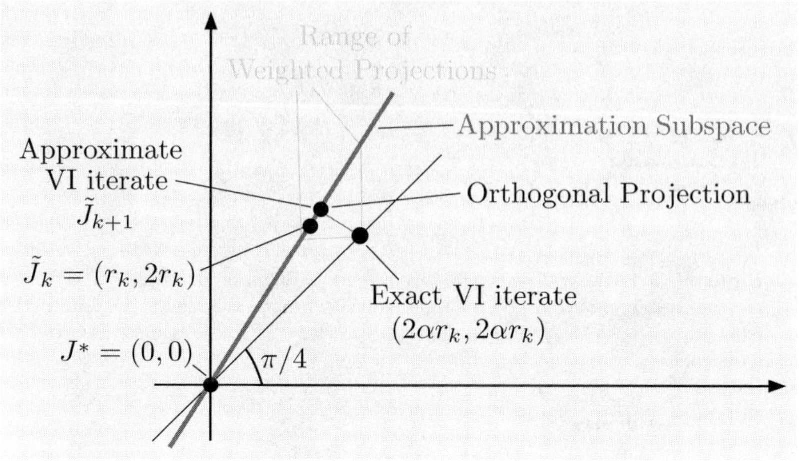

Figure 5.2.2 Illustration of Example 5.2.1. The iterates of approximate VI lie on the approximation subspace, which is the line $\{(r, 2r) \mid r \in \Re\}$. Given an iterate $\tilde{J}_k = (r_k, 2r_k)$, the next exact VI iterate is

$$\big(\alpha\tilde{J}_k(2), \alpha\tilde{J}_k(2)\big) = (2\alpha r_k, 2\alpha r_k).$$

This exact VI iterate is approximated by $\tilde{J}_{k+1}$ obtained by least squares regression, which can be viewed as weighted projection onto the line $\{(r, 2r) \mid r \in \Re\}$, and depends on the weights (ξ_1, ξ_2). The range of weighted projections as the weights vary is shown in the figure. For the natural choice $\xi_1 = \xi_2 = 1$ and α sufficiently close to 1, $\tilde{J}_{k+1}$ is further away from $J^* = (0, 0)$ than $\tilde{J}_k$. The difficulty here is that the mapping that consists of an exact VI followed by weighted projection onto the line $\{(r, 2r) \mid r \in \Re\}$ need not be a contraction.

is obtained by the minimization

$$r_{k+1} \in \arg\min_r \sum_{s=1}^{q} \big(\tilde{J}(i^s, r) - \beta^s\big)^2, \tag{5.20}$$

where (i^s, β^s), $s = 1, \ldots, q$, is a training set with each β^s being the value iterate at the state i^s:

$$\beta^s = (T\tilde{J}_k)(i^s).$$

The critical question now is how to select the sample states i^s, $s = 1, \ldots, q$, to guarantee that the iterates r_k remain bounded, so that a condition of the form (5.18) is satisfied and the instability illustrated with Example 5.2.1 is avoided. It turns out that there is no known general method to guarantee this in infinite horizon problems. However, some practical methods have been developed. One such method is to weigh each state according to its "long-term importance," i.e., proportionally to the number of its occurrences over a long trajectory under a "good" heuristic policy.† To implement this, we may run the system with the heuristic policy starting from a number of representative states, wait for some time for the system to approach steady-state, and record the generated states i^s, $s = 1, \ldots, q$, to be used in the regression scheme (5.20). There is no theoretical guarantee for the stability of this scheme in the absence of additional conditions: it has been used with success in several reported case studies, although its rationale has only a tenuous basis in analysis; see the discussion of a related algorithm, called *projected value iteration*, in the book [Ber12], Section 6.3, and the end-of-chapter references.

We finally note that in the absence of special modifications, optimistic PI with approximations is also subject to the error amplification phenomenon illustrated in Example 5.2.1. Indeed approximate VI is a special case of an optimistic PI method, where each policy evaluation is done with a single VI, and then approximated by least squares regression.

5.3 SIMULATION-BASED POLICY ITERATION WITH PARAMETRIC APPROXIMATION

In this section we will discuss PI methods where the policy evaluation and policy improvement steps are carried out with the use of parametric approximations and Monte-Carlo simulation. We will focus on the discounted problem, but similar methods can be used for the SSP problem.

5.3.1 Self-Learning and Actor-Critic Methods

The name "self-learning" in RL usually refers to some form of PI method that involves the use of simulation for approximate policy evaluation and/or approximate policy improvement. The algorithm that performs the policy evaluation is usually called a *critic*, and if a neural network is used as the parametric architecture, it is called a *critic network*. The algorithm that performs the policy improvement is usually called an *actor*, and if a neural network is involved, it is called an *actor network*.

A PI-type method is often called an *actor-critic method* if it involves approximations in both its actor and its critic portions. The term "actor-critic" actually applies more generally, to methods beyond PI-type, such as for example policy gradient methods to be discussed later.

The two operations needed at each policy iteration are as follows:

---

† In the preceding Example 5.2.1, weighing the two states according to their "long-term importance" would choose ξ_2 to be much larger than ξ_1, since state 2 is "much more important," in the sense that it occurs almost exclusively in system trajectories. Indeed, from Eq. (5.19) it can be seen that when the ratio ξ_1/ξ_2 is close enough to 0, the scalar ζ is close enough to 1, making the scalar $\alpha\zeta$ strictly less than 1, and guaranteeing convergence of $\tilde{J}_k$ to J^*.

(a) *Evaluate the current policy μ^k (critic)*: Here algorithm, system, and simulator are merged in one, and the system "observes itself" by generating simulation cost samples under the policy μ^k. It then combines these samples to "learn" an approximate policy evaluation $\tilde{J}_{\mu^k}$. Usually this is done through some kind of incremental method that involves a least squares minimization/regression using cost samples, and either a linear feature-based architecture or a neural network. Note that *the evaluation may be optimistic*, in the sense that the number of samples used between policy updates may be limited, even very small, with cost function approximation used to correct for the small number of samples. In effect, the policy evaluation of the preceding policy is used as the starting iterate for an optimistic evaluation of the current policy (cf. Section 4.6.2).

(b) *Improve the current policy μ^k (actor)*: Given the approximate policy evaluation $\tilde{J}_{\mu^k}$, we generate or "learn" the new policy μ^{k+1} through the lookahead minimization

$$\mu^{k+1}(i) \in \arg \min_{u \in U(i)} \sum_{j=1}^{n} p_{ij}(u)\big(g(i,u,j) + \alpha \tilde{J}_{\mu^k}(j)\big), \qquad i = 1,\ldots,n, \tag{5.21}$$

(or its multistep version). Alternatively we compute the minimizing control u^s at a set of sample states i^s, $s = 1,\ldots,q$, through

$$u^s \in \arg \min_{u \in U(i^s)} \sum_{j=1}^{n} p_{i^s j}(u)\big(g(i^s,u,j) + \alpha \tilde{J}_{\mu^k}(j)\big).$$

These are the sample values of the improved policy μ^{k+1} at the sample states i^s. They are generalized to "learn" a complete policy μ^{k+1} by using some parametric approximation in policy space scheme (cf. Section 2.1.5). This scheme *may be incremental* in the sense that the improved policy may be updated using very few sample pairs (i^s, u^s).

The preceding two operations are sequentially performed up to the point where some form of "convergence" occurs. To summarize, actor-critic PI methods can be viewed as a repeated application of a two-step process:†

(a) *Critic step*: A form of least squares regression based on (many or few) cost samples, and aiming to evaluate better the cost function of the current policy (while keeping that policy unchanged).

(b) *Actor step*: The improved policy is computed when needed at states of interest via the lookahead minimization (5.21) (or its multistep version). Alternatively, a form of incremental least squares regression is

† Actor-critic methods also arise in other, non-PI contexts, which combine parametric approximation in both policy and value space, and involve gradient descent-type of parameter adjustments. We will briefly discuss such methods in Section 5.7.

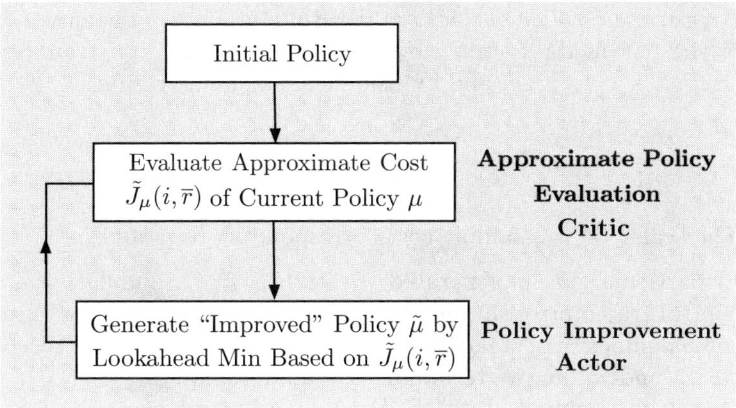

Figure 5.3.1 Block diagram of model-based approximate PI for cost functions.

performed based on (many or few) state-control sample pairs, aiming to improve a parametric form of the current policy (while keeping the policy evaluation unchanged).

Note that while approximate PI schemes of the type described above have been widely reported and advocated, they offer very little in the way of performance guarantees (at least when either costs or policies, or both, are represented with approximation architectures). Moreover, while in these schemes the system learns by itself, it does not learn itself, in the sense that *it does not construct a mathematical model for itself*. An alternative is to adopt instead a two-phase approach: first use system identification and simulation to construct a mathematical model of the system, and then use a model-based PI method (cf. Section 1.3.8). However, we will not discuss system identification and the construction of models in this book.

In what follows in this section, in Section 5.4 (Q-learning, SARSA, and DQN), and in Section 5.5 (temporal differences), we will focus on *critic-only* PI methods, where the policy improvement steps are "exact" in the sense that they are performed with Eq. (5.21). Actor-critic methods, as well as *actor-only* methods that involve approximations only in the actor step, will be discussed further in Section 5.7.

5.3.2 Model-Based Variant of a Critic-Only Method

We will first provide an example of a model-based approximate PI implementation and then discuss its model-free version in Section 5.3.3. In particular, we assume that the cost per stage $g(i,u,j)$ and transition probabilities $p_{ij}(u)$ are available, and that the cost function J_μ of any given policy μ is approximated using a parametric architecture $\tilde{J}_\mu(i,r)$.

We recall that given any policy μ, the exact PI algorithm for costs generates the new policy $\tilde{\mu}$ with a policy evaluation/policy improvement process. We approximate this process as follows; see Fig. 5.3.1.

(a) *Approximate policy evaluation*: To evaluate μ, we determine the value of the parameter vector r by generating a number of training pairs (i^s, β^s), $s = 1, \dots, q$, and by using least squares training:

$$\overline{r} \in \arg\min_{r} \sum_{s=1}^{q} \big(\tilde{J}_\mu(i^s, r) - \beta^s\big)^2. \tag{5.22}$$

The scalar β^s is a sample cost corresponding to i^s and μ.

In particular, β^s is generated by starting at i^s, simulating a state-control trajectory using μ and the known transition probabilities for some number N of stages, accumulating the corresponding discounted costs, and adding a terminal cost approximation $\alpha^N \hat{J}(i_N)$, where i_N is the terminal state of the N-stage trajectory and $\hat{J}$ is some initial guess of J_μ. The guess $\hat{J}$ may be obtained with additional training or some other means, such as using the result of the policy evaluation of the policy preceding μ. Note that *this is similar to the cost function approximation implicitly used in optimistic policy iteration*; cf. Section 4.6.2. It is also possible to simplify the method by using $\hat{J}(i_N) = 0$, or obtaining $\hat{J}$ via a problem approximation process.†

The approximate policy evaluation problem of Eq. (5.22) can be solved with the incremental methods of Section 3.1.3. In particular, the incremental gradient method for this problem is given by

$$r^{k+1} = r^k - \gamma^k \nabla \tilde{J}(i^{s_k}, r^k)\big(\tilde{J}(i^{s_k}, r^k) - \beta^{s_k}\big),$$

where γ^k is a stepsize, (i^{s_k}, β^{s_k}) is the state-cost sample pair that is used at the kth iteration, and r^0 is an initial parameter guess. Here the architecture $\tilde{J}(i, r)$ may be linear or may be nonlinear and differentiable. In the case of a linear architecture, problem (5.22) may also be solved in closed form, i.e., by using exact least squares.

(b) *Policy improvement*: Having solved the approximate policy evaluation problem (5.22), the new "improved" policy $\tilde{\mu}$ is obtained (at the states needed, when needed) by the policy improvement operation

$$\tilde{\mu}(i) \in \arg\min_{u \in U(i)} \sum_{j=1}^{n} p_{ij}(u)\big(g(i, u, j) + \alpha \tilde{J}(j, \overline{r})\big), \tag{5.23}$$

where $\overline{r}$ is the parameter vector obtained from the policy evaluation (5.22). [The transition probabilities $p_{ij}(u)$ are required for this.]

---

† The choice of the terminal cost function approximation $\hat{J}$ may pose difficulties in SSP problems, particularly when most of the cost is incurred upon termination. Dealing with this issue may be problem-dependent. We will not discuss it further in this book.

Trajectory Reuse and Bias-Variance Tradeoff

As we have noted, to each training pair (i^s, β^s) there corresponds an N-stage trajectory over which the sample cost β^s is accumulated, but the length of the trajectory may depend on s. This allows sampling effort economies based on *trajectory reuse.* In particular, suppose that starting at some state i_0 we generate a long trajectory $(i_0, i_1, \ldots, i_N)$ using the policy μ. Then we can obtain the state-cost sample that corresponds to i_0, as discussed above, but we can also obtain additional cost samples for the subsequent states $i_1, i_2, \ldots$, by using the tail portions of the trajectory $(i_0, i_1, \ldots, i_N)$ that start at these states.

Clearly, it is necessary to truncate the sample trajectories to some number of stages N, since we cannot simulate an infinite length trajectory in practice. If N is large, then because of the discount factor, the error for neglecting the stage costs beyond stage N will be small. However, there are other important concerns when choosing the trajectory lengths N.

In particular, a short length reduces the sampling effort, but is also a source of inaccuracy. The reason is that the cost of the tail portion of the trajectory (from stage N to infinity) is approximated by $\alpha^N \hat{J}(i_N)$, where i_N is the terminal state of the N-stage trajectory and $\hat{J}$ is the initial guess of J_μ. This terminal cost compensates for the costs of the neglected stages in the spirit of optimistic PI, but adds an error to the cost samples β^s, which becomes larger as the trajectory length N becomes smaller.

We note two additional benefits of using many training trajectories, each with a relatively short trajectory length:

(1) The cost samples β^s are less noisy, as they correspond to summation of fewer random stage costs. This leads to the so-called *bias-variance tradeoff*: short trajectories lead to larger bias but smaller variance of the cost samples.

(2) With more starting states i_0, there is better opportunity for *exploration of the state space*. By this we mean adequate representation of all possible initial trajectory states in the sample set. This is a major issue in approximate PI, as we will discuss in Section 5.3.4.

Let us also note that the bias-variance tradeoff underlies the motivation for a number of alternative policy evaluation methods such as TD(λ), LSTD(λ), and LSPE(λ), which we will summarize in Section 5.5; see Section 6.3 of the book [Ber12] and other approximate DP/RL books referenced earlier. The papers [Ber13b], [YuB12], and the book [Ber12], Section 6.4, discuss a broad range of short trajectory sampling methods.

5.3.3 Model-Free Variant of a Critic-Only Method

We recall the exact PI algorithm for Q-factors (cf. Section 4.6.3), which given any policy μ, generates the new policy $\tilde{\mu}$ with a policy evalua-

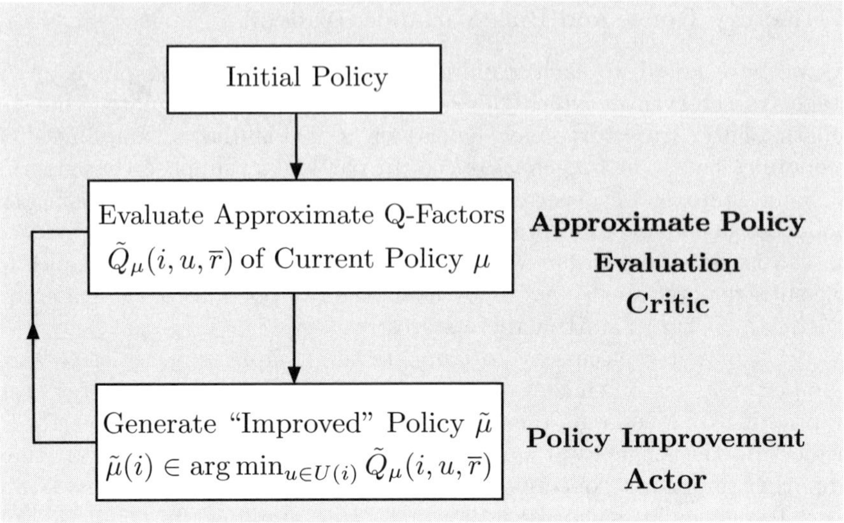

Figure 5.3.2 Block diagram of model-free approximate PI for Q-factors.

tion/policy improvement process. We will provide an approximate version of this algorithm, which involves a model-free simulation-based implementation. To this end, we will use Q-factors to bypass the need for transition probabilities $p_{ij}(u)$ in Eq. (5.23).

We introduce a parametric architecture $\tilde{Q}_\mu(i,u,r)$ for the Q-factors of μ. This architecture may be nonlinear, involving for example a neural network. It may also be a feature-based linear architecture, with a feature vector that depends either on both the state and the control or on just the state. In the former case, the architecture has the form

$$\tilde{Q}_\mu(i,u,r) = r'\phi(i,u),$$

where r is a weight vector that is independent of u. In the latter case, the architecture has the form

$$\tilde{Q}_\mu(i,u,r) = r(u)'\phi(i),$$

where $r(u)$ is a separate weight vector for each control u. This architecture is suitable for problems with a relatively small number of control options at each stage.

We approximate the PI process as follows; see Fig. 5.3.2.

(a) *Approximate policy evaluation*: Here, given a policy μ, we determine the value of the parameter vector r by generating (using a simulator of the system) a number of training triplets (i^s,u^s,β^s), $s = 1,\ldots,q$, and by using a least squares fit:

$$\overline{r} \in \arg\min_r \sum_{s=1}^{q} \big(\tilde{Q}_\mu(i^s,u^s,r) - \beta^s\big)^2. \tag{5.24}$$

In particular, for a given pair (i^s, u^s), the scalar β^s is a sample Q-factor corresponding to (i^s, u^s). It is generated by starting at i^s, using u^s at the first stage, simulating a trajectory of states and controls using μ for a total of N stages, accumulating the corresponding discounted costs, and adding a terminal cost approximation, similar to Section 5.3.2. Thus, β^s is a sample of

$$\sum_{j=1}^{n} p_{i^s j}(u^s)\big(g(i^s, u^s, j) + \alpha J_\mu^{N-1}(j) + \alpha^N \hat{J}(i_N)\big),$$

where $J_\mu^{N-1}(j)$ is the $(N-1)$-stages cost of μ starting at j, and $\alpha^N \hat{J}(i_N)$ is the terminal cost approximation. Note that β^s is computed similar to the scalar β^s of Eq. (5.22), except that *the first stage control is* u^s *rather than* $\mu(i^s)$. Similar to Section 5.3.2, the number of stages N in the sample trajectories may be different for different samples, and can be either large, or fairly small. Again an incremental method may be used to solve the training problem (5.24).

(b) *Policy improvement*: Here we compute the new policy $\tilde{\mu}$ (at the states needed, when needed) according to

$$\tilde{\mu}(i) \in \arg\min_{u \in U(i)} \tilde{Q}_\mu(i, u, \overline{r}), \qquad i = 1, \ldots, n, \tag{5.25}$$

where $\overline{r}$ is the parameter vector obtained from the policy evaluation operation (5.24). [Contrary to Eq. (5.23), this does not require the use of the transition probabilities $p_{ij}(u)$.]

Unfortunately, trajectory reuse is more problematic in Q-factor evaluation than in cost evaluation, because each trajectory generates state-control pairs of the special form $\big(i, \mu(i)\big)$ at every stage after the first, so *pairs* (i, u) *with* $u \neq \mu(i)$ *are not adequately explored*; cf. the discussion in Section 5.3.2. For this reason, it is necessary to make an effort to include in the samples a rich enough set of trajectories that start at pairs (i, u) with $u \neq \mu(i)$. We discuss this issue in Section 5.3.4.

An important alternative to the preceding procedure is a two-stage process for policy evaluation: first compute in model-free fashion a cost function approximation $\tilde{J}_\mu(j, \overline{r})$, using the regression (5.22), and then use a *second sampling process and regression* to approximate further the (already approximate) Q-factor of μ

$$\sum_{j=1}^{n} p_{ij}(u)\big(g(i, u, j) + \alpha \tilde{J}_\mu(j, \overline{r})\big),$$

with some $\tilde{Q}_\mu(i, u, \overline{r})$, possibly obtained with a model-free Q-factor approximation process such as the one of Section 2.1.4. In view of the two-fold approximation needed to obtain $\tilde{Q}_\mu(i, u, \overline{r})$, this scheme is more complex, but allows trajectory reuse and thus deals better with the exploration issue.

5.3.4 Implementation Issues of Parametric Policy Iteration

Approximate PI in its various forms has been the subject of extensive research, both theoretical and applied. Let us provide a few comments, focusing on the preceding critic-only parametric PI schemes.

Architectural Issues and Cost Shaping

The choice of architectures for costs $\tilde J_\mu(i,r)$ and Q-factors $\tilde Q_\mu(i,u,r)$ is critical for the success of parametric approximation schemes for PI. These architectures may involve the use of features, and they can be linear, or they can be nonlinear such as a neural network. A major advantage of a linear feature-based architecture is that the corresponding policy evaluations (5.22) and (5.24) involve linear least squares problems, which admit a closed-form solution. Moreover, when linear architectures are used, there is a broader variety of approximate policy evaluation methods with solid theoretical performance guarantees, such as TD(λ), LSTD(λ), and LSPE(λ), which will be summarized in Section 5.5, and are described in detail in several textbook sources.

Another interesting possibility for architecture choice has to do with *cost shaping*, which we discussed in Section 4.2. This possibility involves a modified cost per stage

$$\hat g(i,u,j) = g(i,u,j) + V(j) - V(i), \qquad i = 1,\ldots,n,$$

for SSP problems, where V can be any approximation to J^* with $V(t) = 0$. The corresponding formula for discounted problems is

$$\hat g(i,u,j) = g(i,u,j) + \alpha V(j) - V(i), \qquad i = 1,\ldots,n.$$

As noted in Section 4.2, cost shaping may change significantly the suboptimal policies produced by approximate DP methods and approximate PI in particular. Generally, V should be chosen close (at least in terms of "shape") to J^* or to the current policy cost function J_{μ^k}, so that the difference $J^* - V$ or $J_{\mu^k} - V$, respectively, can be approximated by an architecture that matches well the characteristics of the problem. It is possible to approximate V with a parametric architecture or some other approximation method, depending on the problem at hand. Moreover, in the context of approximate PI, the choice of V may change from one policy evaluation to the next.

The papers referenced for Section 4.2 provide some applications of cost shaping (or reward shaping as it is sometimes called in RL contexts). An interesting possibility is to use complementary approximations for V and for J^* or J_{μ^k}. For example V may be approximated by a neural network-based approach that aims to discover the general form of J^* or J_{μ^k}, and then a different method may be applied to provide a local correction to V in order to refine the approximation. We will also illustrate this idea within the context of biased aggregation in Section 6.5.

Exploration Issues

Generating an appropriate training set at the policy evaluation step of approximate PI poses considerable challenges, and the literature contains several related proposals. We will discuss the generic issue of *inadequate exploration*, which we noted earlier in connection with model-based approximate PI. This is indeed the Achilles' heel of approximate simulation-based PI schemes, and is often not addressed seriously enough in practice.

We recall that in the PI variant of Section 5.3.2, in order to evaluate a policy μ, we determine the value of the parameter vector r by generating a large number of training pairs (i^s, β^s), $s = 1, \dots, q$, where β^s is a sample cost corresponding to i^s and μ. We then use least squares training:

$$\overline{r} \in \arg\min_r \sum_{s=1}^{q} \big(\tilde{J}_\mu(i^s, r) - \beta^s\big)^2.$$

Each sample cost β^s is generated by starting at i^s, simulating a trajectory of states and controls using μ and the known transition probabilities for some number N_s of stages, accumulating the corresponding discounted costs, and adding a terminal cost approximation

$$\alpha^{N_s} \hat{J}(i_{N_s}),$$

where i_{N_s} is the terminal state of the N_s-stage trajectory and $\hat{J}$ is some initial guess of J_μ. In this context, we also discussed trajectory reuse, which aims at reducing the sampling effort, by using the tail portions of any generated trajectory.

Thus with trajectory reuse, we will be generating many cost or Q-factor samples that start from states frequently visited by μ, and this may bias the simulation by underrepresenting states that are unlikely to occur under μ. As a result, the cost estimates of these underrepresented states may be highly inaccurate, causing potentially serious errors in the calculation of the improved policy $\tilde{\mu}$ via the policy improvement operation.

One possibility to improve the exploration of the state space is to use a large number of initial states. It may then be necessary to use relatively short trajectories to keep the cost of the simulation low. To compensate for the short length of the trajectories it will then be important to introduce a terminal cost function approximation in the policy evaluation step in order to make the cost sample β^s more accurate, as noted earlier. Moreover, when selecting the initial states of these trajectories, we should make sure that they are a representative sample of the portion of the state space most visited not just under the current policy and also under other policies.

A simple plausible scheme when evaluating policy μ^k, is to use *randomization over a memory buffer*, which stores candidate initial states. In particular, we may use a set of initial states that consists of the union of multiple subsets $I, I_0, \ldots, I_{k-1}$, where I is a starting set of initial states, and $I_0, \ldots, I_{k-1}$ are generated while evaluating the previous policies $\mu^0, \ldots, \mu^{k-1}$. Each subset I_m should consist of states generated while evaluating policy μ^m. We evaluate μ^k, by using short trajectories whose starting states are chosen randomly from the subsets $I, I_0, \ldots, I_{k-1}$ with probabilities that may reflect a bias toward more recent policies (i.e., the probability of selecting an initial state from set I_m should increase with m). At the same time, while evaluating policy μ^k, we generate the subset of states I_k to use when training the next policy μ^{k+1}. The initial subset I is special and should be carefully chosen so that it includes not only a subset of representative states, but also successor states under a variety of control choices. The fine details of such a scheme are likely to be settled by trial and error.

Thus in the preceding scheme, all state transitions and associated transition costs used to generate the costs β^s of the training set are generated by using the current policy μ^k. However, the set of initial states of the trajectories used to produce the cost samples is quite diverse: it is obtained by randomization over the union of the starting set I and the set $I_0 \cup \cdots \cup I_{k-1}$, which corresponds to states visited by the preceding policies.

A potential weakness of a scheme of the preceding type is the need for a good terminal cost function approximation. Such a cost may not be easily available, particularly for SSP problems involving substantial "late costs;" for example when most of the cost is incurred upon reaching the destination. Such problems generally require "deep exploration," i.e., long trajectories that reach or get close to the destination.

Exploration schemes like the one just described may also be used in the context of model-free variants of approximate PI for Q-factors; cf. Section 5.3.3. In fact, as we discussed in that section, the need for exploration in the space of state-control pairs is more acute. Again we may generate short trajectories that start with state-control pairs chosen randomly within a set of pairs that are representative of several policies.

Let us also note the possibility to partition the state space, and select a rich enough representative set of initial states for each set of the partition. This allows a better control of exploration, and also ushers the possibility of using a distributed policy evaluation algorithm for each set of the partition; see our earlier discussion and the papers [BeY10], [BeY12], [YuB13a].

There have also been other related approaches to improve exploration, particularly in connection with the temporal difference methods to be discussed in Section 5.5. In some of these approaches, trajectories are generated through a mix of two policies: the policy being evaluated, sometimes called the *target policy*, to distinguish from the other policy, a probabilistically used policy, called the *behavior policy* that introduces enhanced exploration; see the end-of-chapter references. In the RL literature (see e.g., the book [SuB18]), methods that use a behavior policy are called *off-policy* methods, while methods that do not are called *on-policy* methods. A commonly suggested scheme is to use as an off-policy one that is "ϵ-greedy," i.e., that attains the minimum in the policy improvement phase within some threshold $\epsilon > 0$, which is experimentally chosen.

It is important to note, however, that using a behavior policy biases the cost samples towards that policy. Special modifications are needed, which can eliminate this source of bias and indeed can work exclusively with a behavior policy. Such modifications were introduced by Bertsekas and Yu [BeY09] (Section 3) to deal with the solution of general linear systems of equations that do not involve transition probabilities; see also the book [Ber12], Sections 6.4.2 and 7.3.1.

Generally, efficient sampling, and the issue of balancing exploration and the choice of promising controls (the exploration-exploitation tradeoff) is a subject of continuing research. Many authors have suggested practical schemes that may work in given contexts, including on-line situations. For some recent work, see the paper by Russo and Van Roy [RuV16], and the monograph [RVK18]. The paper by Osband, Van Roy, Russo, and Wen [OVR19] develops special schemes to deal with issues of deep exploration.

5.3.5 Convergence Issues of Approximate Policy Iteration - Oscillations

We will now consider the sequence of policies generated by approximate PI. Contrary to exact PI, which is guaranteed to yield an optimal policy, approximate PI produces a sequence of policies, which are only guaranteed to lie asymptotically within a certain error bound from the optimal; cf. Prop. 5.1.4. Moreover, the generated policies may oscillate. By this we mean that after a few iterations, policies tend to repeat in cycles.

This oscillation phenomenon, first described by the author in a 1996 conference [Ber96b], occurs systematically in the absence of special conditions, for both optimistic and nonoptimistic PI methods. It can be observed even in simple examples (see [BeT96], Section 6.4.2, [Ber12], Section 6.4.3).

To describe a generic mechanism that may cause policy oscillations in approximate PI, we focus on the discounted problem, and we introduce the so called *greedy partition*. For a given approximation architecture $\tilde J(\cdot, r)$, this is a partition of the space $\Re^s$ of parameter vectors r into subsets R_μ, each subset corresponding to a stationary policy μ, and defined by †

$$R_\mu = \big\{r \mid T_\mu\big(\tilde J(\cdot, r)\big) = T\big(\tilde J(\cdot, r)\big)\big\}$$

or equivalently

$$R_\mu = \left\{ r \;\middle|\; \mu(i) \in \arg \min_{u\in U(i)} \sum_{j=1}^{n} p_{ij}(u)\big(g(i,u,j) + \alpha \tilde J(j, r)\big),\ i = 1, \ldots, n \right\}.$$

† Since we are now dealing with multiple policies, our notation reflects the dependence of the stage costs and transition probabilities on the control. Also T_μ is the Bellman operator corresponding to a policy μ, while T is defined in terms of minimization of T_μ over μ.

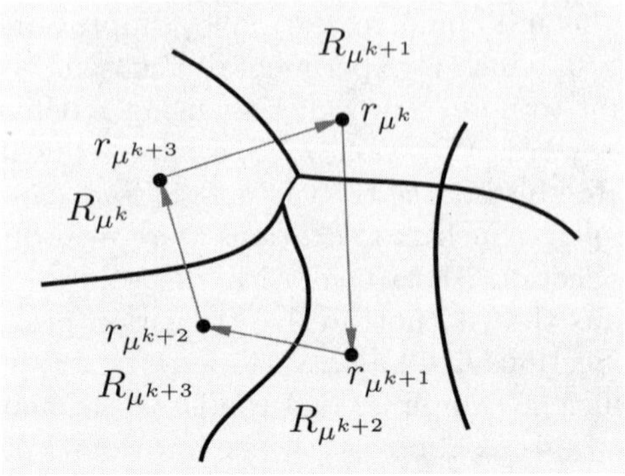

Figure 5.3.3 Greedy partition and cycle of policies generated by nonoptimistic PI with cost function approximation. In particular, μ yields $\overline{\mu}$ by policy improvement if and only if $r_\mu \in R_{\overline{\mu}}$. In this figure, the method cycles between four policies and the corresponding parameters $r_{\mu^k}, r_{\mu^{k+1}}, r_{\mu^{k+2}}, r_{\mu^{k+3}}$.

Thus, R_μ is the set of parameter vectors r for which μ is "greedy" with respect to $\tilde{J}(\cdot, r)$. Note that *the greedy partition depends only on the approximation architecture* $\tilde{J}(\cdot, r)$ (which can be arbitrary, e.g., nonlinear), and does not depend on the method used for policy evaluation.

We will first consider the nonoptimistic version of approximate PI; a similar phenomenon occurs for optimistic PI, as we discuss later. For simplicity, let us assume that we use a policy evaluation method that for each given μ produces a unique parameter vector denoted r_μ. The method starts with a policy μ^0 and generates r_{μ^0} by using the given policy evaluation method. It then finds a policy μ^1 such that $r_{\mu^0} \in R_{\mu^1}$. It then repeats the process with μ^1 replacing μ^0. If some policy μ^k satisfying $r_{\mu^k} \in R_{\mu^k}$ is encountered, the method converges to μ^k (it keeps generating μ^k). This is the necessary and sufficient condition for policy convergence in the nonoptimistic PI method.

In the case of an exact cost function representation where the parameter vector r_μ is equal to the cost-to-go vector J_μ, the condition $r_{\mu^k} \in R_{\mu^k}$ is equivalent to $r_{\mu^k} = Tr_{\mu^k}$, and is satisfied if and only if μ^k is optimal. When there is cost function approximation, however, this condition need not be satisfied for any policy. Since there is a finite number of possible vectors r_μ, one generated from another in a deterministic way, the algorithm ends up repeating some cycle of policies $\mu^k, \mu^{k+1}, \ldots, \mu^{k+m}$ with

$$r_{\mu^k} \in R_{\mu^{k+1}},\ r_{\mu^{k+1}} \in R_{\mu^{k+2}}, \ldots, r_{\mu^{k+m-1}} \in R_{\mu^{k+m}},\ r_{\mu^{k+m}} \in R_{\mu^k};$$

(see Fig. 5.3.3). Furthermore, there may be many different cycles of this type, and the method may end up converging to any one of them. The actual cycle obtained depends on the initial policy μ^0. This is similar to gradient methods applied to minimization of functions with multiple local minima, where the limit of convergence depends on the starting point.

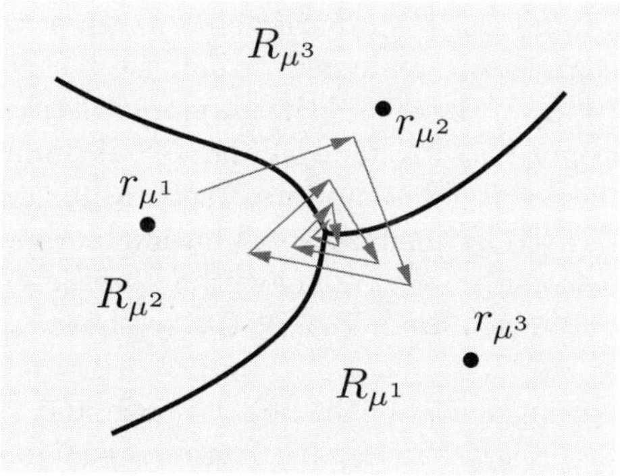

Figure 5.3.4 Illustration of a trajectory of optimistic policy iteration with cost function approximation. The algorithm settles into an oscillation between policies μ^1, μ^2, μ^3 with

$$r_{\mu^1} \in R_{\mu^2}, \qquad r_{\mu^2} \in R_{\mu^3}, \qquad r_{\mu^3} \in R_{\mu^1}.$$

The parameter vectors converge to the common boundary of R_{μ^1}, R_{μ^2}, R_{μ^3}.

Oscillations can in principle be quite damaging, because there is no guarantee that the oscillating policies are "good" policies, and there is often no way to verify how well they perform relative to the optimal. However, we will later give an argument (from Section 6.4.2 of the book [BeT96]), which suggests that oscillations may not degrade significantly the approximate PI performance for many types of problems. Moreover, we note that oscillations can be avoided and convergence can be shown under special conditions, which arise in particular when an aggregation approach is used; see Chapter 6 and the approximate PI survey [Ber11a]. Also, when policies converge, there is a more favorable error bound, cf. Prop. 5.1.5.

Oscillations and Chattering in Optimistic PI Methods

We now consider policy oscillations in the context of optimistic approximate PI. Then the trajectory of the method is less predictable and depends on the fine details of the policy evaluation, such as the frequency of the policy updates.

Generally, given the current policy μ, optimistic PI will move towards the corresponding "target" parameter r_μ, for as long as μ continues to be greedy with respect to the current cost-to-go approximation $\tilde{J}(\cdot, r)$, that is, for as long as the current parameter vector r belongs to the set R_μ. Once, however, the parameter r crosses into another set, say $R_{\overline{\mu}}$, the policy $\overline{\mu}$ becomes greedy, and r changes course and starts moving towards the new "target" $r_{\overline{\mu}}$. Thus, the "targets" r_μ of the method, and the corresponding policies μ and sets R_μ may keep changing, similar to nonoptimistic policy iteration. Simultaneously, the parameter vector r will move near the boundaries that separate the regions R_μ that the method visits, thus following reduced versions of the cycles that nonoptimistic PI may follow (see Fig. 5.3.3). Furthermore, as Fig. 5.3.4 indicates, if diminishing parameter changes are made between policy updates (such as for example when a diminishing stepsize is used by the policy evaluation method) and the method eventually cycles between several policies, the parameter vectors will tend to converge to the common boundary of the regions R_μ corresponding to these policies.

The simultaneous oscillation in policy space and convergence in parameter space in optimistic PI is called *chattering*. The literature reflects some confusion regarding the nature of this phenomenon, and in fact it is often erroneously assumed that if the sequence of generated parameters converges, the same is true for the sequence of generated policies.

An interesting observation is that the choice of the approximate policy evaluation method and exploration scheme may not be crucial for the quality of the final policy obtained. Using a different method changes the targets r_μ somewhat, but leaves the greedy partition unchanged. As a result, different methods "fish in the same waters" and tend to yield similar ultimate cycles of policies.

Note that when chattering occurs, the limit of optimistic PI tends to be on a common boundary of several subsets of the greedy partition and *may not meaningfully represent a cost approximation for any of the policies involved in the oscillation.* Thus, the parameter limit of the method cannot always be used to construct an approximation of the cost-to-go of any policy or the optimal cost-to-go. As a result, at the end of optimistic PI one may need to go back and perform a screening process; that is, evaluate the many policies generated by the method and select the most promising.

Still even with chattering, it is possible that the cost functions of the different policies involved in oscillation may not be "too different," so the cost functions of the generated policies may appear to be "nearly converging" (which is not the same as saying that they are all "good" policies). For an explanation, suppose that we have convergence to a parameter vector $\bar{r}$ and that there is a steady-state policy oscillation involving a collection of policies $\mathcal{M}$. Then, all the policies in $\mathcal{M}$ are greedy with respect to $\tilde{J}(\cdot,\bar{r})$, which implies that there is a subset of states i such that there are at least two different controls $\mu_1(i)$ and $\mu_2(i)$ satisfying

$$
\begin{aligned}
\min_{u\in U(i)} \sum_j p_{ij}(u)\big(g(i,u,j)+\alpha\tilde{J}(j,\bar{r})\big)
&= \sum_j p_{ij}\big(\mu_1(i)\big)\Big(g\big(i,\mu_1(i),j\big)+\alpha\tilde{J}(j,\bar{r})\Big) \\
&= \sum_j p_{ij}\big(\mu_2(i)\big)\Big(g\big(i,\mu_2(i),j\big)+\alpha\tilde{J}(j,\bar{r})\Big).
\end{aligned}
$$

Each equation of this type can be viewed as a constraining relation on the parameter vector $\overline{r}$. Thus, excluding singular situations, there will be at most s relations of the preceding form holding, where s is the dimension of $\overline{r}$. This implies that there will be at most s "ambiguous" states where more than one control is greedy with respect to $\tilde{J}(\cdot, \overline{r})$.

Now assume that we have a problem where the total number of states is much larger than s, and in addition there are no "critical" states; that is, the cost consequences of changing a policy in just a small number of states (say, of the order of s) is relatively small. It then follows that all policies in the set $\mathcal{M}$ involved in chattering have roughly the same cost function (although these policies are not necessarily "good" policies). Furthermore, one may argue that the cost approximation $\tilde{J}(\cdot, \overline{r})$ is close to the cost approximation $\tilde{J}(\cdot, r_\mu)$ that would be generated for any of the policies $\mu \in \mathcal{M}$. Note, however, that the assumption of "no critical states" may be hard to quantify and check, and may also not be true.

5.4 Q-LEARNING

In this section we will focus on discounted problems and discuss various Q-learning algorithms, which can be implemented in model-free fashion (if this is desirable). The original method of this type is related to the VI algorithm for Q-factors, described in Sections 4.2 and 4.3. Instead of approximating the cost functions of successive policies as in the PI method, it aims to obtain the optimal Q-factors, thereby avoiding the multiple policy evaluation steps of PI. We will consider Q-learning as well as a variety of related methods with the shared characteristic that they involve exact or approximate Q-factors. Some of these methods are also related to (highly) optimistic PI methods that are based on the use of Q-factors (cf. the methods of Section 5.3.3).

We first discuss the original form of Q-learning for discounted problems; the books [BeT96] and [Ber12], and the papers [Tsi94] and [YuB13b] contain discussions of Q-learning for SSP problems. Then we consider PI algorithms for Q-factors, including optimistic asynchronous versions, with Q-factor approximation.

In the discounted problem, the optimal Q-factors are defined for all pairs (i, u) with $u \in U(i)$, by

$$Q^*(i, u) = \sum_{j=1}^{n} p_{ij}(u)\big(g(i, u, j) + \alpha J^*(j)\big).$$

As discussed in Section 4.3, these Q-factors satisfy for all (i, u),

$$Q^*(i, u) = \sum_{j=1}^{n} p_{ij}(u)\left(g(i, u, j) + \alpha \min_{v \in U(j)} Q^*(j, v)\right),$$

and are the unique solution of this set of equations. Moreover the optimal Q-factors can be obtained by the VI algorithm $Q_{k+1} = FQ_k$, where F is the Bellman operator for Q-factors defined by

$$(FQ)(i,u) = \sum_{j=1}^{n} p_{ij}(u)\left(g(i,u,j) + \alpha \min_{v\in U(j)} Q(j,v)\right), \qquad \text{for all } (i,u). \tag{5.26}$$

It is straightforward to show that F is a contraction with modulus α, similar to the Bellman operator for costs [cf. Eq. (5.3)]. Thus the algorithm $Q_{k+1} = FQ_k$ converges to Q^* from every starting point Q_0.

The original and most widely known Q-learning algorithm ([Wat89]) is a stochastic version of VI, whereby the expected value in Eq. (5.26) is suitably approximated by sampling and simulation. In particular, an infinitely long sequence of state-control pairs $\{(i_k,u_k)\}$ is generated according to some probabilistic mechanism. For each pair (i_k,u_k), a state j_k is generated according to the probabilities $p_{i_kj}(u_k)$. Then the Q-factor of (i_k,u_k) is updated using a stepsize $\gamma^k \in (0,1]$ while all other Q-factors are left unchanged:

$$Q_{k+1}(i,u) = (1-\gamma^k)Q_k(i,u) + \gamma^k(F_kQ_k)(i,u), \qquad \text{for all } (i,u), \tag{5.27}$$

where

$$(F_kQ_k)(i,u) = \begin{cases} g(i_k,u_k,j_k) + \alpha \min_{v\in U(j_k)} Q_k(j_k,v) & \text{if } (i,u) = (i_k,u_k), \\ Q_k(i,u) & \text{if } (i,u) \neq (i_k,u_k). \end{cases} \tag{5.28}$$

Note that $(F_kQ_k)(i_k,u_k)$ is a single sample approximation of the expected value defining $(FQ_k)(i_k,u_k)$ in Eq. (5.26).

To guarantee the convergence of the algorithm (5.27)-(5.28) to the optimal Q-factors, some conditions must be satisfied. Chief among these are that all state-control pairs (i,u) must be generated infinitely often within the infinitely long sequence $\{(i_k,u_k)\}$, and that the successor states j must be independently sampled at each occurrence of a given state-control pair. Furthermore, the stepsize γ^k should satisfy the conditions

$$\gamma^k > 0, \quad \text{for all } k, \qquad \sum_{k=0}^{\infty} \gamma^k = \infty, \quad \sum_{k=0}^{\infty} (\gamma^k)^2 < \infty,$$

which are typical of stochastic approximation methods (see e.g, the books [BeT96], [Ber12], Section 6.1.4), as for example when $\gamma^k = c_1/(k+c_2)$, where c_1 and c_2 are some positive constants. In addition some other technical conditions should hold. A mathematically rigorous convergence proof was given in the paper [Tsi94], which embeds Q-learning within a broad class of asynchronous stochastic approximation algorithms. This proof (also reproduced in [BeT96]) combines the theory of stochastic approximation algorithms with the convergence theory of asynchronous DP and asynchronous iterative methods; cf. the papers [Ber82], [Ber83], and the book [BeT89].

In practice, Q-learning has some drawbacks, the most important of which is that the number of Q-factors/state-control pairs (i, u) may be excessive. To alleviate this difficulty, we may introduce a Q-factor approximation architecture. One of these possibilities will be considered next.

5.4.1 Optimistic Policy Iteration with Parametric Q-Factor Approximation - SARSA and DQN

We have discussed so far Q-learning algorithms with an exact representation of Q-factors. We will now consider Q-learning with Q-factor approximation. As we noted earlier, we may view Q-factors as optimal costs of a certain discounted DP problem, whose states are the state-control pairs (i, u) in addition to the original states. We may thus apply the approximate PI methods discussed earlier. For this, we need to introduce a parametric architecture $\tilde{Q}(i, u, r)$. This architecture could be linear feature-based, or nonlinear such as one that uses a neural network.

We have already discussed in Section 5.3.3 a model-free approximate PI method that is based on Q-factors and least squares training/regression. There are also optimistic approximate PI methods, which use a policy for a limited number of stages with cost function approximation for the remaining states, and/or a few samples in between policy updates.

As an example, let us consider an extreme version that uses a single sample between policy updates. At the start of iteration k, we have the current parameter vector r^k, we are at some state i^k, and we have chosen a control u^k. Then:

(1) We simulate the next transition (i^k, i^{k+1}) using $p_{i^k j}(u^k)$.

(2) We generate the control u^{k+1} with the minimization

$$u^{k+1} \in \arg \min_{u \in U(i^{k+1})} \tilde{Q}(i^{k+1}, u, r^k).$$

[In some schemes, to enhance exploration, u^{k+1} is chosen with a small probability to be a random element of $U(i^{k+1})$ or one that is "ϵ-greedy," i.e., attains within some ϵ the minimum above.]

(3) We update the parameter vector via

$$r^{k+1} = r^k - \gamma^k \nabla \tilde{Q}(i^k, u^k, r^k) \cdot \big(\tilde{Q}(i^k, u^k, r^k) - g(i^k, u^k, i^{k+1}) - \alpha \tilde{Q}(i^{k+1}, u^{k+1}, r^k)\big), \tag{5.29}$$

where γ^k is a positive stepsize, and $\nabla(\cdot)$ denotes gradient with respect to r evaluated at the current parameter vector r^k. To get a sense for the rationale of this iteration, note that if $\tilde{Q}$ is a linear feature-based architecture, $\tilde{Q}(i,u,r) = \phi(i,u)'r$, then $\nabla\tilde{Q}(i^k,u^k,r^k)$ is just the feature vector $\phi(i^k,u^k)$, and iteration (5.29) becomes

$$r^{k+1} = r^k - \gamma^k \phi(i^k,u^k)\big(\phi(i^k,u^k)'r^k - g(i^k,u^k,i^{k+1}) - \alpha\phi(i^{k+1},u^{k+1})'r^k\big).$$

Thus r^k *is changed in an incremental gradient direction*: the one opposite to the gradient (with respect to r) of the incremental error

$$\big(\phi(i^k,u^k)'r - g(i^k,u^k,i^{k+1}) - \alpha\phi(i^{k+1},u^{k+1})'r^k\big)^2,$$

evaluated at the current iterate r^k.

The process is now repeated with r^{k+1}, i^{k+1}, and u^{k+1} replacing r^k, i^k, and u^k, respectively.

Extreme optimistic PI schemes of the type just described have been used in practice, and are often referred to as SARSA (State-Action-Reward-State-Action); see e.g., the books [BeT96], [BBD10], [SuB18]. When Q-factor approximation is used, their behavior is very complex, their theoretical convergence properties are unclear, and there are no associated performance bounds in the literature. The method just described is more commonly used in a less extreme/optimistic form, whereby several (perhaps many) state-control-transition cost-next state samples are batched together and suitably averaged before updating the vector r^k.

Other variants of the method attempt to save in sampling effort by storing the generated samples in a buffer and reusing them in some randomized fashion in subsequent iterations (cf. our discussion of exploration in Section 5.3.4). This is also called sometimes *experience replay*, an idea that has been used since the early days of RL, both to save in sampling effort and to enhance exploration. The DQN (Deep Q Network) scheme, championed by DeepMind (see Mnih et al. [MKS15]), is based on this idea (the term "Deep" is a reference to DeepMind's affinity for deep neural networks, but experience replay does not depend on the use of a deep neural network architecture).

5.5 ADDITIONAL METHODS - TEMPORAL DIFFERENCES

In this section, we summarize a few additional methods for approximation in value space in infinite horizon problems. We focus on *policy evaluation with a linear parametric architecture*, and on the simulation-based temporal difference methods. One of the aims of these methods is to address a bias-variance tradeoff that is similar in nature to the one discussed in Section 5.3.2. Our presentation is brief, somewhat abstract, and makes use of linear algebra mathematics. It may be skipped without loss of continuity. This is only a summary; it is meant to provide a connection to other material in this chapter, and orientation for further reading into both the optimization and artificial intelligence literature on the subject.

Approximate Policy Evaluation Using Projections

Our main concern in policy evaluation is to solve approximately the Bellman equation corresponding to a given policy μ. Thus, for discounted problems, we are interested in solving the linear system of equations

$$J_\mu(i) = \sum_{i=1}^{n} p_{ij}\big(\mu(i)\big)\Big(g\big(i,\mu(i),j\big) + \alpha J_\mu(j)\Big), \qquad i = 1,\ldots,n,$$

or in shorthand,

$$J_\mu = T_\mu J_\mu, \tag{5.30}$$

where T_μ is the Bellman operator for μ, given by

$$(T_\mu J)(i) = \sum_{i=1}^{n} p_{ij}\big(\mu(i)\big)\Big(g\big(i,\mu(i),j\big) + \alpha J(j)\Big), \qquad i = 1,\ldots,n. \tag{5.31}$$

Let us consider the approximate solution of this equation by parametric approximation (cf. Section 5.3). This amounts to replacing J_μ with some vector that lies within the manifold represented by the approximation architecture

$$\mathcal{M} = \Big\{\big(\tilde J(1,r),\ldots,\tilde J(n,r)\big) \mid \text{all parameter vectors } r\Big\}. \tag{5.32}$$

The approximate solution of systems of equations within an approximation manifold of the form (5.32) has a long history in scientific computation, particularly when the manifold is linear. A central approach involves the use of projections with respect to a weighted quadratic norm

$$\|J\|^2 = \sum_{i=1}^{n} \xi_i \big(J(i)\big)^2, \tag{5.33}$$

where $J(i)$ are the components of the vector J and ξ_i are some positive weights. The projection of a vector J onto the manifold $\mathcal{M}$ is denoted by $\Pi(J)$. Thus

$$\Pi(J) \in \arg\min_{V\in\mathcal{M}} \|J - V\|^2.$$

Note that for a nonlinear parametric architecture, such as a neural network, the projection may not exist and may not be unique. However, in the case of a linear architecture, where the approximation manifold $\mathcal{M}$ is a subspace, the projection does exist and is unique; this is a consequence of the fundamental orthogonal projection theorem of calculus and real analysis.

We may consider three general approaches for approximation of J_μ.

(a) Project J_μ onto $\mathcal{M}$ to obtain $\Pi(J_\mu)$, which we use as an approximation of J_μ.

(b) Start with some approximation $\hat{J}$ of J_μ, perform N value iterations to obtain $T_\mu^N \hat{J}$, and project onto $\mathcal{M}$ to obtain $\Pi(T_\mu^N \hat{J})$. We then use $\Pi(T_\mu^N \hat{J})$ as an approximation to J_μ.

(c) Solve a projected version $J_\mu = \Pi(T_\mu J_\mu)$ of the Bellman Eq. (5.30), and use the solution of this projected equation as an approximation to J_μ. We will also discuss related projected versions that involve other operators in place of T_μ.

The preceding three approaches are hard to implement exactly; for example, (a) is impossible since we do not know the values of J_μ. However, it turns out that it is possible to implement approximately these approaches by using a Monte Carlo simulation methodology that is suitable for large problems. To explain this methodology we first discuss the implementation of the projection operation through sampling for the case where the parametric architecture is linear and $\mathcal{M}$ is a subspace.

Projection by Monte Carlo Simulation

Let us focus on the case where the manifold $\mathcal{M}$ is a subspace of the form

$$\mathcal{M} = \{\Phi r \mid r \in \Re^m\}, \tag{5.34}$$

where $\Re^m$ is the space of m-dimensional vectors, and Φ is an $n \times m$ matrix with rows denoted by $\phi(i)'$, $i = 1, \ldots, n$. Here we use the notational convention that all vectors are column vectors, and prime denotes transposition, so $\phi(i)'$ is an m-dimensional row vector, and the subspace $\mathcal{M}$ may be viewed as the space spanned by the n-dimensional columns of Φ.

We consider projection with respect to the weighted Euclidean norm of Eq. (5.33), so $\Pi(J)$ is of the form Φr^*, where

$$r^* \in \arg\min_{r\in\Re^m} \|\Phi r - J\|_\xi^2 = \arg\min_{r\in\Re^m} \sum_{i=1}^n \xi_i \big(\phi(i)'r - J(i)\big)^2. \tag{5.35}$$

We can perform the above minimization by setting to 0 the gradient of the cost function,

$$2\sum_{i=1}^n \xi_i \phi(i)\big(\phi(i)'r^* - J(i)\big) = 0.$$

We then obtain the solution in closed form,

$$r^* = \left(\sum_{i=1}^n \xi_i \phi(i)\phi(i)'\right)^{-1} \sum_{i=1}^n \xi_i \phi(i) J(i), \tag{5.36}$$

assuming that the inverse above exists. One of the difficulties here is that when n is very large, the matrix-vector calculations in this formula can be very time-consuming.

On the other hand, assuming (by normalizing ξ if necessary) that $\xi = (\xi_1, \ldots, \xi_n)$ is a probability distribution, we may view the two terms in the right side of Eq. (5.36) as expected values with respect to ξ, and approximate them by Monte Carlo simulation. In particular, suppose that we generate a set of state index samples i^s, $s = 1, \ldots, q$, according to the distribution ξ, and form the Monte Carlo estimates

$$\frac{1}{q}\sum_{s=1}^{q} \phi(i^s)\phi(i^s)' \approx \sum_{i=1}^{n} \xi_i \phi(i)\phi(i)', \qquad \frac{1}{q}\sum_{s=1}^{q} \phi(i^s)\beta^s \approx \sum_{i=1}^{n} \xi_i \phi(i) J(i), \tag{5.37}$$

where β^s is a simulation-based approximation to the exact value $J(i^s)$. It may be modeled as

$$\beta^s = J(i^s) + n(i^s),$$

where $n(i^s)$ is a random variable ("noise") resulting from the randomness of the simulation. For the Monte Carlo estimate in the right side of Eq. (5.37) to be asymptotically correct [in the sense that it converges to $\sum_{i=1}^{n} \xi_i \phi(i)\phi(i)'$ as $q \to \infty$], we must have

$$\frac{1}{q}\sum_{s=1}^{q} \phi(i^s) n(i^s) \approx 0, \tag{5.38}$$

which is implied by a zero sample mean condition for the noise.†

Given the Monte Carlo approximations (5.37) of the two terms in Eq. (5.36), we can estimate r^* with

$$\overline{r} = \left(\sum_{s=1}^{q} \phi(i^s)\phi(i^s)'\right)^{-1} \sum_{s=1}^{q} \phi(i^s)\beta^s, \tag{5.40}$$

---

† A suitable zero mean condition for the noise $n(i^s)$ has the form

$$\lim_{q\to\infty} \frac{\sum_{s=1}^{q} \delta(i^s = i) n(i^s)}{\sum_{s=1}^{q} \delta(i^s = i)} = 0, \qquad \text{for all } i = 1, \ldots, n, \tag{5.39}$$

where $\delta(i^s = i) = 1$ if $i^s = i$ and $\delta(i^s = i) = 0$ if $i^s \neq i$. It states that the Monte Carlo averages of the noise terms corresponding to every state i are zero. Then the expression in Eq. (5.38) has the form

$$\begin{aligned} \frac{1}{q}\sum_{s=1}^{q} \phi(i^s) n(i^s) &= \frac{1}{q}\sum_{i=1}^{n} \phi(i) \sum_{s=1}^{q} \delta(i^s = i) n(i^s) \\ &= \frac{1}{q}\sum_{i=1}^{n} \phi(i) \left(\sum_{s'=1}^{q} \delta(i^{s'} = i)\right) \frac{\sum_{s=1}^{q} \delta(i^s = i) n(i^s)}{\sum_{s=1}^{q} \delta(i^s = i)}, \end{aligned}$$

and converges to 0 as $q \to \infty$, assuming Eq. (5.39) [and also that each index i is sampled infinitely often so that Eq. (5.39) makes sense].

(assuming sufficiently many samples are obtained to ensure the existence of the inverse above).† This is also equivalent to estimating r^* by approximating the least squares minimization (5.35) with the following least squares training problem

$$\overline{r} \in \arg\min_{r \in \Re^m} \sum_{s=1}^{q} \big(\phi(i^s)'r - \beta^s\big)^2. \tag{5.41}$$

Thus simulation-based projection can be implemented in two equivalent ways:

(a) Replacing the expected values in the exact projection formula (5.36) with the simulation-based estimates (5.37) [cf. Eq. (5.40)].

(b) Replacing the exact least squares problem (5.35) with the simulation-based least squares approximation (5.41).

These dual possibilities of implementing projection by simulation can be used interchangeably. In particular, *the least squares training problems for approximation in value space considered in this book may be viewed as simulation-based approximate projection calculations.*

Generally, we wish that the estimate $\overline{r}$ converges to r^* as the number of samples q increases. An important point is that it is not necessary that the simulation produces independent samples. Instead it is sufficient that the long term empirical frequencies by which the indices i appear in the simulation sequence are consistent with the probabilities ξ_i of the projection norm, i.e.,

$$\xi_i = \lim_{k\to\infty} \frac{1}{q} \sum_{s=1}^{q} \delta(i^s = i), \qquad i = 1, \ldots, n, \tag{5.42}$$

where $\delta(i^s = i) = 1$ if $i^s = i$ and $\delta(i^s = i) = 0$ if $i^s \neq i$.

Another important point is that the probabilities ξ_i need not be predetermined. In fact, often the exact values of ξ_i do not matter much, and *one may wish to first specify a reasonable and convenient sampling scheme, and let ξ_i be implicitly specified via Eq. (5.42).*

Projected Equation View of Approximate Policy Evaluation

Let us now discuss the approximate policy evaluation method for costs of Section 5.3.2 [cf. Eq. (5.22)]. It can be interpreted in terms of a projected equation, written abstractly as

$$\tilde{J}_\mu \approx \Pi(T_\mu^N \hat{J}), \tag{5.43}$$

---

† The preceding derivation and the formula (5.40) actually make sense even if $\xi = (\xi_1, \ldots, \xi_n)$ has some zero components, as long as the inverses in Eqs. (5.36) and (5.40) exist. This is related to the concept of *seminorm projection*; see the paper by Yu and Bertsekas [YuB12] for a discussion related to approximate DP.

where:†

(a) $\hat{J}$ is some initial guess of J_μ (the terminal cost function approximation discussed in Section 5.3.2), and $\tilde{J}_\mu$ is the vector

$$\tilde{J}_\mu = \big(\tilde{J}(1,\overline{r}),\ldots,\tilde{J}(n,\overline{r})\big),$$

which is the approximate policy evaluation of μ, used in the policy improvement operation (5.23). Here $\overline{r}$ is the solution of the training problem (5.22).

(b) T_μ is the Bellman operator corresponding to μ, which maps a vector $J = \big(J(1),\ldots,J(n)\big)$ into the vector $T_\mu J$ of Eq. (5.31).

(c) T_μ^N denotes the N-fold application of the operator T_μ, where N is the length of the sample trajectories used in the least squares regression problem (5.22). In particular, $(T_\mu^N \hat{J})(i)$ is the cost of starting at i, using μ for N stages, with terminal cost function $\hat{J}$ (or equivalently, the result of an optimistic policy evaluation using N value iterations starting from $\hat{J}$). The sample state-cost pairs (i^s,β^s) are obtained from trajectories corresponding to this N-stage problem.

(d) $\Pi(T_\mu^N \hat{J})$ denotes projection of the vector $T_\mu^N \hat{J}$ on the manifold of possible approximating vectors $\mathcal{M}$ with respect to a weighted norm, where each weight ξ_i represents the relative frequency of the state i as initial state of a training trajectory. This projection is approximated by the least squares regression (5.22). In particular, the cost samples β^s of the training set are noisy samples of the values $(T_\mu^N \hat{J})(i^s)$, and the projection is approximated with a least squares minimization, to yield the function $\tilde{J}_\mu$ of Eq. (5.43).

Suppose now that $T_\mu^N \hat{J}$ is close to J_μ (which happens if either N is large or $\hat{J}$ is close to J_μ, or both) and the number of samples q is large (so that the simulation-based regression approximates well the projection operation Π). Then from Eq. (5.43), the approximate evaluation $\tilde{J}_\mu$ of μ approaches the projection of J_μ on the approximation manifold (5.32), which can be viewed as the best possible approximation of J_μ (at least relative to the distance metric defined by the weighted projection norm). This provides an abstract formal rationale for the parametric PI method of Section 5.3.2, which is based on Eq. (5.43).

TD(λ), LSTD(λ), and LSPE(λ)

Projected equations also fundamentally underlie *temporal difference methods* (TD for short), a prominent class of simulation-based methods for ap-

† The equation (5.43) assumes that all trajectories have equal length N, and thus does not allow trajectory reuse. If trajectories of different lengths are allowed, the term T_μ^N in the equation should be replaced by a more complicated weighted sum of powers of T_μ; see the paper [YuB12] for related ideas.

proximate evaluation of a policy. Examples of such methods are TD(λ), LSTD(λ), and LSPE(λ), where λ is a scalar with $0 \leq \lambda < 1$.†

These three methods require a linear parametric approximation architecture $\tilde{J}_\mu = \Phi r$, and all aim at the same problem. This is the problem of solving a projected equation of the form

$$\Phi r = \Pi\big(T_\mu^{(\lambda)} \Phi r\big), \tag{5.44}$$

where T_μ is the operator (5.31), $T_\mu^{(\lambda)} J$ is defined by

$$(T_\mu^{(\lambda)} J)(i) = (1 - \lambda) \sum_{\ell=0}^{\infty} \lambda^\ell \, (T_\mu^{\ell+1} J)(i), \qquad i = 1, \ldots, n, \tag{5.45}$$

and Π is projection on the approximation subspace

$$\mathcal{M} = \{\Phi r \mid r \in \Re^m\},$$

with respect to some weighted projection norm. One interpretation of the equation $J = T_\mu^{(\lambda)} J$ is as a *multistep version of Bellman's equation*. It has the same solution, J_μ, as the "one-step" Bellman equation $J = T_\mu J$, which corresponds to $\lambda = 0$.

Of course the projected equation (5.44) cannot be solved exactly when the number of states n is large, since the projection is a high dimensional operation that requires computations of order n. Instead the key idea is to *replace the projection by a simulation-based approximate projection*, of the type discussed earlier. This leads to the problem of finding a vector $\overline{r}$ such that

$$\Phi \overline{r} = \tilde{\Pi}\big(T_\mu^{(\lambda)} \Phi \overline{r}\big), \tag{5.46}$$

where $\tilde{\Pi}$ is an approximate projection implemented by sampling.

The meaning of the above equation is to *find a vector $\Phi \overline{r} \in \mathcal{M}$, which when transformed with $T_\mu^{(\lambda)}$ and then projected approximately (using $\tilde{\Pi}$) back onto $\mathcal{M}$, yields itself*. To implement the approximate projection, suppose that for q samples of initial states i^s, $s = 1, \ldots, q$, we were able compute the corresponding samples of $(T_\mu^{(\lambda)} \Phi \overline{r})(i^s)$ (this is hypothetical since $\overline{r}$ is unavailable). Then the parameter vector $\overline{r}$ would be obtained as

$$\overline{r} \in \arg\min_r \sum_{s=1}^{q} \big(\phi(i^s)' r - \text{sample of } (T_\mu^{(\lambda)} \Phi \overline{r})(i^s)\big)^2, \tag{5.47}$$

where $\phi(i)'$ is the ith row of Φ, so that the inner product $\phi(i)'r$ denotes the ith component of the vector Φr. This least squares problem implements the

† TD stands for "temporal difference," LSTD stands for "least squares temporal difference," and LSPE stands for "least squares policy evaluation."

approximate projection $\tilde{\Pi}$ of Eq. (5.46) [i.e., $\bar{r}$, the solution of Eq. (5.46) minimizes over r the quadratic expression in Eq. (5.47) (which also involves $\bar{r}$!)].

A key fact now is that the optimality condition for the problem in Eq. (5.47) can be written as a closed form equation that can be solved for $\bar{r}$, and indeed LSTD(λ) does exactly that. By contrast LSPE(λ) and TD(λ) solve this optimality condition iteratively and incrementally, in the spirit of the methods of Section 3.1.3. We will first give a high level description of the three methods, and then focus on the simpler case where $\lambda = 0$.

(a) The LSTD(λ) method, after the q samples have been collected, writes the optimality condition for Eq. (5.47) as a linear equation of the form

$$C_\lambda \bar{r} = d_\lambda, \tag{5.48}$$

where C_λ is some $m \times m$ square matrix, and d_λ is an m-dimensional vector (which depend on λ). The components of C_λ and d_λ are explicitly computed, and LSTD(λ) produces the approximate cost function $\tilde{J}_\mu(i) = \Phi\bar{r}$, where $\bar{r} = C_\lambda^{-1} d_\lambda$ is the solution of Eq. (5.48).

(b) The LSPE(λ) method solves the projected equation (5.44) by using a simulation-based *projected value iteration*,

$$J_{k+1} = \tilde{\Pi}\big(T_\mu^{(\lambda)} J_k\big). \tag{5.49}$$

Here the projection is implemented iteratively, with sampling-based least squares regression, in a manner that resembles the incremental aggregated gradient method of Section 3.1.3.

(c) The TD(λ) method is a simpler iterative stochastic approximation method for solving the linear equation (5.48), which resembles the incremental gradient method of Section 3.1.3. It can also be viewed as a stochastic version of the proximal algorithm for solving this linear equation (the parameter λ is related to the penalty parameter of the proximal algorithm; see the author's papers [Ber16c] and [Ber18d]).

An interesting question is how to select λ and what is its role. There is a bias-variance tradeoff here, similar to the one we discussed in Section 5.3.2. We will address this issue later in this section.

TD(0), LSTD(0), and LSPE(0)

Let us describe in more detail LSTD(0) for evaluation of a given policy μ. We assume that the simulation generates a sample sequence of q transitions using μ:

$$(i^1, i^2), (i^2, i^3), \dots, (i^q, i^{q+1}),$$

with corresponding transition costs

$$g(i^1, i^2), g(i^2, i^3), \dots, g(i^q, i^{q+1}).$$

Here, to simplify notation, we do not show the dependence of the transition costs on the control applied by μ. As earlier, let the ith row of the matrix Φ be the m-dimensional row vector $\phi(i)'$, so that the cost $J_\mu(i)$ is approximated as the inner product $\phi(i)'r$:

$$J_\mu(i) \approx \phi(i)'r.$$

Since $\lambda = 0$, we have $T_\mu^{(\lambda)} = T_\mu$, the samples of $T_\mu \Phi \overline{r}$ in Eq. (5.47) are

$$g(i^s, i^{s+1}) + \alpha\phi(i^{s+1})'\overline{r}, \qquad s = 1, \ldots, q,$$

and the least squares problem in Eq. (5.47) has the form

$$\overline{r} \in \arg\min_r \sum_{s=1}^{q} \big(\phi(i^s)'r - g(i^s, i^{s+1}) - \alpha\phi(i^{s+1})'\overline{r}\big)^2. \tag{5.50}$$

(Note that this is an unusual optimization problem because its solution $\overline{r}$ appears in its quadratic cost function.) By setting the gradient of the minimized expression above to zero, we obtain the condition for $\overline{r}$ to attain the above minimum:

$$\sum_{s=1}^{q} \phi(i^s)\big(\phi(i^s)'\overline{r} - g(i^s, i^{s+1}) - \alpha\phi(i^{s+1})'\overline{r}\big) = 0. \tag{5.51}$$

Solving this equation for $\overline{r}$ yields the LSTD(0) solution:

$$\overline{r} = \left(\sum_{s=1}^{q} \phi(i^s)\big(\phi(i^s) - \alpha\phi(i^{s+1})\big)'\right)^{-1} \sum_{s=1}^{q} \phi(i^s) g(i^s, i^{s+1}). \tag{5.52}$$

Note that the inverse in the preceding equation must exist for the method to be well-defined; otherwise the iteration has to be modified. A modification is also needed when the matrix inverted is nearly singular; in this case the simulation noise may introduce serious numerical problems. Various methods have been developed to deal with the near singularity issue using regularization/stabilization ideas; see Wang and Bertsekas [WaB13a], [WaB13b], and the DP textbook [Ber12], Section 7.3.

The expression

$$d^s(\overline{r}) = \phi(i^s)'\overline{r} - g(i^s, i^{s+1}) - \alpha\phi(i^{s+1})'\overline{r} \tag{5.53}$$

that appears in Eq. (5.51) is referred to as the *temporal difference associated with the sth transition and parameter vector $\overline{r}$*. In the artificial intelligence literature, temporal differences are viewed as fundamental to learning and are accordingly interpreted, but we will not go further in this direction; see the RL textbooks that we have cited.

The LSPE(0) method is similarly derived. It consists of a simulation-based approximation of the projected value iteration method

$$J_{k+1} = \tilde{\Pi}(T_\mu J_k),$$

[cf. Eq. (5.49)]. At the kth iteration, it uses only the samples $s = 1, \ldots, k$, and updates the parameter vector according to

$$r^{k+1} = r^k - \left(\sum_{s=1}^{k} \phi(i^s)\phi(i^s)' \right)^{-1} \sum_{s=1}^{k} \phi(i^s) d^s(r^k), \qquad k = 1, 2, \ldots, \quad (5.54)$$

where $d^s(r^k)$ is the temporal difference of Eq. (5.53), evaluated at the iterate r^k; the form of this iteration is derived similar to the case of LSTD(0). The vector r^q obtained after q iterations, when all the samples have been processed, is the one used for the approximate evaluation of J_μ. Note that the inverse in Eq. (5.54) can be updated economically from one iteration to the next, using fast linear algebra operations (cf. the use of the Sherman Morrison formula in the incremental Newton method in Section 3.1.3).

Overall, it can be shown that LSTD(0) and LSPE(0) [with efficient matrix inversion in Eq. (5.54)] require essentially identical amount of work to process the q samples associated with the current policy μ [this is also true for the LSTD(λ) and LSPE(λ) methods; see [Ber12], Section 6.3]. An advantage offered by LSPE(0) is that because it is iterative, it allows carrying over the final parameter vector r^q, as a "hot start" when passing from one policy evaluation to the next, in the context of an approximate PI scheme.

The TD(0) method has the form

$$r^{k+1} = r^k - \gamma^k \phi(i^k) d^k(r^k), \qquad k = 1, 2, \ldots, \quad (5.55)$$

where γ^k is a diminishing stepsize parameter and $d^k(r^k)$ is the temporal difference, cf. Eq. (5.53). It can be seen that TD(0) resembles an *incremental gradient* iteration for solving the least squares training problem (5.50), but with $\bar{r}$ replaced by the current iterate r^k. The reason is that the gradient of the typical kth term in the least squares sum of Eq. (5.50) is the vector $\phi(i^k)d^k(r^k)$ that appears in the TD(0) iteration (5.55) (cf. Section 3.1.3). Thus at each iteration, TD(0) uses only one sample, and changes r^k in the opposite direction to the corresponding incremental gradient using a stepsize γ^k that must be carefully controlled.

By contrast the LSPE(0) iteration (5.54) uses the full sum

$$\sum_{s=1}^{k} \phi(i^s) d^s(r^k),$$

which may be viewed as an *aggregated incremental* method, with scaling provided by the matrix $\left(\sum_{s=1}^{k} \phi(i^s)\phi(i^s)' \right)^{-1}$. This explains why TD(0) is generally much slower and more fragile than LSPE(0). While TD(0) does not require a matrix inversion, the same is true for LSPE(0), once modified to replace the scaling matrix with its diagonal approximation, similar to the incremental Newton method (cf. Example 3.1.9). On the other hand there is no way to avoid matrix inversion in the implementation of LSTD(0).

The properties, the analysis, and the implementation of TD methods in the context of approximate PI are quite complicated. In particular, the issue of exploration is important and must be addressed. Moreover there are convergence, oscillation, and reliability issues to contend with. LSTD(λ) relies on matrix inversion and not on iteration, so it does not have a serious convergence issue, but the system (5.48) may be singular or nearly singular, in which case very accurate simulation is needed to approximate the matrix C well enough for its inversion to be reliable, as we have noted earlier (see [WaB13a], [WaB13b], and [Ber12], Section 7.3). LSPE(λ) has a convergence issue because the mapping $\Pi T_\mu^{(\lambda)}$ may not be a contraction mapping (even though T_μ is) and the projected value iteration (5.49) may not be convergent. It turns out that the mapping $\Pi T_\mu^{(\lambda)}$ is guaranteed to be a contraction for λ sufficiently close to 1, so the convergence difficulty may be circumvented by suitably increasing λ (see [Ber12], Section 6.3.6).

Direct and Indirect Policy Evaluation Methods

In trying to compare the approximate policy evaluation methods discussed in this section, we may draw a distinction between *direct methods*, which aim to compute approximately the projection $\Pi(J_\mu)$, and *indirect methods*, which try to solve the projected equation (5.44).

The method of Section 5.3.2 is direct and is based on Eq. (5.43). In particular, as $N \to \infty$ and $q \to \infty$, it yields the approximate evaluation $\Pi(J_\mu)$. The TD methods are indirect, and aim at computing the solution of the projected equation (5.44). The solution of this equation is of the form Φr_λ^*, where the parameter vector r_λ^* depends on λ. In particular the projected equation solution Φr_λ^* is different from $\Pi(J_\mu)$. It can be shown that it satisfies the error bound

$$\|J_\mu - \Phi r_\lambda^*\|_\xi \leq \frac{1}{\sqrt{1-\alpha_\lambda^2}} \|J_\mu - \Pi(J_\mu)\|_\xi, \tag{5.56}$$

where

$$\alpha_\lambda = \frac{\alpha(1-\lambda)}{1-\alpha\lambda}$$

and $\|\cdot\|_\xi$ is a special projection norm of the form (5.33), where ξ is the steady-state probability distribution of the controlled system Markov chain under policy μ. Moreover as $\lambda \to 1$ the projected equation solution Φr_λ^* approaches $\Pi(J_\mu)$. Based on this fact, methods which aim to compute $\Pi(J_\mu)$, such as the direct method of Section 5.3.2 are sometimes called

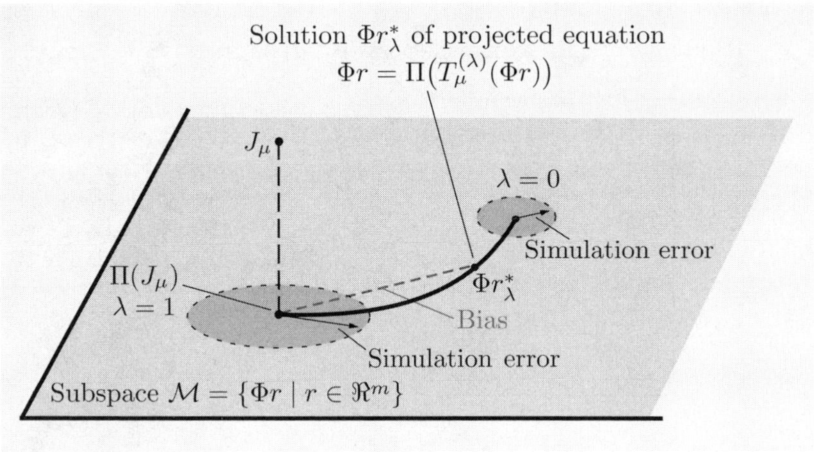

Figure 5.5.1 Illustration of the bias-variance tradeoff in estimating the solution of the projected equation for different values of λ. As λ increases from $\lambda = 0$ towards $\lambda = 1$, the solution Φr_λ^* of the projected equation

$$\Phi r = \Pi\big(T^{(\lambda)}(\Phi r)\big)$$

approaches the projection $\Pi(J_\mu)$. The difference

$$\Phi r_\lambda^* - \Pi(J_\mu)$$

is the bias, and it decreases to 0 as λ approaches 1, while the simulation error variance increases.

TD(1). We refer to [Ber12], Section 6.3.6, for an account of this analysis, which is beyond the scope of this book.

The difference $\Phi r_\lambda^* - \Pi(J_\mu)$ is commonly referred to as the *bias* and is illustrated in Figure 5.5.1. As indicated in this figure and as the estimate (5.56) suggests, there is a *bias-variance tradeoff*. As λ is decreased, the solution of the projected equation (5.44) changes and more bias is introduced relative to the "ideal" approximation ΠJ_μ (this bias can be embarrassingly large as shown by examples in the paper [Ber95]). At the same time, however, the simulation samples of $T_\mu^{(\lambda)} J$ contain less noise as λ is decreased [cf. Eq. (5.45)]. This provides another view of the bias-variance tradeoff, which we discussed in Section 5.3.2 in connection with the use of short trajectories.

5.6 EXACT AND APPROXIMATE LINEAR PROGRAMMING

Another method for exact solution of infinite horizon DP problems is based on the use of linear programming. In particular, J^* can be shown to be the unique optimal solution of a certain linear program. Focusing

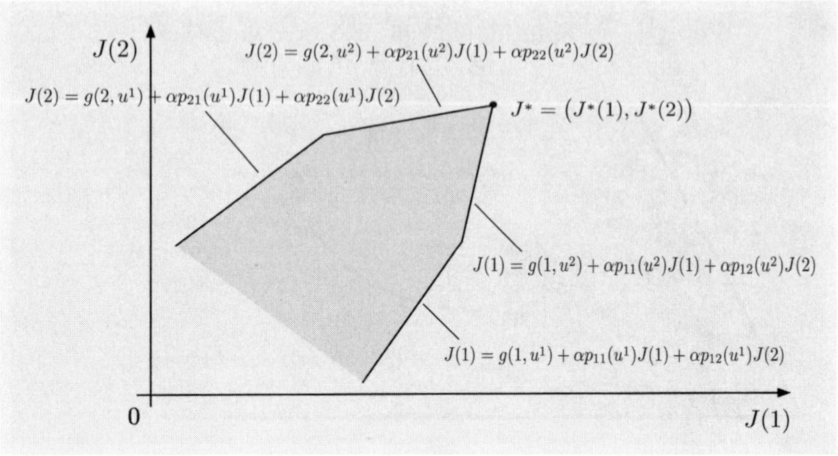

Figure 5.6.1 A linear program associated with a two-state discounted problem. The constraint set is shaded, and the objective to maximize is $J(1) + J(2)$. Note that because we have $J(i) \leq J^*(i)$ for all i and vectors J in the constraint set, the vector J^* maximizes any linear cost function of the form $\sum_{i=1}^n \beta_i J(i)$, where $\beta_i \geq 0$ for all i. If $\beta_i > 0$ for all i, then J^* is the unique optimal solution of the corresponding linear program.

on α-discounted problems, the key idea is that J^* is the "largest" (on a component-by-component basis) vector J that satisfies the constraint

$$J(i) \leq \sum_{j=1}^n p_{ij}(u)\big(g(i,u,j) + \alpha J(j)\big), \qquad \text{for all } i = 1,\ldots,n \text{ and } u \in U(i), \tag{5.57}$$

so that $J^*(1),\ldots,J^*(n)$ solve the linear program

$$\begin{aligned} &\text{maximize } \sum_{i=1}^n J(i) \\ &\text{subject to the constraint (5.57),} \end{aligned} \tag{5.58}$$

(see Fig. 5.6.1). The constraint (5.57) is sometimes called the *Bellman inequality*, and is used in a variety of DP contexts besides the framework of this section.

To verify this, let us use the VI algorithm to generate a sequence of vectors $J_k = \big(J_k(1),\ldots,J_k(n)\big)$ starting with an initial condition vector $J_0 = \big(J_0(1),\ldots,J_0(n)\big)$ such that

$$J_0(i) \leq \min_{u \in U(i)} \sum_{j=1}^n p_{ij}(u)\big(g(i,u,j) + \alpha J_0(j)\big) = J_1(i), \qquad \text{for all } i.$$

This inequality and the monotonicity property of the Bellman operator can be used to show that

$$J_0(i) \le J_1(i) \le \min_{u \in U(i)} \sum_{j=1}^{n} p_{ij}(u)\big(g(i,u,j) + \alpha J_1(j)\big) = J_2(i), \qquad \text{for all } i,$$

and similarly

$$J(i) = J_0(i) \le J_k(i) \le J_{k+1}(i), \qquad \text{for all } i \text{ and } k.$$

Since $J_k(i)$ converges to $J^*(i)$ as $k \to \infty$, it follows that we will also have

$$J(i) = J_0(i) \le J^*(i), \qquad \text{for all } i.$$

Thus out of all J satisfying the constraint (5.57), J^* is the largest on a component-by-component basis.

Unfortunately, for large n the dimension of the linear program (5.58) can be very large and its solution can be impractical, particularly in the absence of special structure. In this case, we may consider finding an approximation to J^*, which can be used in turn to obtain a (suboptimal) policy through approximation in value space.

One possibility is to approximate $J^*(i)$ with a linear feature-based architecture

$$\tilde J(i,r) = \sum_{\ell=1}^{m} r_\ell \phi_\ell(i),$$

where $r = (r_1, \ldots, r_m)$ is a vector of parameters, and for each state i, $\phi_\ell(i)$ are some features. It is then possible to determine r by using $\tilde J(i,r)$ in place of J^* in the preceding linear programming approach. In particular, we compute r as the solution of the program

$$\begin{aligned} &\text{maximize} \quad \sum_{i \in \tilde I} \tilde J(i,r) \\ &\text{subject to} \quad \tilde J(i,r) \le \sum_{i=1}^{n} p_{ij}(u)\big(g(i,u,j) + \alpha \tilde J(j,r)\big), \quad i \in \tilde I, \; u \in \tilde U(i), \end{aligned}$$

where $\tilde I$ is either the state space $I = \{1, \ldots, n\}$ or a suitably chosen subset of I, and $\tilde U(i)$ is either $U(i)$ or a suitably chosen subset of $U(i)$. This is a linear program because $\tilde J(i,r)$ is assumed linear in the parameter vector r.

The major difficulty with this approximation approach is that while the dimension of r may be moderate, the number of constraints can be extremely large. It can be as large as nm, where n is the number of states and m is the maximum number of elements of the control constraint sets $U(i)$. Thus for a large problem it is essential to reduce drastically the number of constraints. Random sampling methods may be used to select a suitable subset of the constraints to enforce (perhaps using some known suboptimal policies), and progressively enrich the subset as necessary. With such constraint sampling schemes, the linear programming approach may be practical even for problems with a very large number of states. Its application, however, may require considerable sophistication and computation (see de Farias and Van Roy [DFV03], [DFV04], [DeF04]).

We finally mention the possibility of using linear programming to evaluate approximately the cost function J_μ of a stationary policy μ in the context of approximate PI. The motivation for this is that the linear program to evaluate μ involves fewer constraints (a single constraint for each state).

5.7 APPROXIMATION IN POLICY SPACE

We will now consider an alternative to approximation in value space: approximation within the space of policies, restricting ourselves to α-discounted problems.† In particular, we parametrize stationary policies with a parameter vector r and denote them by $\tilde{\mu}(r)$, with components

$$\tilde{\mu}(i, r), \qquad i = 1, \ldots, n.$$

The parametrization may involve features and/or a neural network. The idea is then to optimize some measure of performance with respect to r. Thus in marked difference to approximation in value space, the point of departure for approximation in policy space is not the DP framework, and the algorithmic ideas of VI and PI. Instead it is general purpose optimization methodologies, such as gradient descent and random search.

Note that it is possible for a suboptimal control scheme to employ both types of approximation: in policy space and in value space, with a distinct architecture for each case (examples of such schemes have been discussed briefly in Sections 2.1.5 and 5.3.3). When neural networks are used, this is known as the simultaneous use of a "policy network" (or "actor network") and a "value network" (or "critic network"), each with its own set of parameters (see e.g., the following discussion on expert training).

Let us provide some examples where policy parametrization is natural and/or has been successful in practice.

Example 5.7.1: (Supply Chain Parametrization)

There are many problems where the general structure of an optimal or near-optimal policy is known through analysis or insight into the problem's struc-

† The SSP case is somewhat more complicated because of the need to focus on proper policies, i.e., policies that are guaranteed to terminate starting from all initial states. However, the techniques of Sections 5.7.1 and 5.7.2 apply, with modest modifications, to finite horizon problems.

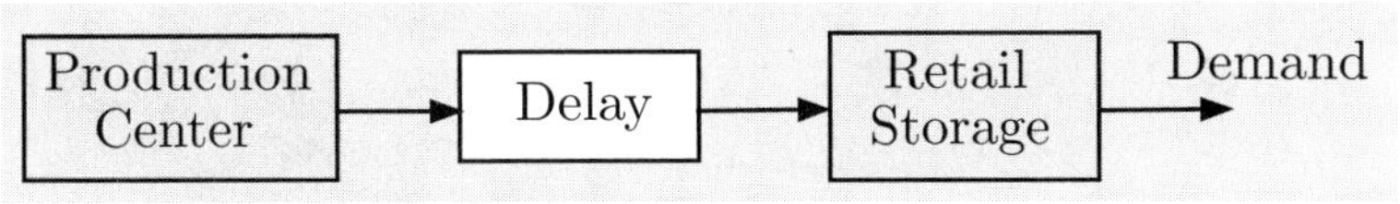

Figure 5.7.1. Illustration of a simple supply chain system.

ture. An important case are supply chain systems involving production, inventory, and retail centers that are connected with transportation links. A simple example is illustrated in Fig. 5.7.1.

Here a retail center places orders to the production center, depending on current stock. There may be orders in transit, and stochastic demand and delays. Such a problem can be formulated by DP but can be very difficult to solve exactly. However, intuitively, there is a near-optimal policy that has a simple form: when the retail inventory goes below some critical level r_1, order an amount to bring the inventory to a target level r_2. Here the policy is specified by the parameter vector $r = (r_1, r_2)$, and can be trained by one of the methods of this section. This type of approach readily extends to the case of a complex network of production/retail centers, multiple products, etc.

Example 5.7.2: (PID Control)

A time-honored and very popular scheme for control system design is the PID (Proportional-Integral-Derivative) controller. It is widely used to maintain the output of a single-input single-output dynamic system around a set point (or to follow a sequence of set points). The internal description of the system is not assumed known, but the error e_k between the output and the set point at time k can be measured; see Fig. 5.7.2.†

The input/control u_k applied at time k is the sum of three terms that depend on the observed errors $e_0, \ldots, e_k$ up to time k. The first term, called *proportional*, is $r_p e_k$ where r_p is some constant. The second term, called *integral*, is

$$r_i \sum_{m=0}^{k} e_m$$

(i.e., it is proportional to a running sum of the errors), where r_i is another constant. The third term, called *derivative*, is $r_d d_k$, where r_d is a third constant and d_k is the most recent error difference,

$$d_k = e_k - e_{k-1},$$

† Note that PID control by its nature applies to problems with continuous state and control spaces, and it may not be readily applicable to the discounted and SSP problems of the present chapter. We discuss it here because it has a strong connection to the approximation in policy space material of this section, so it may become the starting point for extensions of the methodology given here.

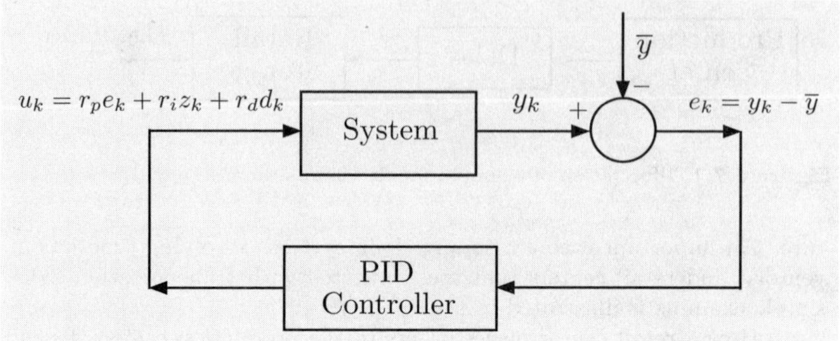

Figure 5.7.2. Illustration of PID control for a system with a single control input u_k and a single output y_k. The objective is to keep the output near a "set point" $\overline{y}$. The controller observes the error $e_k = y_k - \overline{y}$ and applies control

$$u_k = r_p e_k + r_i z_k + r_d d_k,$$

where r_p, r_i, and r_d are scalar parameters to be determined, z_k is the sum of all errors up to time k,

$$z_k = \sum_{m=0}^{k} e_m,$$

generated by

$$z_k = z_{k-1} + e_k,$$

and d_k is a damped version of the error difference $e_k - e_{k-1}$, generated by

$$d_k = (1 - \beta) d_{k-1} + \beta(e_k - e_{k-1}),$$

where β is a damping factor with $0 < \beta < 1$. A mathematical model of the system need not be known. The three terms comprising the controller, $r_p e_k$, $r_i z_k$, $r_d d_k$ are the proportional, integral, and derivative terms, respectively.

or a damped version thereof generated by a recursion such as

$$d_k = (1 - \beta) d_{k-1} + \beta(e_k - e_{k-1}),$$

where β is damping factor with $0 < \beta < 1$ (this is to mitigate the effects of noise in the error e_k).

The three constants (r_p, r_i, r_d) can be viewed as a parametrization of the controller, and they can be tuned to achieve good performance (typically a stable closed-loop system, zero error in steady-state, and satisfactory transient behavior). The proportional term serves to drive the error towards 0. The integral term guarantees that for a stable closed-loop system, the error vanishes asymptotically, and among others, is very important to neutralize the effects of nonzero mean disturbances in the system's dynamics. The derivative term is often not needed, and when it is, its purpose is mainly to improve the transient behavior of the closed-loop system by "anticipating" changes in the error e_k.

Note that, *contrary to the MPC controllers discussed in Section 2.5.1, PID control is a model-free scheme*: a mathematical model of the system need not be known. Moreover a single set of (r_p, r_i, r_d) values may be sufficient to provide good performance over a wide range of operating conditions for the system.

There are many variations and extensions of the PID scheme, and many practical methods for tuning the parameters have been developed over the years, some of them manual/heuristic; see e.g., the books by Astrom and Hagglund [AsH95], [AsH06]. The approach of optimizing the parameters of a controller by using a cost function that encodes the steady-state and transient system performance is known as *extremum seeking control*; see e.g., the books by Ariyur and Krstic [ArK03], and by Zhang and Ordonez [ZhO11]. Its application to the PID case is described in several sources; see e.g., the paper by Killingsworth and Krstic [KiK06]. For recent related work, see Frihauf, Krstic, and Basar [FKB13], Radenkovic and Krstic [RaK18], and the references quoted there. Another related optimization-based method that has been applied to PID control is *iterative feedback tuning*; see Lequin et al. [LGM03], and the references quoted there. Collectively these methods address reasonably well the tuning of PID-type schemes as applied to problems where the model parameters change slowly.

Our main purpose in this example is to point out that PID controller design can also be viewed within the context of approximation in policy space. This brings to bear the methodology of this section for the purpose of determining the controller parameters. We note, however, that PID control and other related schemes are often intended for use in real-time adaptive control of systems with frequently and unpredictably changing parameters (cf. Section 1.3.8). This is a challenging application context that RL methods are not able to fully address with the current state of the art.

Example 5.7.3: (Policy Parametrization Through Cost Parametrization)

In an important approach for parametrization of policies we start with a parametric cost function approximation $\tilde{J}(j, r)$. We then define a policy parametrization through the one-step lookahead minimization

$$\tilde{\mu}(i, r) \in \arg \min_{u \in U(i)} \sum_{j=1}^{n} p_{ij}(u)\big(g(i, u, j) + \alpha \tilde{J}(j, r)\big), \tag{5.59}$$

where $\tilde{J}$ is a function of a given form that depends on r.† For example, $\tilde{J}$ may be a linear feature-based architecture, with features possibly obtained through a separately trained neural network. The policies $\tilde{\mu}(r)$ of Eq. (5.59) form a class of one-step lookahead policies parametrized by r. We may then determine r through some form of policy training method if this is convenient.

† The policy parametrization (5.59) may not be differentiable with respect to r, and this may make it unsuitable for use in some methods that rely on gradients. In such cases one may approximate the policy $\tilde{\mu}(i, r)$ of Eq. (5.59) with

As an illustration, we note that this type of scheme has been used for learning a high-scoring strategy in the game of tetris; see Szita and Lorinz [SzL06], and Thiery and Scherrer [ThS09]. The parametrization in policy space was derived through a feature-based parametrization in value space (cf. Example 3.1.3). The algorithm used for training was the cross-entropy method, a random search algorithm to be discussed in Section 5.7.1.

Example 5.7.4: (Policy Parametrization Through Feature Parametrization)

An interesting special case of policy parametrization is based on state features, so that $\tilde{\mu}$ depends on the state i through some feature vector $\phi(i)$, i.e.,

$$\tilde{\mu}(i, r) = \hat{\mu}\big(\phi(i), r\big),$$

for some function $\hat{\mu}$. Then to implement the policy, we only need to know $\phi(i)$ rather than i. Moreover, to train the policy with most of the simulation-based methods of this section, we only need a cost simulator [to generate samples of $J_{\tilde{\mu}}(i)$], and a feature simulator [to generate feature values $\phi(i)$]. In particular, we do not need the simulated values of i. An example of feature-based policy parametrization is provided by the cost parametrization approach of the preceding example, in the case where the cost function approximation $\tilde{J}$ is based on features.

Another context where feature parametrization may be useful is in partial state information problems (POMDP), where we use as state at a given time k the entire observation and control history I_k up to k (the information record considered in Example 3.1.6). When the horizon is infinite, the size of I_k increases without bound, thus complicating the implementation of methods based on approximation in policy space. On the other hand, convenient features for policy parametrization may be obtained from sufficient statistics of the information record I_k, such as those discussed in Example 3.1.6. Some possibilities include a partial control-observation history (a subset of I_k), or some estimate of the state or other "good" features of I_k.

---

a *randomized* policy that applies at state i a control $u \in U(i)$ with probability

$$p(u \mid i, r) = \frac{e^{-\beta \tilde{Q}(i,u,r)}}{\sum_{v \in U(i)} e^{-\beta \tilde{Q}(i,v,r)}},$$

where $\tilde{Q}(i, u, r)$ is the approximate Q-factor given by

$$\tilde{Q}(i, u, r) = \sum_{j=1}^{n} p_{ij}(u)\big(g(i, u, j) + \alpha \tilde{J}(j, r)\big),$$

and β is a positive scalar, which controls the accuracy of the approximation. This is called the *soft-min* approximation in the literature (or *soft-max* approximation in the case of maximization of reward).

The feature-based policy parametrization approach may also be convenient for rollout algorithms with a base policy that is feature-parametrized (e.g., a finite history controller for POMDP). It may also be useful in the context of the rollout-based approximate PI methods to be given later in this chapter (cf. Section 5.7.3).

Example 5.7.5: (Policy Parametrization, Unconventional Problem Structures, and Multiagent Problems)

An important point to keep in mind is that *approximation in policy space is a more broadly applicable methodology* than approximation in value space. In particular, policy space approximation is not tied to the cost-to-go formalism of DP, which is the essential framework for the development of the approximate VI and PI methodologies. As a result, the idea of policy parametrization applies to problems that share some structure with the finite and infinite horizon DP problems we have discussed so far, but do not admit a formal treatment by DP.

Situations of this type are common in practice. For example, the transition probabilities between states may depend not just on the control, but also on preceding transitions. This dependence may be hard to understand or model mathematically, but may be captured in a simulator that can be used to implement approximation in policy space.

Another important example is multiagent problems, where there is a dynamic system whose state evolves in time, and there are multiple decision makers that do not share the same information about the (collective) system state. Each agent has access to local observations, and may receive some of the other agents' observations (or a summary thereof) with delay: the agent's decision at each time is based on just the information available to him/her at that time. This type of information pattern is unconventional and is not allowed in the DP context (except through the use of approximations; see, e.g., [BBD08], [Ber19c], [KMP13], [QuL19], [ZYL18], for related work and references). Consequently there is no legitimate DP framework and associated Bellman's equation for this type of problem, and the VI and PI methods are not valid anymore. Still, however, while approximation in value space does not apply, it is possible to parametrize the policies of the agents and set up a framework for parameter optimization of the type to be described.

In what follows we will discuss three training approaches for approximation in policy space. The first approach (Section 5.7.1) is based on determining the parameter r by optimization of some measure of cost derived from the given DP problem. The second approach (Section 5.7.2) is less ambitious and is reminiscent of supervised learning. Here we collect state-control data produced by a human or software "expert," and we obtain the parameter r by matching approximately the decision making of the expert through some least squares error minimization. The third approach (Section 5.7.3) uses approximation in policy space for policy improvement within an actor-only PI framework. Here we apply the expert training approach of Section 5.7.2, with rollout being used as a software expert.

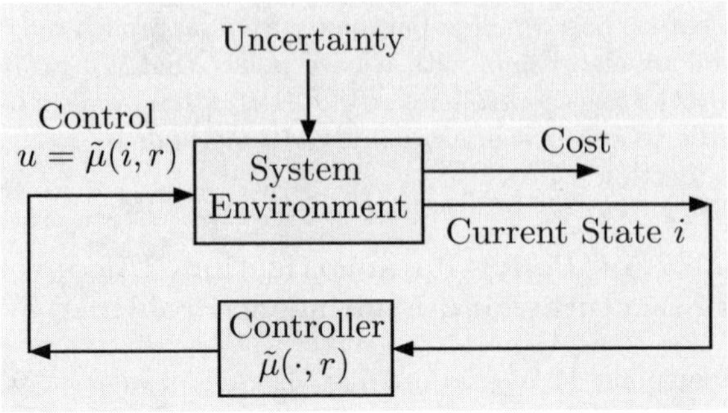

Figure 5.7.3 The optimization framework for approximation in policy space. Here policies are parametrized with a parameter vector r and denoted by $\tilde{\mu}(r)$, with components $\tilde{\mu}(i, r)$, $i = 1, \ldots, n$. Each parameter value r determines a policy $\mu(r)$, and a cost $J_{\tilde{\mu}(r)}(i_0)$ for each initial state i_0, as indicated in the figure. The optimization approach determines r through the minimization

$$\min_r E\big\{J_{\tilde{\mu}(r)}(i_0)\big\},$$

where the expected value above is taken with respect to a suitable probability distribution of i_0.

5.7.1 Training by Cost Optimization - Policy Gradient, Cross-Entropy, and Random Search Methods

In this section we discuss a major training approach for approximation in policy space. It is based on controller parameter optimization: we parametrize the policies by a vector r, and we optimize the corresponding expected cost over r. In particular, we determine r through the minimization

$$\min_r E\big\{J_{\tilde{\mu}(r)}(i_0)\big\}, \tag{5.60}$$

where $J_{\tilde{\mu}(r)}(i_0)$ is the cost of the policy $\tilde{\mu}(r)$ starting from the initial state i_0, and the expected value above is taken with respect to a suitable probability distribution of the initial state i_0 (cf. Fig. 5.7.3).

Note that in the case where the initial state i_0 is known and fixed, the method involves just minimization of $J_{\tilde{\mu}(r)}(i_0)$ over r. This simplifies a great deal the minimization, particularly when the problem is deterministic.

Gradient Methods for Cost Optimization

Let us first consider methods that perform the minimization (5.60) by using a gradient method, and for simplicity let us assume that the initial condition i_0 is known. Thus the aim is to minimize $J_{\tilde{\mu}(r)}(i_0)$ over r by using the gradient method

$$r^{k+1} = r^k - \gamma^k \nabla J_{\tilde{\mu}(r^k)}(i_0), \qquad k = 0, 1, \ldots, \tag{5.61}$$

assuming that $J_{\tilde{\mu}(r)}(i_0)$ is differentiable with respect to r. Here γ^k is a positive stepsize parameter, and $\nabla(\cdot)$ denotes gradient with respect to r evaluated at the current iterate r^k.

The difficulty with this method is that the gradients $\nabla J_{\tilde{\mu}(r^k)}(i_0)$ may not be explicitly available. In this case, the gradients can be approximated by finite differences of cost function values $J_{\tilde{\mu}(r^k)}(i_0)$. Unfortunately, when the problem is stochastic, the cost function values may be computable only through Monte Carlo simulation. This may introduce a large amount of noise, so it is likely that many samples will need to be averaged in order to obtain sufficiently accurate gradients, thereby making the method inefficient. On the other hand, when the problem is deterministic, this difficulty does not appear, and the use of the gradient method (5.61) or other methods that do not rely on the use of gradients (such as coordinate descent) is facilitated.

In this section we will focus on alternative and typically more efficient gradient-like methods for stochastic problems, which are based on sampling. Some popular methods of this type are based on incremental gradient ideas (cf. Section 3.1.3) and the use of randomized policies [i.e., policies that map a state i to a probability distribution over the set of controls $U(i)$, rather than mapping onto a single control].† We discuss these gradient-like methods next.

Incremental Gradient Methods Based on Randomization

To get a sense of the general principle underlying the incremental gradient approach that uses randomization and sampling, let us digress from the DP context of this chapter, and consider the generic optimization problem

$$\min_{z \in Z} F(z), \tag{5.62}$$

where Z is a subset of the m-dimensional space $\Re^m$, and F is some real-valued function over $\Re^m$.

We will take the unusual step of converting this problem to the *stochastic* optimization problem

$$\min_{p \in \mathcal{P}_Z} E_p\{F(z)\}, \tag{5.63}$$

† The AlphaGo and AlphaZero programs (Silver et al. [SHM16], [SHS17]) also use randomized policies, and a policy adjustment scheme that involves incremental changes along "directions of improvement." However, these changes are implemented through the MCTS algorithm used by these programs, without the explicit use of a gradient (see the discussion in Section 2.4.2). Thus it may be said that the AlphaGo and AlphaZero programs involve a form of approximation in policy space (as well as approximation in value space), which bears resemblance but cannot be classified as a policy gradient method.

where z is viewed as a random variable, $\mathcal{P}_Z$ is the set of probability distributions over Z, p denotes the generic distribution in $\mathcal{P}_Z$, and $E_p\{\cdot\}$ denotes expected value with respect to p. Of course this enlarges the search space from Z to $\mathcal{P}_Z$, but it enhances the use of randomization schemes and simulation-based methods, even if the original problem is deterministic. Moreover, the stochastic optimization problem (5.63) may have some nice differentiability properties that are lacking in the original deterministic version (5.62); see the paper [Ber73] for an analysis of this differentiability issue under convexity assumptions on F.

At this point it is not clear how the stochastic optimization problem (5.63) relates to our stochastic DP context of this chapter. We will return to this question later, but for the purpose of orientation, we note that to obtain a problem of the form (5.63), we must take another unusual step: enlarge the set of policies to include *randomized policies*, mapping a state i into a probability distribution over the set of controls $U(i)$.

Suppose now that we restrict attention to a subset $\tilde{\mathcal{P}}_Z \subset \mathcal{P}_Z$ of probability distributions $p(z;r)$ that are parametrized by some continuous parameter r, e.g., a finite-dimensional vector in some Euclidean space.† In other words, we approximate the stochastic optimization problem (5.63) with the restricted problem

$$\min_r E_{p(z;r)}\big\{F(z)\big\}.$$

Then we may use a gradient method for solving this problem, such as

$$r^{k+1} = r^k - \gamma^k \nabla\Big(E_{p(z;r^k)}\big\{F(z)\big\}\Big), \qquad k = 0, 1, \ldots, \tag{5.64}$$

where $\nabla(\cdot)$ denotes gradient with respect to r of the function in parentheses, evaluated at the current iterate r^k.

Likelihood-Ratio Policy Gradient Methods

We will first consider an incremental version of the gradient method (5.64). This method requires that $p(z;r)$ is differentiable with respect to r. It relies on a convenient gradient formula, sometimes referred to as the *log-likelihood trick*, which involves the natural logarithm of the sampling distribution.

This formula is obtained by the following calculation, which is based on interchanging gradient and expected value, and using the gradient formula $\nabla(\log p) = \nabla p / p$. We have

$$\nabla\Big(E_{p(z;r)}\big\{F(z)\big\}\Big) = \nabla\left(\sum_{z\in Z} p(z;r)F(z)\right)$$

† To be on safe mathematical ground, we assume that $p(z;r)$ is a discrete distribution in what follows in this section.

$$= \sum_{z \in Z} \nabla p(z;r) F(z)$$

$$= \sum_{z \in Z} p(z;r) \frac{\nabla p(z;r)}{p(z;r)} F(z)$$

$$= \sum_{z \in Z} p(z;r) \nabla \Big(\log \big(p(z;r)\big) \Big) F(z),$$

and finally

$$\nabla \Big(E_{p(z;r)} \{F(z)\} \Big) = E_{p(z;r)} \Big\{ \nabla \Big(\log \big(p(z;r)\big) \Big) F(z) \Big\}, \tag{5.65}$$

where for any given z, $\nabla\Big(\log\big(p(z;r)\big)\Big)$ is the gradient with respect to r of the function $\log\big(p(z;\cdot)\big)$, evaluated at r (the gradient is assumed to exist).

The preceding formula suggests an incremental implementation of the gradient iteration (5.64) that approximates the expected value in the right side in Eq. (5.65) with a single sample (cf. Section 1.3). The typical iteration of this method is as follows.

Sample-Based Gradient Method for Parametric Approximation of $\min_{z \in Z} F(z)$

Let r^k be the current parameter vector.

(a) Obtain a sample z^k according to the distribution $p(z;r^k)$.

(b) Compute the gradient $\nabla\Big(\log\big(p(z^k;r^k)\big)\Big)$.

(c) Iterate according to

$$r^{k+1} = r^k - \gamma^k \nabla \Big(\log \big(p(z^k;r^k)\big) \Big) F(z^k). \tag{5.66}$$

The advantage of the preceding sample-based method is its simplicity and generality. It allows the use of parametric approximation for *any* minimization problem (well beyond DP), as long as the logarithm of the sampling distribution $p(z;r)$ can be conveniently differentiated with respect to r, and samples of z can be obtained using the distribution $p(z;r)$.

Note that in iteration (5.66) r is adjusted along a random direction. This direction does not involve at all the gradient of F, only the gradient of the logarithm of the sampling distribution! As a result the iteration has a *model-free character*: we don't need to know the form of the function F as long as we have a simulator that produces the cost function value $F(z)$ for any given z. This is also a major advantage offered by many random search methods.

An important issue is the efficient computation of the sampled gradient $\nabla\big(\log\big(p(z^k;r^k)\big)\big)$. In the context of DP, including the SSP and discounted problems that we have been dealing with, there are some specialized procedures and corresponding parametrizations to approximate this gradient conveniently. The following is an example.

Example 5.7.6 (Policy Gradient Method for Discounted DP)

Consider the α-discounted problem and denote by z the infinite horizon state-control trajectory:

$$z = \{i_0, u_0, i_1, u_1, \ldots\}.$$

We consider a parametrization of randomized policies with parameter r, so the control at state i is generated according to a distribution $p(u \mid i; r)$ over $U(i)$. Then for a given r, the state-control trajectory z is a random vector with probability distribution denoted $p(z; r)$. The cost corresponding to the trajectory z is

$$F(z) = \sum_{m=0}^{\infty} \alpha^m g(i_m, u_m, i_{m+1}), \tag{5.67}$$

and the problem is to minimize over r

$$E_{p(z;r)}\big\{F(z)\big\}.$$

To apply the sample-based gradient method (5.66), given the current iterate r^k, we must generate the sample state-control trajectory z^k, according to the distribution $p(z; r^k)$, compute the corresponding cost $F(z^k)$, and also calculate the gradient

$$\nabla\Big(\log\big(p(z^k;r^k)\big)\Big). \tag{5.68}$$

Let us assume that the logarithm of the randomized policy distribution $p(u \mid i; r)$ is differentiable with respect to r (a soft-min policy parametrization is often recommended for this purpose). Then the logarithm that is differentiated in Eq. (5.68) can be written as

$$\begin{aligned}\log\big(p(z^k;r^k)\big) &= \log \prod_{m=0}^{\infty} p_{i_m i_{m+1}}(u_m)p(u_m \mid i_m; r^k)\\ &= \sum_{m=0}^{\infty} \log\big(p_{i_m i_{m+1}}(u_m)\big) + \sum_{m=0}^{\infty} \log\big(p(u_m \mid i_m; r^k)\big),\end{aligned}$$

and its gradient (5.68), which is needed in the iteration (5.66), is given by

$$\nabla\Big(\log\big(p(z^k;r^k)\big)\Big) = \sum_{m=0}^{\infty} \nabla\Big(\log\big(p(u_m \mid i_m; r^k)\big)\Big). \tag{5.69}$$

This gradient involves the current randomized policy, but does not involve the transition probabilities and the costs per stage.

The policy gradient method (5.66) can now be implemented with a finite horizon approximation whereby r^k is changed after a finite number N of time steps [so the infinite cost and gradient sums (5.67) and (5.69) are replaced by finite sums]. The method takes the form

$$r^{k+1} = r^k - \gamma^k \sum_{m=0}^{N-1} \nabla \log \big(p(u_m \mid i_m; r^k)\big) F_N(z_N^k),$$

where $z_N^k = (i_0, u_0, \ldots, i_{N-1}, u_{N-1})$ is the generated N-step trajectory, and $F_N(z_N^k)$ is the corresponding cost. The initial state i_0 of the trajectory is chosen randomly, with due regard to exploration issues.

Policy gradient methods for other types of DP problems can be similarly developed, as well as variations involving a combination of policy and cost function approximations [e.g., replacing $F(z)$ of Eq. (5.67) by a parametrized estimate that has smaller variance]. Cost shaping is also useful in this connection (cf. Section 5.3.4). This leads to a class of actor-critic methods that differ from the PI-type methods that we discussed in Section 5.3.1. A further discussion is beyond our scope, and we refer to the end-of-chapter literature for a variety of specific schemes.

Implementation Issues

There are several issues to consider in the implementation of the sample-based gradient method (5.66). The first of these is that the problem solved is a randomized version of the original. If the method produces a parameter $\overline{r}$ in the limit and the distribution $p(z;\overline{r})$ is not atomic (i.e., it is not concentrated at a single point), then a solution $\overline{z} \in Z$ must be extracted from $p(z;\overline{r})$. In the SSP and discounted problems of this chapter, the subset $\tilde{\mathcal{P}}_Z$ of parametric distributions typically contains the atomic distributions, while it can be shown that minimization over the set of all distributions $\mathcal{P}_Z$ produces the same optimal value as minimization over Z (the use of randomized policies does not improve the optimal cost of the problem), so this difficulty does not arise.

Another issue is how to collect the samples z^k. Different methods must strike a balance between convenient implementation and a reasonable guarantee that the search space Z is sufficiently well explored.

Finally, there is the issue of improving sampling efficiency. To this end, let us note a simple generalization of the gradient method (5.66), which can often improve its performance. It is based on the gradient formula

$$\nabla\Big(E_{p(z;r)}\big\{F(z)\big\}\Big) = E_{p(z;r)}\Big\{\nabla\Big(\log\big(p(z;r)\big)\Big)\big(F(z) - b\big)\Big\}, \tag{5.70}$$

where b is any scalar. This formula generalizes Eq. (5.65), where $b = 0$, and holds in view of the following calculation, which shows that the term multiplying b in Eq. (5.70) is equal to 0:

$$\begin{aligned} E_{p(z;r)}\Big\{\nabla\Big(\log\big(p(z;r)\big)\Big)\Big\} &= E_{p(z;r)}\left\{\frac{\nabla p(z;r)}{p(z;r)}\right\} \\ &= \sum_{z\in Z} p(z;r)\frac{\nabla p(z;r)}{p(z;r)} \\ &= \sum_{z\in Z} \nabla p(z;r) = \nabla\left(\sum_{z\in Z} p(z;r)\right) = 0, \end{aligned}$$

where the last equality holds because $\sum_{z\in Z} p(z;r)$ is identically equal to 1 and hence does not depend on r.

Based on the gradient formula (5.70), we can modify the iteration (5.66) to read as follows:

$$r^{k+1} = r^k - \gamma^k\nabla\Big(\log\big(p(z^k;r^k)\big)\Big)\big(F(z^k) - b\big), \tag{5.71}$$

where b is some fixed scalar, called the *baseline*. Whereas the choice of b does not affect the gradient $\nabla\Big(E_{p(z;r)}\big\{F(z)\big\}\Big)$ [cf. Eq. (5.70)], it affects the incremental gradient

$$\nabla\Big(\log\big(p(z^k;r^k)\big)\Big)\big(F(z^k) - b\big),$$

which is used in Eq. (5.71). Thus, by optimizing the baseline b, empirically or through a calculation (see e.g., [DNP11]), we can improve the performance of the algorithm. Moreover, for discounted and SSP problems, state-dependent baseline functions may be used, whereby $F(z)$ is replaced in Eq. (5.67) by $F(z) + V(i_0)$, where $V(i_0)$ is a suitably obtained estimate of $F(z)$ or $J^*(i_0)$ (cf. the idea of cost shaping of Sections 4.2 and 5.3.4).

Random Direction Methods

We will now consider an alternative class of incremental versions of the policy gradient method (5.64), repeated here for convenience:

$$r^{k+1} = r^k - \gamma^k\nabla\Big(E_{p(z;r^k)}\big\{F(z)\big\}\Big), \qquad k = 0, 1, \ldots. \tag{5.72}$$

These methods are based on the use of a random search direction and only *two sample function values per iteration*. They are generally faster than methods that use a finite difference approximation of the entire cost function gradient; see the book by Spall [Spa03] for a detailed discussion, and the paper by Nesterov and Spokoiny [NeS17] for a more theoretical view. Moreover they do not require the derivative of the sampling distribution or its logarithm.

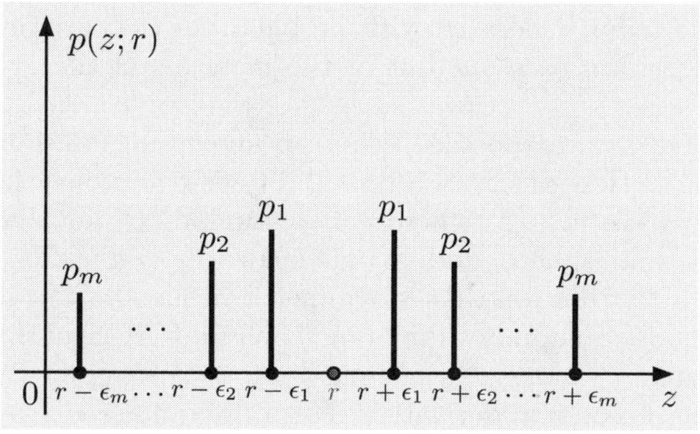

Figure 5.7.4 The distribution $p(z;r)$ used in the gradient iteration (5.73).

We first consider the case where z and r are scalars, and later discuss the multidimensional case. As earlier, to avoid mathematical complications, we assume that $p(z;r)$ is a discrete distribution; the development is similar for more general distributions. In particular, we assume that $p(z;r)$ is symmetric and is concentrated with probabilities p_i at the points $r+\epsilon_i$ and $r-\epsilon_i$, where $\epsilon_1,\dots,\epsilon_m$ are some small positive scalars. Thus we have

$$E_{p(z;r)}\{F(z)\} = \sum_{i=1}^m p_i\big(F(r+\epsilon_i)+F(r-\epsilon_i)\big),$$

and

$$\nabla\Big(E_{p(z;r)}\{F(z)\}\Big) = \sum_{i=1}^m p_i\big(\nabla F(r+\epsilon_i)+\nabla F(r-\epsilon_i)\big).$$

The gradient iteration (5.72) becomes

$$r^{k+1} = r^k - \gamma^k \sum_{i=1}^m p_i\big(\nabla F(r+\epsilon_i)+\nabla F(r-\epsilon_i)\big), \qquad k=0,1,\dots. \tag{5.73}$$

Let us now consider approximation of the gradient by finite differences:

$$\nabla F(r+\epsilon_i) \approx \frac{F(r+\epsilon_i)-F(r)}{\epsilon_i}, \qquad \nabla F(r-\epsilon_i) \approx \frac{F(r)-F(r-\epsilon_i)}{\epsilon_i}.$$

We approximate the gradient iteration (5.73) by

$$r^{k+1} = r^k - \gamma^k \sum_{i=1}^m p_i \frac{F(r^k+\epsilon_i)-F(r^k-\epsilon_i)}{\epsilon_i}, \qquad k=0,1,\dots. \tag{5.74}$$

One possible sample-based/incremental version of this iteration is

$$r^{k+1} = r^k - \gamma^k \frac{F(r^k+\epsilon_{i_k})-F(r^k-\epsilon_{i_k})}{\epsilon_{i_k}}, \tag{5.75}$$

where i^k is an index generated with probabilities that are proportional to p_{i^k}. This algorithm uses one out of the m terms of the gradient in Eq. (5.74).

The extension to the case, where z and r are multidimensional, is straightforward. Here $p(z;r)$ is a probability distribution, whereby z takes values of the form $r+\epsilon d$, where d is a random vector that lies on the surface of the unit sphere, and ϵ (independently of d) takes scalar values according to a distribution that is symmetric around 0. The idea is that at r^k, we first choose randomly a direction d^k on the surface of the unit sphere, and then change r^k along d^k or along $-d^k$, depending on the sign of the corresponding directional derivative. For a finite difference approximation of this iteration, we sample z^k along the line $\{r^k+\epsilon d^k \mid \epsilon \in \Re\}$, and similar to the iteration (5.75), we set

$$r^{k+1} = r^k - \gamma^k \frac{F(r^k + \epsilon^k d^k) - F(r^k - \epsilon^k d^k)}{\epsilon^k} d^k, \qquad (5.76)$$

where ϵ^k is the sampled value of ϵ.

Let us also discuss the case where $p(z;r)$ is a discrete but *nonsymmetric* distribution, i.e., z takes values of the form $r+\epsilon d$, where d is a random vector that lies on the surface of the unit sphere, and ϵ is a zero mean scalar. Then the analog of iteration (5.76) is

$$r^{k+1} = r^k - \gamma^k \frac{F(r^k + \epsilon^k d^k) - F(r^k)}{\epsilon^k} d^k, \qquad (5.77)$$

where ϵ^k is the sampled value of ϵ. Thus in this case, we still require two function values per iteration. Generally, for a symmetric sampling distribution, iteration (5.76) tends to be more accurate than iteration (5.77), and is often preferred.

Algorithms of the form (5.76) and (5.77) are known as *random direction methods*. They use only two cost function values per iteration, and a direction d^k that need not be related to the gradient of F in any way. There is some freedom in selecting d^k, which could potentially be exploited in specific schemes. However, selecting the stepsize γ^k and the sampling distribution for ϵ can be tricky, particularly when the values of F are noisy.

Random Search and Cross-Entropy Methods

The main drawback of the policy gradient methods that we have considered in this section is potential unreliability due to the stochastic uncertainty corrupting the calculation of the gradients, the slow convergence that is typical of gradient methods in many settings, and the presence of local minima. For this reason, methods based on random search have been considered as potentially more reliable alternatives. Viewed from a high level, random search methods are similar to policy gradient methods in that they

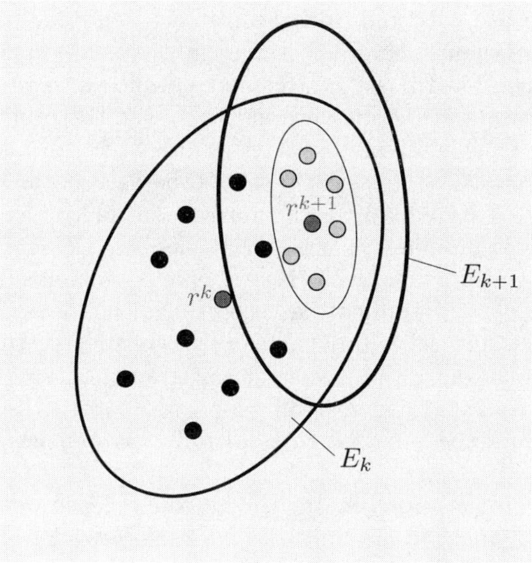

Figure 5.7.5 Schematic illustration of the cross-entropy method. At the current iterate r^k, we construct an ellipsoid E_k centered at r^k. We generate a number of random samples within E_k, and we "accept" a subset of the samples that have "low" cost. We then choose r^{k+1} to be the sample mean of the accepted samples, and construct a sample "covariance" matrix of the accepted samples. We then form the new ellipsoid E_{k+1} using this matrix and a suitably enlarged radius, and continue. Notice the resemblance with a policy gradient method: we move from r^k to r^{k+1} in a direction of cost improvement.

aim at iterative cost improvement through sampling. However, they need not involve randomized policies, they are not subject to cost differentiability restrictions, and they offer some global convergence guarantees, so in principle they are not affected much by local minima.

Let us consider a parametric policy optimization approach based on solving the problem

$$\min_r E\big\{J_{\tilde{\mu}(r)}(i_0)\big\},$$

cf. Eq. (5.60). Random search methods for this problem explore the space of the parameter vector r in some randomized but intelligent fashion. There are several types of such methods for general optimization, and some of them have been suggested for approximate DP. We will briefly describe the *cross-entropy method*, which has gained considerable attention.

The method, when adapted to the approximate DP context, bears resemblance to policy gradient methods, in that it generates a parameter sequence $\{r^k\}$ by changing r^k to r^{k+1} along a direction of "improvement." This direction is obtained by using the policy $\tilde{\mu}(r^k)$ to generate randomly cost samples corresponding to a set of sample parameter values that are concentrated around r^k. The current set of sample parameters are then screened: some are accepted and the rest are rejected, based on a cost improvement criterion. Then r^{k+1} is determined as a "central point" or as the "sample mean" in the set of accepted sample parameters, some more samples are generated randomly around r^{k+1}, and the process is repeated; see Fig. 5.7.5. Thus successive iterates r^k are "central points" of successively better groups of samples, so in some broad sense, the random sample generation process is guided by cost improvement. This idea is shared with evolutionary programming; see e.g., the books [Bac96], [DeJ06].

The cross-entropy method is very simple to implement, does not suffer from the fragility of gradient-based optimization, does not involve randomized policies, and relies on some supportive theory. Importantly, the method does not require the calculation of gradients, and it does not require differentiability of the cost function. Moreover, it does not need a model to compute the required costs of different policies; a simulator is sufficient.

Like all random search methods, the convergence rate guarantees of the cross-entropy method are limited, and its success depends on domain-specific insights and the skilled use of heuristics. However, the method relies on solid ideas and has gained a favorable reputation. In particular, it was used with impressive success in the context of the game of tetris; see Szita and Lorinz [SzL06], and Thiery and Scherrer [ThS09]. There have also been reports of domain-specific successes with alternative but related random search methods; see Salimans et al. [SHC17]. We refer to the end-of-chapter literature for details and examples of implementation.

5.7.2 Expert-Based Supervised Learning

According to the second approach for approximation in policy space, we choose the parameter r by "training" on a large number of sample state-control pairs (i^s, u^s), $s = 1, \ldots, q$, such that for each s, u^s is a "good" control at state i^s. This can be done for example by solving the least squares problem†

$$\min_r \sum_{s=1}^{q} \big\| u^s - \tilde{\mu}(i^s, r) \big\|^2 \tag{5.78}$$

(possibly with added regularization). In particular, we may determine u^s by a human or a software "expert" that can choose "near-optimal" controls at given states, so $\tilde{\mu}$ is trained to match the behavior of the expert. We have also discussed this approach in Section 2.4.3, in the context of finite horizon problems, and in Section 3.5 in the context of classification methods

† It is implicitly assumed here (and in similar situations later) that the controls are members of a Euclidean space so that the distance between two controls can be measured by their normed difference.

for approximation in policy space. In the context of artificial intelligence, it comes within the framework of *supervised learning* methods.†

Another possibility, which we have already discussed in various forms, is to select a large number of sample states i^s, $s = 1, \dots, q$, and generate the controls u^s, $s = 1, \dots, q$, through the one-step lookahead minimization

$$u^s = \arg \min_{u \in U(i^s)} \sum_{j=1}^{n} p_{ij}(u)\big(g(i^s, u, j) + \alpha \tilde{J}(j)\big), \tag{5.79}$$

where $\tilde{J}$ is a suitable one-step lookahead function (multistep lookahead can also be used). Similarly, once a parametric Q-factor approximation architecture $\tilde{Q}(i, u, r)$ is chosen, we can select a large number of sample states i^s, $s = 1, \dots, q$, and then compute the controls u^s, $s = 1, \dots, q$, through the one-step lookahead minimization

$$u^s = \arg \min_{u \in U(i^s)} \tilde{Q}(i^s, u, r). \tag{5.80}$$

In this case, we will be collecting sample state-control pairs (i^s, u^s), $s = 1, \dots, q$, using approximation in value space through Eq. (5.79) or Eq. (5.80), and then applying approximation in policy space through Eq. (5.78).

Note that once the sample state-control pairs (i^s, u^s), $s = 1, \dots, q$, have been collected, an alternative to solving the least squares problem (5.78) is to use *interpolation* (rather than parametric approximation). By this we mean to specify for each $i \notin \{i^1, \dots, i^s\}$ a probability distribution $\{\phi_{i1}, \dots, \phi_{is}\}$, and to use the policy $\tilde{\mu}$ defined by

$$\tilde{\mu}(i) = \sum_{s=1}^{q} \phi_{is} u^s, \qquad i = 1, \dots, n. \tag{5.81}$$

In general, this requires that the control constraint set is a convex subset of a Euclidean space so that the interpolated controls (5.81) are feasible. This is not necessary if all the interpolation probabilities ϕ_{is} are either 0 or 1 (an example in the *nearest neighbor approach*, where for a given state i, we set $\phi_{i\bar{s}} = 1$ for the state $i^{\bar{s}}$ that is "closest" to i in some sense). Interpolation approaches are central to the aggregation methodology of Chapter 6, and will be discussed in greater detail there.

---

† Tesauro [Tes89a], [Tes89b] constructed a backgammon player, trained by a neural network and a supervised learning approach (called "comparison training"), which used examples from human expert play (he was the expert who provided the training samples). However, his subsequent TD-based algorithm [Tes92], [Tes94], [Tes95], performed substantially better, and his rollout-based algorithm [TeG96] performed even better. The Deepchess program by David, Netanyahu, and Wolf [DNW16] provides another example of an expert-based supervised training approach.

Of course in the expert training approach we cannot expect to obtain a policy that performs better than the expert with which it is trained, in the case of Eq. (5.78), or the one-step lookahead policy that is based on the approximation $\tilde{J}$ or $\tilde{Q}$, in the case of Eq. (5.79) or Eq. (5.80), respectively. However, a major advantage is that once the parametrized policy is obtained, its on-line implementation is fast and does not involve extensive calculations such as minimizations of the form (5.79). This advantage is generally shared by schemes that are based on approximation in policy space. Moreover the policy obtained by emulating the expert can be improved through rollout.

5.7.3 Approximate Policy Iteration, Rollout, and Approximation in Policy Space

In this section we revisit approximate PI, but with a view towards combining it with rollout and approximation in policy space. We describe how approximation in policy space can provide the basis for an alternative PI implementation, namely *approximate the generated policies directly instead of approximating their cost functions or Q-factors*. This is an *actor-only* method, as opposed to the approximate PI methods of Sections 5.3.2 and 5.3.3, which involve a critic component.

In particular, the approximate PI methods given in Sections 5.3.2 and 5.3.3 use approximate policy evaluation (approximation in value space to represent the cost function or Q-factors of the current policy) followed by fairly exact policy improvement through one-step or multistep lookahead. By contrast, the methods described in this section use fairly exact policy evaluation through the use of rollout for a sample set of states, followed by approximate policy improvement [representation of the improved (or rollout) policy using a policy architecture]. The idea is to view the PI algorithm as a *perpetual rollout process*, which uses at any one time one out of a parametrized collection of base policies, and occasionally "improves" the base policy using the rollout results and approximation in policy space.

As an example, let us consider a PI algorithm where at the typical iteration we have a policy μ, which we use as the base policy for generating by (possibly truncated) rollout many state-control sample pairs (i^s, u^s), $s = 1, \ldots, q$ (cf. the rollout algorithm of Section 5.1.2).† We then obtain an "improved" policy $\tilde{\mu}(i, \overline{r})$, using an approximation architecture (such

† The approximation of the rollout policy using state control samples is subject to exploration issues, i.e., the choice of states starting from which we generate rollout trajectories with a given base policy. In particular, it is important to include in the sample set of initial states i^s, $s = 1, \ldots, q$, a subset of states that are "favored" by the rollout trajectories; e.g., start from some initial subset of states i^s and selectively add to this subset states that are encountered along the rollout trajectories.

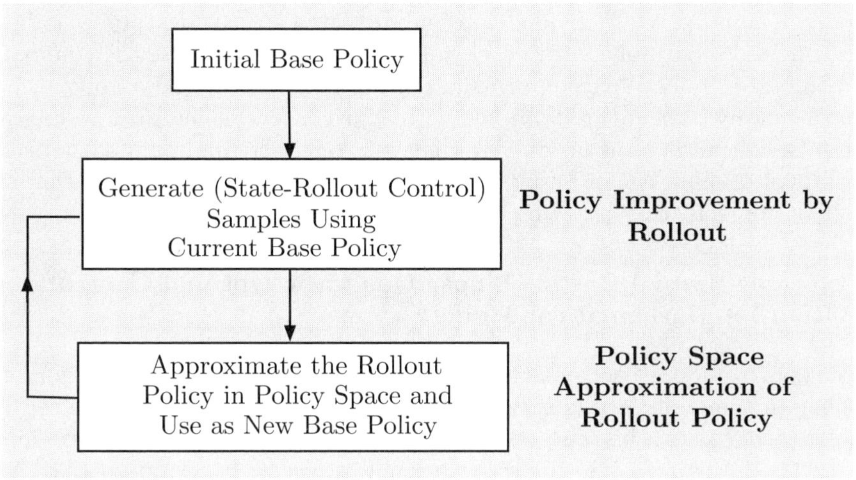

Figure 5.7.6 Block diagram of approximate PI based on policy improvement by rollout and policy approximation in policy space. Here the policy evaluation and policy improvement processes may be exact, but the implementation of the "improved" policy is approximate, using approximation in policy space and regression.

as a neural network), where the parameter $\overline{r}$ is obtained from the least squares/regression minimization

$$\overline{r} \in \arg\min_{r} \sum_{s=1}^{q} \big\| u^s - \tilde{\mu}(i^s, r) \big\|^2 \tag{5.82}$$

(possibly with added regularization); see Fig. 5.7.6. The "improved" policy $\tilde{\mu}(i, \overline{r})$ is then used as a base policy to generate samples of the corresponding rollout policy, which is then approximated in policy space, etc. This is similar to the expert training approach of Section 5.7.2 [cf. Eq. (5.78)]; we just use the rollout policy as the "expert" and emulate its decisions using sampling and supervised learning. The training problem (5.82) relates to the ones solved in classification problems (often with the use of neural networks or other architectures), and can be addressed with similar algorithms: the parameter $\overline{r}$ defines a classifier, which given a state i, classifies i as requiring control $\tilde{\mu}(i, \overline{r})$; cf. Section 3.5.†

Among notable characteristics of the scheme just described, we mention the substantial computation required to generate the rollout policy sample pairs (i^s, u^s), $s = 1, \ldots, q$, particularly for a stochastic problem. This PI process must be performed off-line. However, it can take advantage of parallelization and a partitioned architecture, whereby the state

† A simpler alternative to parametric approximation and the least squares minimization (5.82) is the interpolation scheme (5.81).

space is appropriately divided into subsets, and the rollout policy is approximated separately within each subset. The scheme shares the common advantage of policy space approximation: it yields in the end a policy that can be easily implemented on-line, without the need for one-step or multistep lookahead minimization. Alternatively, once the final policy has been obtained off-line, it can be used on-line as a base policy for generation of rollout controls, thus allowing on-line replanning.

Favorable Special Cases - Linear Quadratic Optimal Control Without a Mathematical Model

The approximate PI scheme of this section can actually produce an *optimal* policy in some interesting special cases. In particular, let $\mathcal{M}$ be the set of policies μ that can be represented by the approximation architecture, i.e., have the form $\mu = \tilde{\mu}(r)$ for some parameter vector r. Suppose that $\mathcal{M}$ has the property that if a policy that belongs to $\mathcal{M}$ is used as a base policy, then the corresponding rollout policy also belongs to $\mathcal{M}$ (i.e., exact policy improvement produces a policy in $\mathcal{M}$). Then it can be seen that the algorithm of this section, starting with a policy within $\mathcal{M}$ and using a large number of samples q for the least squares minimization (5.82) (so that the rollout policy is the same as the one produced by exact policy improvement) produces a sequence of policies in $\mathcal{M}$, which is essentially the same sequence as the one produced by the exact PI algorithm.

It follows that in this favorable case, the algorithm of this section inherits the convergence properties of exact PI, without requiring a mathematical model; a simulator of the system is sufficient. This supports the conjecture that if the set $\mathcal{M}$ "nearly contains" the policies produced by exact PI, then the policies generated by the approximate PI scheme of this section generates a sequence of policies whose performance oscillates at near optimal levels.

An important special case is when the system is linear, the cost function is infinite horizon-discounted quadratic, and $\mathcal{M}$ is the class of control laws that consist of a linear function of the state (so perfect state information is assumed). It can then be shown (under mild assumptions, cf. [Ber12], Section 4.2) that the optimal policy belongs to $\mathcal{M}$.† Moreover, exact PI, starting from a policy in $\mathcal{M}$, generates policies within $\mathcal{M}$ and converges to an optimal policy. Thus the scheme of this section also converges to an optimal policy for this choice of $\mathcal{M}$, and works without requiring a model of the system.

---

† We have not discussed in this book infinite horizon problems with continuous state and control spaces. While such problems can be challenging in general, the linear quadratic case is well behaved, and is supported by a solid VI and PI methodology; see e.g., [Ber12], [Ber17]. In our discussion here, we simply quote results from this methodology without a proof elaboration.

Rollout and Approximate PI for POMDP

We will now discuss the application of the rollout-based approximate PI algorithm to the partially observable version of the α-discounted problem. Here at each stage, instead of observing the current state, we obtain an observation z with probabilities $p(z \mid j, u)$, where j is the current state, and u is the preceding control. We assume that at each stage, the belief state $b = \big(b(1), \ldots, b(n)\big)$ is available, where $b(i)$ is the conditional probability that the state is i, given the past history, and that we can also compute $B(b, u, z)$, the next belief state, given that the current belief state is b, the control applied at b is u, and that the next measurement is z. We can view $B(b, u, z)$ as a *belief state generator*. Finally, we assume that the initial base policy and all the base policies generated by the policy space scheme (5.82) are functions of b.

Let us consider the computation of the rollout control $\tilde{\mu}(b)$ at some fixed b, using the truncated rollout algorithm, with μ being the current base policy (some function of the belief state), and $\tilde{J}$ being some cost function approximation. This rollout control is given by

$$\tilde{\mu}(b) \in \arg\min_{u \in U} \sum_{i=1}^{n} b(i) \tilde{F}_\mu(i, b, u), \tag{5.83}$$

where $\tilde{F}_\mu(i, b, u)$ can be viewed as an approximate Q-factor, obtained by starting from state i, applying control u, then applying the policy μ for some number m of stages, and finally approximating the cost of the remaining stages using the function $\tilde{J}$. [A multistep lookahead version of Eq. (5.83) is also possible.] For given values of b and u, the expression minimized in Eq. (5.83) can be approximated by simulation. Here is a possibility, which involves *simultaneous simulation of the states of the original system, as well as the belief states obtained by the belief state generator*; the corresponding simulator is illustrated in Fig. 5.7.7.

The state space is sampled according to the distribution b to obtain a dataset of states $\mathcal{I} = \{i^m \mid m = 1, \ldots, M\}$ (if n is relatively small, we can take $\mathcal{I} = \{1, \ldots, n\}$). Then for each state in the dataset, say i^m, and each control $u \in U$, we do the following: We obtain the next state j_1 according to the transition probabilities $p_{i^m j}(u)$, and we record the cost $g(i^m, u, j_1)$. Then we obtain a random observation sample z_1 according to $p(z_1 \mid j_1, u)$, and compute the next belief state $b_1 = B(b, u, z_1)$ [which allows us to compute the next control $\mu(b_1)$ for the next simulation step]. In a similar manner, we run a simulation trajectory starting from j_1 for m stages, using policy μ, while obtaining a sequence of m successive control-next state-observation-belief state quadruplets

$$\big(\mu(b_1), j_2, z_2, b_2\big), \ldots, \big(\mu(b_m), j_{m+1}, z_{m+1}, b_{m+1}\big);$$

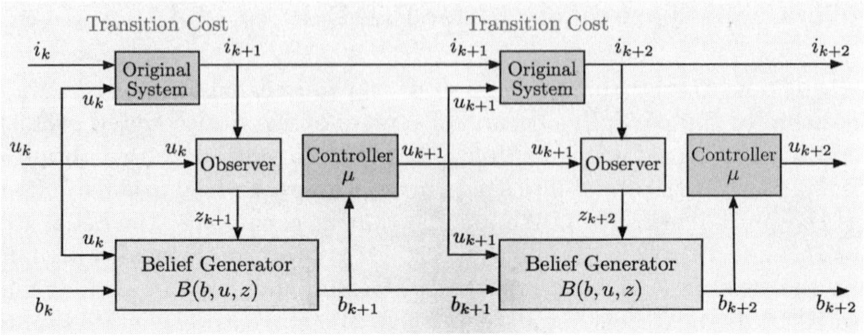

Figure 5.7.7 Block diagram of simultaneous simulation of the states of the original system, and the belief states obtained by the belief state generator, under a base policy $\mu(b)$.

cf. Fig. 5.7.7. Simultaneously, we obtain the corresponding sum of stage costs plus the terminal (approximate) cost

$$g(i^m, u, j_1) + \alpha g\big(j_1, \mu(b_1), j_2\big) + \cdots + \alpha^m g\big(j_m, \mu(b_m), j_{m+1}\big) + \alpha^{m+1} \tilde{J}(b_{m+1}).$$

In this way we obtain a single sample of $\tilde{F}_\mu(i^m, b, u)$ that corresponds to the state-control pair (i^m, u).

By averaging many such random samples for every state $i^m \in \mathcal{I}$, we compute an estimate of $\sum_{i=1}^n b(i)\tilde{F}_\mu(i, b, u)$ for each $u \in U$. By minimizing this estimate over u, we obtain the rollout control $\tilde{\mu}(b)$ at b; cf. Eq. (5.83). After collecting q such sample belief-rollout control pairs $\big(b^s, \tilde{\mu}(b^s)\big)$, $s = 1, \ldots, q$, we can generate a new base policy by using approximation in policy space [cf. Eq. (5.82)]. The computations are similar to the perfect state information case, except that obtaining each belief-rollout control pair through the minimization of Eq. (5.83) requires one additional operation: sampling the state space using the current belief state b and carrying out the computations for each pair (i^m, u) with $i^m \in \mathcal{I}$ and $u \in U$, rather than a single state-control pair. Still, a lot more computation may be needed relative to the perfect state information case, depending on the number of states i such that $b(i) > 0$. This argues for small lengths of lookahead minimization and rollout, and greater emphasis on the terminal cost function approximation $\tilde{J}$.

Variations

There are many variations of the actor-only approximate PI scheme of this section. Basically, all the variants of rollout (multistep lookahead, rollout truncation, and terminal cost function approximation) are applicable. Moreover, optimistic variants of PI can be used, whereby only a small number of samples q are generated between changes in the parameter vector and the attendant change of base policy. To implement such an optimistic variant, one may use an incremental gradient or Newton method (cf. Section 3.1.3) to solve the regression minimization (5.82) while adding new terms to the least squares sum as new samples (i^s, u^s) become available, with u^s being the control obtained with the current policy (the one that corresponds to the current value of r). Naturally, as in all approximate PI methods, exploration is an important issue to address, using a judicious choice of the sampled states i^s.

Here is an example of an extreme case of optimistic incremental gradient-type algorithm, where the parameter r is updated after each sample of state control pair (i^s, u^s) is obtained. The algorithm parallels the SARSA Q-learning algorithm given in Section 5.4.1.

At the start of iteration k, we have the current parameter vector r^k, and the corresponding policy $\mu^k = \tilde{\mu}(\cdot, r^k)$. Then:

(1) We select a state i^k (with due regard to exploration).

(2) We compute the rollout control u^k at i^k, using $\mu^k = \tilde{\mu}(\cdot, r^k)$ as base policy, i.e.,

$$u^k \in \arg \min_{u \in U(i^k)} \sum_{j=1}^{n} p_{i^k j}(u)\big(g(i^k, u, j) + \alpha J_{\mu^k}(j)\big).$$

(3) We update the parameter vector via

$$r^{k+1} = r^k - \gamma^k \nabla \tilde{\mu}(i^k, r^k)\big(\tilde{\mu}(i^k, r^k) - u^k\big),$$

where γ^k is a positive stepsize, and $\nabla \tilde{\mu}(i^k, r^k)$ denotes the gradient matrix of $\tilde{\mu}(i^k, \cdot)$ evaluated at the current parameter vector r^k.

As in the case of SARSA, there are also less optimistic variants of the preceding algorithm, whereby several states and their rollout controls are computed before updating the parameter vector r.

5.8 NOTES AND SOURCES

Section 5.1: The performance bound of Props. 5.1.1 for multistep lookahead with a terminal cost function approximation is well known; see e.g., Prop. 6.1.1 in the author's DP textbook [Ber17] (and earlier editions), as well as [Ber18a], Section 2.2. The bound of Prop. 5.1.3 for multistep lookahead, truncated rollout, and terminal cost function approximation is new.

The bounds of Section 5.1.3 for approximate PI were first given in the neuro-dynamic book by Bertsekas and Tsitsiklis [BeT96], Sections 6.2.2 and 6.2.3, for both discounted and SSP problems. These bounds, together with related bounds for approximate optimistic PI, due to Thiery and Scherrer [ThS10], are quite fundamental. While they are often conservative, they correctly describe the qualitative behavior of approximate PI and its variations, and they differentiate the behavior of these methods from their VI counterparts, which are generally unstable in the absence of a suitable sampling policy (cf. Example 5.2.1). Extensions of these bounds to approximate PI for general spaces abstract DP problems under contraction and monotonicity assumptions can be found in the author's abstract DP book [Ber18a], Sections 2.4.1 and 2.5.2.

Section 5.2: Fitted VI algorithms have been used for finite horizon problems since the early days of DP. They are conceptually simple and easily implementable, and they are used widely for approximation of either optimal costs or Q-factors (see e.g., the papers by Gordon [Gor95], Longstaff and Schwartz [LoS01], Ormoneit and Sen [OrS02], Ernst, Geurts, and Wehenkel [EGW06], Antos, Munos, and Szepesvari [AMS07], and Munos and Szepesvari [MuS08]). Approximations to VI may also be implemented by relaxing the constraints of the iteration and/or introducing bounding operations that simplify the calculations; see e.g., Lincoln and Rantzer [LiR06], and Spaan and Vlassis [SpV05] in the context of POMDP.

Section 5.3: Approximate PI methods of the type considered in Section 5.3.3 has been proposed by Fern, Yoon, and Givan [FYG06], and variants have also been discussed and analyzed by several other authors. The method (with some variations) has been used to train a tetris playing computer program that performs impressively better than programs that are based on other variants of approximate PI, and various other methods; see Scherrer [Sch13], Scherrer et al. [SGG15], and Gabillon, Ghavamzadeh, and Scherrer [GGS13], who also provide an analysis of the method.

Section 5.4: Q-learning was proposed by Watkins [Wat89], and had a major impact in the development of the RL field. A rigorous convergence proof of Q-learning was given by Tsitsiklis [Tsi94], in a more general framework that combined several ideas from stochastic approximation theory and the theory of distributed asynchronous computation. This proof covered discounted problems, and SSP problems where all policies are proper. It also covered SSP problems with improper policies, assuming that the Q-learning iterates are either nonnegative or bounded. Convergence for SSP problems without the nonnegativity or the boundedness assumption was shown by Yu and Bertsekas [YuB13b]. Q-learning for optimal stopping problems was analyzed by Tsitsiklis and Van Roy [TsV99b], with followup work by Yu and Bertsekas [YuB07].

Optimistic asynchronous versions of PI based on Q-learning, which have solid convergence properties, are given by Bertsekas and Yu [BeY10], [BeY12], [YuB13a]. The distinctive feature of the Q-learning methods of [BeY12], [YuB13a] is that the policy evaluation process aims towards the solution of an optimal stopping problem rather than towards the solution of the Bellman equation associated with the policy; this is needed to avoid the pathological behavior identified by Williams and Baird [WiB93], and noted earlier in Section 4.6.3.

The original proposal of SARSA is attributed to Rummery and Niranjan [RuN94], with related work reported in the papers by Peng and Williams [PeW96], and Wiering and Schmidhuber [WiS98]. The ideas of the DQN algorithm attracted much attention following the paper by Mnih et al. [MKS15], which reported impressive test results on a suite of 49 classic Atari 2600 games.

The advantage updating idea, which was noted in the context of finite horizon problems in Section 3.4, can be readily extended to infinite horizon problems. In this context, it was proposed by Baird [Bai93], [Bai94]; see [BeT96], Section 6.6. A related variant of approximate PI and Q-learning, called *differential training* and aiming to approximate cost-to-go differences rather than cost-to-go values, has been proposed by the author in [Ber97b]; see also Weaver and Baxter [WeB99].

Section 5.5: Projected equations underlie Galerkin methods, which have a long history in scientific computation. These methods are used widely for approximate solution of many types of large-scale problems, including linear systems arising from discretization of partial differential and integral equations. The connection of the approximate DP context of policy evaluation based on projected equations with Galerkin methods was first discussed by Yu and Bertsekas [YuB10], and Bertsekas [Ber11b]. However, the Monte Carlo simulation ideas that are central in approximate DP differentiate the projected equation methods of the present chapter from the Galerkin methodology. On the other hand, Galerkin methods apply to a wide range of problems, far beyond DP, and the simulation-based ideas of approximate DP can consequently be extended to apply more broadly (see the papers by Bertsekas and Yu [BeY09], Wang and Bertsekas [WaB13a], [WaB13b], and the author's textbook [Ber12], Section 7.3, which discuss the application of TD methods to the solution of general linear systems of equations).

Temporal difference ideas were introduced in the works of Samuel [Sam59], [Sam67] on a checkers-playing program. The work by Sutton [Sut88], following earlier work by Barto, Sutton, and Anderson [BSA83], formalized the use of temporal differences and proposed the TD(λ) method. This was a major development, and motivated a lot of research in RL and simulation-based DP, particularly following an impressive early success with the backgammon playing program of Tesauro [Tes92], [Tes94]. The relation of TD(λ) and the solution of projected equations was clarified in the works of Tsitsiklis and Van Roy [TsV97], [TsV99b], and was also described in the neuro-dynamic programming book [BeT96].

The three methods TD(λ), LSTD(λ), and LSPE(λ) are discussed in detail in the journal and textbook RL literature. For a discussion that extends our presentation of Section 5.5, see Chapters 6 and 7 of the book

[Ber12].

The convergence of TD(λ) was proved by Tsitsiklis and Van Roy [TsV97], with extensions in [TsV99a] and [TsV99b]. The author's papers [Ber16c], [Ber18d] describe the connection of TD and proximal algorithms, a central methodology in convex optimization. In particular, in these works, TD(λ) is shown to be a stochastic version of the proximal algorithm for solving linear systems of equations, and extensions of TD(λ) for solving nonlinear systems of equations are described (the parameter λ is related to the penalty parameter of the proximal algorithm).

The LSTD(λ) algorithm was first proposed by Bradtke and Barto [BrB96] for $\lambda = 0$, and was extended for $\lambda > 0$ later by Boyan [Boy02]. Convergence analyses of LSTD(λ) under assumptions of increasing generality were given by Nedić, Bertsekas, and Yu [NeB03], [BeY09], [Yu12].

The LSPE(λ) algorithm was first proposed by Bertsekas and Ioffe [BeI96] under the name *λ-policy iteration*, and it was used to train a tetris playing program using the feature-based linear architecture described in Example 3.1.3. The motivation was to provide a better alternative to TD(λ)-based PI, which failed within the tetris context. LSPE(λ) was also given in the book [BeT96], Section 2.3.1, with subsequent contributions, including versions that use diagonal scaling to avoid matrix inversion, by Nedić, Borkar, Yu, Scherrer, and the author [NeB03], [BBN04], [YuB07], [BeY09], [YuB09a], [Ber11a], [Ber11b], [Yu12], [Sch13], [Ber18a].

In our discussion here, we did not go much into the sampling implementation details of TD(λ), LSTD(λ), and LSPE(λ); see the approximate DP/RL textbooks cited earlier, and the paper by Bertsekas and Yu [BeY09], which extends the TD simulation-based methodology to the solution of large systems of general linear equations.

Section 5.6: The linear programming approach for exact infinite horizon DP was proposed by D'Epenoux [D'Ep60]. Approximation methods using basis functions and linear programming were suggested with little analysis by Schweitzer and Seidman [ScS85], and have been further developed by de Farias and Van Roy [DFV03], [DFV04], [DeF04]. For a challenging application to pricing of network services, see Paschalidis and Tsitsiklis [PaT00]. For more recent work based on variants of the linear programming formulation, see Cogill et al. [CRV06], Desai, Farias, and Moallemi [DFM12], [DFM13], Wang, O'Donoghue, and Boyd [WOB15], and Beuchat, Warrington, and Lygeros [BWL19], and the references quoted there. Approximation based on linear programming, while not discussed at length in this book, is a promising approach that deserves further attention.

Section 5.7: Our coverage of approximation in policy space, policy gradient, and random search methods has been limited, and aimed to provide an entry point into the field. For a detailed discussion and references on policy gradient methods, we refer to the book by Sutton and Barto [SuB18], the monographs by Deisenroth, Neumann, and Peters [DNP11], and the surveys by Peters and Schaal [PeS08], and Grondman et al. [GBL12]. An influential paper in this context is the one by Williams [Wil92], who proposed among others the likelihood-ratio policy gradient method given here. The methods of [Wil92] are commonly referred to as REINFORCE in the literature (see e.g., [SuB18], Ch. 13).

There are several related early works on search along randomly chosen directions (Rastrigin [Ras63], Matyas [Mat65], Aleksandrov, Sysoyev, and Shemeneva [ASS68], Rubinstein [Rub69]); see also Spall [Spa92], [Spa03], Duchi, Jordan, Wainwright, and Wibisono [DJW12], [DJW15], and Nesterov and Spokoiny [NeS17], for more modern related works. For early works on simulation-based policy gradient schemes for various DP problems, see Glynn [Gly87], [Gly90], L'Ecuyer [L'Ec91], Fu and Hu [FuH94], Jaakkola, Singh, and Jordan [JSJ95], Cao and Chen [CaC97], Cao and Wan [CaW98].

The challenge in the successful implementation of policy gradient methods is twofold: the difficulties with slow convergence and local minima that are inherent in gradient optimization, and the detrimental effects of simulation noise. Much work has been directed towards variations that address these difficulties, including the use of a baseline and variance reduction methods (Greensmith, Bartlett, and Baxter [GBB04], Greensmith [Gre05]), and scaling based on the so-called *natural gradient* (Kakade [Kak02]) or second order information (see Wang and Paschalidis [WaP17], and the references quoted there). For an interesting discussion of the connections between PID control, model-free approximation in policy space, and policy gradient methods, we refer to the paper and blog of B. Recht [Rec18a], [Rec18b].

We have not covered actor-critic methods within the policy gradient context. Such methods were introduced in the paper by Barto, Sutton, and Anderson [BSA83]. The more recent works of Baxter and Bartlett [BaB01], Konda and Tsitsiklis [KoT99], [KoT03], Marbach and Tsitsiklis [MaT01], [MaT03], and Sutton et al. [SMS99] have been influential. Actor-critic algorithms that are suitable for POMDP and involve gradient estimation have been given by Yu [Yu05], and Estanjini, Li, and Paschalidis [ELP12].

The cross-entropy method was initially developed in the context of rare event simulation and was later adapted for use in optimization. For textbook accounts, see Rubinstein and Kroese [RuK04], [RuK13], [RuK16], and Busoniu et al. [BBD10], and for surveys see de Boer et al. [BKM05], and Kroese et al. [KRC13]. The method was proposed for policy search in an approximate DP context by Mannor, Rubinstein, and Gat [MRG03]. It was applied with success to the game of tetris by Szita and Lorinz [SzL06], and Thiery and Scherrer [ThS09]. For recent analysis, see Joseph and Bhatnagar [JoB16], [JoB18].

The expert training methods of Section 5.7.2 are similar to the comparison training method discussed in Section 2.4.3, which was proposed by Tesauro [Tes89a], [Tes89b], [Tes01]. Methods that use learning from data generated by an expert are often referred in the literature by the names *imitation learning* and *apprenticeship learning*; see e.g., Abbeel and Ng [AbN04], Neu and Szepesvari [NeS12], and Schaal [Sch99].

Training using a human expert has generated considerable interest in robotics (where it is often referred to as *learning from demonstration*); see Argall et al. [ACV09], and for some recent work, see Ben Amor et al. [BVE13] and Lee [Lee17]. For a recent analysis of related possibilities, see Hanawal et al. [HLZ18].

The actor-only PI/rollout methods of Section 5.7.3, depending on their implementation, are similar to several methods proposed in the literature, first by Lagoudakis and Parr [LaP03], and then by other authors (see e.g., Dimitrakakis and Lagoudakis [DiL08], Lazaric, Ghavamzadeh, and Munos [LGM10], Gabillon et al. [GLG11], Liu and Wei [LiW14], Farahmand et al. [FPB15], and the references quoted there). The paper by Yan, Diaconis, Rusmevichientong, and Van Roy [YDR04] developed a different type of PI/rollout method based on recursive application of rollout policies for a version of solitaire.

Our discussion of the application to POMDP is new in the form given here. Generally POMDP are quite challenging for the current RL methodology, both theoretically and practically, because of the continuous nature of the belief space as well as the excessive accumulation of potentially useful information over time.

The main strength of the actor-only PI/rollout methods is that they rely on the reliability and robustness of the rollout approach, which has been demonstrated by many practical studies, combined with the use of mature and well-developed classification algorithms. Note also that the generation of samples of the rollout policies, while computationally intensive, can benefit greatly from the availability of parallel computation.

We have noted in Section 5.7.3 the application of PI with approximation in policy space for adaptive control in problems with a linear-quadratic structure and perfect state information. An alternative simulation-based PI method based on approximation in value space has been proposed by Bradtke, Ydstie, and Barto [BYB94]. In its idealized form (i.e., with perfect state information, and an infinite number of simulation samples), it obtains the optimal policy without knowledge of the linear system equation parameters. Related methods for adaptive control of discrete and continuous-time systems have been discussed in the books by Vrabie, Vamvoudakis, and Lewis [VVL13], Jiang and Jiang [JiJ17], and Liu et al. [LWW17].

5.9 APPENDIX: MATHEMATICAL ANALYSIS

In this appendix we provide proofs of the mathematical results stated in this chapter. We also prove some supplementary results that are described in the chapter without formal statement.

We will use extensively the triangle inequality $\|J + J'\| \leq \|J\| + \|J'\|$, which holds for every norm $\|\cdot\|$. We will also use the Bellman operators T and T_μ for the discounted problem:

$$(TJ)(i) = \min_{u \in U(i)} \sum_{j=1}^{n} p_{ij}(u)\big(g(i,u,j) + \alpha J(j)\big), \qquad i = 1, \ldots, n,$$

$$(T_\mu J)(i) = \sum_{i=1}^{n} p_{ij}\big(\mu(i)\big)\Big(g\big(i, \mu(i), j\big) + \alpha J(j)\Big), \qquad i = 1, \ldots, n.$$

A key property for our analysis is that these operators are contractions, i.e., for all J, J', and μ, we have

$$\|TJ - TJ'\| \leq \alpha\|J - J'\|, \qquad \|T_\mu J - T_\mu J'\| \leq \alpha\|J - J'\|,$$

where $\|J\|$ is the maximum norm $\|J\| = \max_{i=1,\ldots,n} \big|J(i)\big|$, cf. Prop. 4.3.5. Another key property is the monotonicity of these operators, i.e.,

$$TJ \geq TJ', \quad T_\mu J \geq T_\mu J', \qquad \text{for all } J \text{ and } J' \text{ with } J \geq J'.$$

Moreover, we have the "constant shift" property, which states that if the functions J is increased uniformly by a constant c, then the functions TJ and $T_\mu J$ are also increased uniformly by the constant αc.

5.9.1 Performance Bounds for Multistep Lookahead

We first prove the basic performance bounds for ℓ-step lookahead schemes and discounted problems.

Proposition 5.1.1: (Limited Lookahead Performance Bounds)

(a) Let $\tilde{\mu}$ be the ℓ-step lookahead policy corresponding to $\tilde{J}$. Then

$$\|J_{\tilde{\mu}} - J^*\| \leq \frac{2\alpha^\ell}{1-\alpha}\|\tilde{J} - J^*\|, \tag{5.84}$$

where $\|\cdot\|$ denotes the maximum norm $\|J\| = \max_{i=1,\ldots,n} \big|J(i)\big|$.

(b) Define

$$\hat{J}(i) = \min_{u \in \overline{U}(i)} \sum_{j=1}^{n} p_{ij}(u)\big(g(i,u,j) + \alpha\tilde{J}(j)\big), \qquad i = 1, \ldots, n, \tag{5.85}$$

where $\overline{U}(i) \subset U(i)$ for all $i = 1, \ldots, n$, and let $\tilde{\mu}$ be the one-step lookahead policy obtained by minimization in the right side of this equation. Then we have

$$J_{\tilde{\mu}}(i) \leq \tilde{J}(i) + \frac{c}{1-\alpha}, \qquad i = 1, \ldots, n, \tag{5.86}$$

where

$$c = \max_{i=1,\ldots,n} \big(\hat{J}(i) - \tilde{J}(i)\big). \tag{5.87}$$

Proof: (a) The line of proof of this part is to show Eq. (5.84) for the case of one-step lookahead ($\ell = 1$), and then generalize the argument to the case $\ell > 1$ by replacing $\tilde{J}$ with $T^{\ell-1}\tilde{J}$. In the course of the proof, we will use the contraction property of T and T_μ (cf. Prop. 4.3.5).

We first prove a preliminary relation. Using the fact

$$\|T_{\tilde{\mu}}^m J^* - T_{\tilde{\mu}}^{m-1} J^*\| = \|T_{\tilde{\mu}}^{m-1}(T_{\tilde{\mu}} J^*) - T_{\tilde{\mu}}^{m-1} J^*\| \leq \alpha^{m-1} \|T_{\tilde{\mu}} J^* - J^*\|,$$

and the triangle inequality, we have for every k,

$$\|T_{\tilde{\mu}}^k J^* - J^*\| \leq \sum_{m=1}^{k} \|T_{\tilde{\mu}}^m J^* - T_{\tilde{\mu}}^{m-1} J^*\| \leq \sum_{m=1}^{k} \alpha^{m-1} \|T_{\tilde{\mu}} J^* - J^*\|.$$

By taking the limit as $k \to \infty$ and using the fact $T_{\tilde{\mu}}^k J^* \to J_{\tilde{\mu}}$, we obtain

$$\|J_{\tilde{\mu}} - J^*\| \leq \frac{1}{1-\alpha} \|T_{\tilde{\mu}} J^* - J^*\|. \tag{5.88}$$

Denote now $\bar{J} = T^{\ell-1}\tilde{J}$. By using the triangle inequality and the definition of $\tilde{\mu}$, which implies that $T_{\tilde{\mu}}\bar{J} = T\bar{J}$, the rightmost expression of Eq. (5.88) is estimated as follows:

$$\begin{aligned}
\|T_{\tilde{\mu}} J^* - J^*\| &\leq \|T_{\tilde{\mu}} J^* - T_{\tilde{\mu}} \bar{J}\| + \|T_{\tilde{\mu}} \bar{J} - T\bar{J}\| + \|T\bar{J} - J^*\| \\
&= \|T_{\tilde{\mu}} J^* - T_{\tilde{\mu}} \bar{J}\| + \|T\bar{J} - TJ^*\| \\
&\leq 2\alpha \|\bar{J} - J^*\| \\
&= 2\alpha \|T^{\ell-1}\tilde{J} - T^{\ell-1} J^*\| \\
&\leq 2\alpha^\ell \|\tilde{J} - J^*\|.
\end{aligned}$$

By combining the preceding two relations, we obtain Eq. (5.84).

(b) Let us denote by e the unit vector whose components are all equal to 1. Then by the definition (5.87) of c, we have

$$T_{\tilde{\mu}}\tilde{J} = \hat{J} \leq \tilde{J} + ce.$$

Applying $T_{\tilde{\mu}}$ to both sides of this relation, and using the monotonicity and constant shift property of $T_{\tilde{\mu}}$, we obtain

$$T_{\tilde{\mu}}^2 \tilde{J} \leq T_{\tilde{\mu}}\tilde{J} + \alpha ce.$$

Continuing similarly, we have

$$T_{\tilde{\mu}}^{k+1}\tilde{J} \leq T_{\tilde{\mu}}^{k}\tilde{J} + \alpha^k ce, \qquad k = 0, 1, \ldots.$$

Adding the first $k+1$ of these relations, we obtain

$$T_{\tilde{\mu}}^{k+1}\tilde{J} \leq \tilde{J} + (1 + \alpha + \cdots + \alpha^k)ce, \qquad k = 0, 1, \ldots.$$

Taking the limit as $k \to \infty$, and using the fact $T_{\tilde{\mu}}^{k+1}\tilde{J} \to J_{\tilde{\mu}}$, we obtain the desired inequality (5.86). **Q.E.D.**

5.9.2 Performance Bounds for Rollout

We next show the basic cost improvement property of rollout.

Proposition 5.1.2: (Cost Improvement by Rollout) Let $\tilde{\mu}$ be the rollout policy obtained by the one-step lookahead minimization

$$\min_{u \in \overline{U}(i)} \sum_{j=1}^{n} p_{ij}(u)\big(g(i,u,j) + \alpha J_\mu(j)\big),$$

where μ is a base policy [cf. Eq. (5.85) with $\tilde{J} = J_\mu$] and we assume that $\mu(i) \in \overline{U}(i) \subset U(i)$ for all $i = 1, \ldots, n$. Then $J_{\tilde{\mu}} \leq J_\mu$.

Proof: Let us denote

$$\hat{J}(i) = \min_{u \in \overline{U}(i)} \sum_{j=1}^{n} p_{ij}(u)\big(g(i,u,j) + \alpha J_\mu(j)\big).$$

We have for all $i = 1, \ldots, n$,

$$\hat{J}(i) \leq \sum_{j=1}^{n} p_{ij}(u)\big(g\big(i,\mu(i),j\big) + \alpha J_\mu(j)\big) = J_\mu(i),$$

where the equality on the right holds by Bellman's equation. Thus we have $c = \max_{i=1,\ldots,n}\big(\hat{J}(i) - J_\mu(i)\big) \leq 0$, and the result follows from Prop. 5.1.1(b) with $\tilde{J} = J_\mu$ [cf. Eq. (5.86)]. **Q.E.D.**

We finally show the following performance bound for the truncated rollout algorithm with cost function approximation.

Proposition 5.1.3: (Performance Bound of Truncated Rollout with Terminal Cost Function Approximation) Let ℓ and m be positive integers, let μ be a policy, and let $\tilde{J}$ be a function of the state. Consider a truncated rollout scheme consisting of ℓ-step lookahead, followed by rollout with a policy μ for m steps, and a terminal cost function approximation $\tilde{J}$ at the end of the m steps. Let $\tilde{\mu}$ be the policy generated by this scheme.

(a) We have

$$\|J_{\tilde{\mu}} - J^*\| \leq \frac{2\alpha^\ell}{1-\alpha}\|T_\mu^m \tilde{J} - J^*\|,$$

where T_μ is the Bellman operator of Eq. (5.4), and $\|\cdot\|$ denotes the maximum norm $\|J\| = \max_{i=1,\ldots,n} \big|J(i)\big|$.

(b) We have

$$J_{\tilde{\mu}}(i) \leq \tilde{J}(i) + \frac{c}{1-\alpha}, \qquad i = 1,\ldots,n, \tag{5.89}$$

where

$$c = \max_{i=1,\ldots,n} \big((T_\mu \tilde{J})(i) - \tilde{J}(i)\big).$$

(c) We have

$$J_{\tilde{\mu}}(i) \leq J_\mu(i) + \frac{2}{1-\alpha}\|\tilde{J} - J_\mu\|, \qquad i = 1,\ldots,n. \tag{5.90}$$

Proof: (a) This is simply Prop. 5.1.1(a) adapted to the truncated rollout scheme [$\tilde{J}$ is replaced in Eq. (5.84) by $T_\mu^m \tilde{J}$].

(b) We first assume that $c = 0$ and prove the result for this case. If $c = 0$, by the definition of c, we have $\tilde{J} \geq T_\mu \tilde{J}$, which together with the monotonicity of T and T_μ, implies that

$$\tilde{J} \geq T_\mu^m \tilde{J} \geq T_\mu^{m+1} \tilde{J} \geq T T_\mu^m \tilde{J} \geq T^{\ell-1} T_\mu^m \tilde{J} \geq T^\ell T_\mu^m \tilde{J} = T_{\tilde{\mu}} T^{\ell-1} T_\mu^m \tilde{J}. \tag{5.91}$$

By writing this inequality for just the first, fifth, and last expressions above, we have

$$\tilde{J} \geq T^{\ell-1} T_\mu^m \tilde{J} \geq T_{\tilde{\mu}} T^{\ell-1} T_\mu^m \tilde{J}.$$

The right side of the preceding relation and the monotonicity of $T_{\tilde{\mu}}$, imply that the sequence $\{T_{\tilde{\mu}}^k T^{\ell-1} T_\mu^m \tilde{J}\}$ is monotonically nonincreasing as k increases, while the left side shows that the sequence is bounded above by $\tilde{J}$.

Since by the convergence property of VI, this sequence converges to $J_{\tilde{\mu}}$ as $k \to \infty$, we obtain $\tilde{J} \geq J_{\tilde{\mu}}$. Thus the result follows for the case $c = 0$.

To prove the result for general c, we introduce the function J' given by

$$J' = \tilde{J} + \frac{c}{1-\alpha} e,$$

where e is the unit vector whose components are all equal to 1. Applying T_μ to this equation, we have

$$T_\mu J' = T_\mu \tilde{J} + \frac{\alpha c}{1-\alpha} e,$$

while from the definition of c, it follows that

$$T_\mu \tilde{J} \leq \tilde{J} + ce.$$

Combining the preceding two relations, we obtain

$$T_\mu J' \leq \tilde{J} + ce + \frac{\alpha c}{1-\alpha} e = \tilde{J} + \frac{c}{1-\alpha} e = J',$$

or $T_\mu J' \leq J'$. The rollout policy $\tilde{\mu}$ does not change if we replace $\tilde{J}$ with J', since these two functions differ by a constant. Thus by using the version of the result already proved for $c = 0$ and the fact $T_\mu J' \leq J'$, we have $J_{\tilde{\mu}} \leq J'$, which is equivalent to the desired Eq. (5.89).

(c) Let $c = \|\tilde{J} - J_\mu\|$, so that $J_\mu \leq \tilde{J} + ce$ and $T_\mu \tilde{J} \leq T_\mu J_\mu + \alpha ce = J_\mu + \alpha ce$. Using these relations, it follows that

$$T_\mu \tilde{J} \leq J_\mu + \alpha ce \leq \tilde{J} + ce + \alpha ce.$$

By using part (b) and the fact $\tilde{J} \leq J_\mu + ce$ (which follows from the definition of c), we obtain

$$J_{\tilde{\mu}} \leq \tilde{J} + \frac{c + \alpha c}{1-\alpha} e \leq J_\mu + ce + \frac{c + \alpha c}{1-\alpha} e = J_\mu + \frac{2c}{1-\alpha} e.$$

Q.E.D.

A version of part (b) for the case where $m = 0$ is that if $T\tilde{J} = T_{\tilde{\mu}} \tilde{J} \leq \tilde{J} + ce$, then $J_{\tilde{\mu}} \leq \tilde{J} + \frac{c}{1-\alpha} e$. This follows by using induction to show that for all k we have

$$T_{\tilde{\mu}}^k \tilde{J} \leq \tilde{J} + (1 + \alpha + \cdots + \alpha^{k-1}) ce,$$

and by taking the limit as $k \to \infty$.

5.9.3 Performance Bounds for Approximate Policy Iteration

To prove the performance bound of Prop. 5.1.4, we focus on the discounted problem, and we make use of the contraction property of T_μ. We want to prove the following performance bound.

Proposition 5.1.4: (Performance Bound for Approximate PI) Consider the discounted problem, and let $\{\mu^k\}$ be the sequence generated by the approximate PI algorithm defined by the approximate policy evaluation (5.11) and the approximate policy improvement (5.12). Then we have

$$\limsup_{k\to\infty} \|J_{\mu^k} - J^*\| \le \frac{\epsilon + 2\alpha\delta}{(1-\alpha)^2}.$$

The essence of the proof is contained in the following lemma, which quantifies the amount of approximate policy improvement at each iteration. Note that this lemma relates to rollout, i.e., a single policy iteration that produces a rollout policy $\tilde{\mu}$ starting from a base policy μ.

Lemma 5.9.1: (Error Bound for Rollout with Approximations) Consider the discounted problem, and let J, $\tilde{\mu}$, and μ satisfy

$$\|J - J_\mu\| \le \delta, \qquad \|T_{\tilde{\mu}}J - TJ\| \le \epsilon, \tag{5.92}$$

where δ and ϵ are some scalars. Then we have

$$\|J_{\tilde{\mu}} - J^*\| \le \alpha\|J_\mu - J^*\| + \frac{\epsilon + 2\alpha\delta}{1-\alpha}. \tag{5.93}$$

Proof: The contraction property of T and $T_{\tilde{\mu}}$ implies that

$$\|T_{\tilde{\mu}}J_\mu - T_{\tilde{\mu}}J\| \le \alpha\delta, \qquad \|TJ - TJ_\mu\| \le \alpha\delta,$$

and hence

$$T_{\tilde{\mu}}J_\mu \le T_{\tilde{\mu}}J + \alpha\delta e, \qquad TJ \le TJ_\mu + \alpha\delta e,$$

where e is the unit vector whose components are all equal to 1. Using also Eq. (5.92), we have

$$T_{\tilde{\mu}}J_\mu \le T_{\tilde{\mu}}J + \alpha\delta\, e \le TJ + (\epsilon + \alpha\delta)e \le TJ_\mu + (\epsilon + 2\alpha\delta)e. \tag{5.94}$$

Combining this inequality with $TJ_\mu \le T_\mu J_\mu = J_\mu$, we obtain

$$T_{\tilde{\mu}}J_\mu \le J_\mu + (\epsilon + 2\alpha\delta)e. \tag{5.95}$$

We will show that this relation implies that

$$J_{\tilde{\mu}} \leq J_\mu + \frac{\epsilon + 2\alpha\delta}{1-\alpha} e. \tag{5.96}$$

Indeed, by applying $T_{\tilde{\mu}}$ to both sides of Eq. (5.95), it follows that

$$T_{\tilde{\mu}}^2 J_\mu \leq T_{\tilde{\mu}} J_\mu + \alpha(\epsilon + 2\alpha\delta)e \leq J_\mu + (1+\alpha)(\epsilon + 2\alpha\delta)e.$$

Applying $T_{\tilde{\mu}}$ again to both sides of this relation, and continuing similarly, we have for all k,

$$T_{\tilde{\mu}}^k J_\mu \leq J_\mu + (1 + \alpha + \cdots + \alpha^{k-1})(\epsilon + 2\alpha\delta)e.$$

By taking the limit as $k \to \infty$, and by using the VI convergence property $T_{\tilde{\mu}}^k J_\mu \to J_{\tilde{\mu}}$, we obtain Eq. (5.96).

By applying $T_{\tilde{\mu}}$ to Eq. (5.96), it follows that

$$T_{\tilde{\mu}} J_{\tilde{\mu}} \leq T_{\tilde{\mu}} J_\mu + \frac{\alpha(\epsilon + 2\alpha\delta)}{1-\alpha} e,$$

so using the fact $J_{\tilde{\mu}} = T_{\tilde{\mu}} J_{\tilde{\mu}}$, we have

$$J_{\tilde{\mu}} \leq T_{\tilde{\mu}} J_\mu + \frac{\alpha(\epsilon + 2\alpha\delta)}{1-\alpha} e.$$

Subtracting J^* from both sides, we obtain

$$J_{\tilde{\mu}} - J^* \leq T_{\tilde{\mu}} J_\mu - J^* + \frac{\alpha(\epsilon + 2\alpha\delta)}{1-\alpha} e. \tag{5.97}$$

Also from the contraction property of T,

$$TJ_\mu - J^* = TJ_\mu - TJ^* \leq \alpha \|J_\mu - J^*\| e$$

which, in conjunction with Eq. (5.94), yields

$$T_{\tilde{\mu}} J_\mu - J^* \leq TJ_\mu - J^* + (\epsilon + 2\alpha\delta)e \leq \alpha \|J_\mu - J^*\|\, e + (\epsilon + 2\alpha\delta)e.$$

Combining this relation with Eq. (5.97), we obtain

$$J_{\tilde{\mu}} - J^* \leq \alpha \|J_\mu - J^*\|\, e + \frac{\alpha(\epsilon + 2\alpha\delta)}{1-\alpha} e + (\epsilon + 2\alpha\delta)e = \alpha \|J_\mu - J^*\|\, e + \frac{\epsilon + 2\alpha\delta}{1-\alpha} e,$$

which is equivalent to the desired relation (5.93), since $J_{\tilde{\mu}} \geq J^*$. **Q.E.D.**

Proof of Prop. 5.1.4: Applying Lemma 5.9.1, we have

$$\|J_{\mu^{k+1}} - J^*\| \leq \alpha \|J_{\mu^k} - J^*\| + \frac{\epsilon + 2\alpha\delta}{1-\alpha},$$

which by taking the lim sup of both sides as $k \to \infty$ yields the desired result. **Q.E.D.**

We next prove the performance bound for approximate PI, assuming that the generated policy sequence is convergent.

Proposition 5.1.5: (Performance Bound for Approximate PI when Policies Converge) Let $\tilde{\mu}$ be a policy generated by the approximate PI algorithm under conditions (5.11), (5.12), and (5.13). Then we have

$$\|J^* - J_{\tilde{\mu}}\| \leq \frac{\epsilon + 2\alpha\delta}{1-\alpha}.$$

Proof: Let $\bar{J}$ be the cost vector obtained by approximate policy evaluation of $\tilde{\mu}$. Then in view of Eqs. (5.11), (5.12), we have

$$\|\bar{J} - J_{\tilde{\mu}}\| \leq \delta, \qquad \|T_{\tilde{\mu}}\bar{J} - T\bar{J}\| \leq \epsilon.$$

From this relation, the fact $J_{\tilde{\mu}} = T_{\tilde{\mu}} J_{\tilde{\mu}}$, and the triangle inequality, we have

$$\begin{aligned} \|TJ_{\tilde{\mu}} - J_{\tilde{\mu}}\| &\leq \|TJ_{\tilde{\mu}} - T\bar{J}\| + \|T\bar{J} - T_{\tilde{\mu}}\bar{J}\| + \|T_{\tilde{\mu}}\bar{J} - J_{\tilde{\mu}}\| \\ &= \|TJ_{\tilde{\mu}} - T\bar{J}\| + \|T\bar{J} - T_{\tilde{\mu}}\bar{J}\| + \|T_{\tilde{\mu}}\bar{J} - T_{\tilde{\mu}}J_{\tilde{\mu}}\| \\ &\leq \alpha\|J_{\tilde{\mu}} - \bar{J}\| + \epsilon + \alpha\|\bar{J} - J_{\tilde{\mu}}\| \\ &\leq \epsilon + 2\alpha\delta. \end{aligned} \tag{5.98}$$

For every k, by using repeatedly the triangle inequality and the contraction property of T, we have

$$\|T^k J_{\tilde{\mu}} - J_{\tilde{\mu}}\| \leq \sum_{\ell=1}^{k} \|T^\ell J_{\tilde{\mu}} - T^{\ell-1} J_{\tilde{\mu}}\| \leq \sum_{\ell=1}^{k} \alpha^{\ell-1} \|TJ_{\tilde{\mu}} - J_{\tilde{\mu}}\|,$$

and by taking the limit as $k \to \infty$,

$$\|J^* - J_{\tilde{\mu}}\| \leq \frac{1}{1-\alpha} \|TJ_{\tilde{\mu}} - J_{\tilde{\mu}}\|.$$

Combining this relation with Eq. (5.98), we obtain the desired performance bound. **Q.E.D.**

6

Aggregation

Contents

In this chapter we consider approximation in value space using a problem approximation approach that is based on aggregation. In particular, we construct a simpler and more tractable "aggregate" problem by creating special "groups" of multiple states, which we view as "aggregate states." We solve the aggregate problem exactly by DP, and we use its optimal cost-to-go function in a one-step or multistep lookahead approximation scheme for the original problem.

In addition to problem approximation, aggregation is related to feature-based parametric approximation. For example, it often produces a piecewise constant cost function approximation, which may be viewed as a linear feature-based parametrization, where the features are 0-1 membership functions; see Example 3.1.1. Aggregation can also be combined with other approximation schemes, to add a local correction to a cost function approximation $\tilde{J}$, which is already available, possibly through an earlier use of a neural network approximation.

Aggregation can be applied to both finite horizon and infinite horizon problems. In this chapter, we will focus primarily on the discounted infinite horizon problem. We will introduce aggregation in a simple intuitive form in Section 6.1, and then generalize to a more sophisticated form of aggregation in Section 6.2.

6.1 AGGREGATION WITH REPRESENTATIVE STATES

In this section we focus on a relatively simple form of aggregation, which involves a special subset of states, called *representative*. Our approach is to view these states as the states of a smaller optimal control problem, called *aggregate problem*, which we will formulate and solve exactly in place of the original. We will then use the optimal aggregate costs of the representative states to approximate the optimal costs of the original problem states by interpolation. Let us describe a classical example.

Example 6.1.1 (Coarse Grid Approximation)

Consider a discounted problem where the state space is a grid of points $i = 1, \ldots, n$ on the plane. We introduce a coarser grid that consists of a subset $\mathcal{A}$ of the states/points, which we call representative and denote by x; see Fig. 6.1.1. We now wish to formulate a lower-dimensional DP problem just on the coarse grid of states. The difficulty here is that there may be positive transition probabilities $p_{xj}(u)$ from some representative states x to some non-representative states j. To deal with this difficulty, we introduce artificial transition probabilities ϕ_{jy} from non-representative states j to representative states y, which we call *aggregation probabilities*. In particular, a transition from representative state x to a nonrepresentative state j, is followed by a transition from j to some other representative state y with probability ϕ_{jy}; see Fig. 6.1.2.

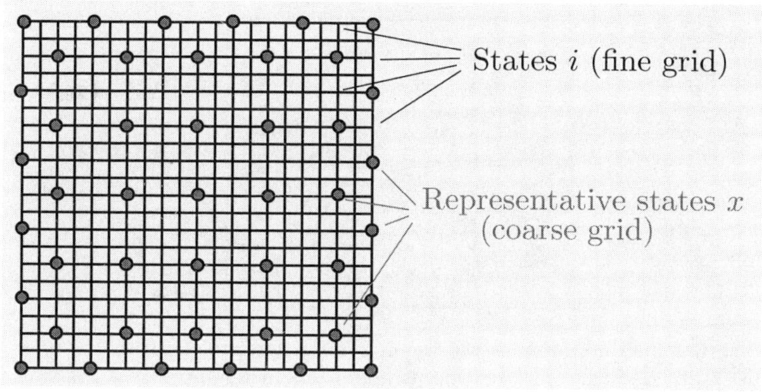

Figure 6.1.1 Illustration of aggregation with representative states; cf. Example 6.1.1. A relatively small number of states are viewed as representative. We define transition probabilities between pairs of aggregate states and we also define the associated expected transition costs; cf. Eq. (6.1). These specify a smaller DP problem, called the aggregate problem, which is solved exactly. The optimal cost function J^* of the original problem is approximated by interpolation from the optimal costs of the representative states r_y^* in the aggregate problem:

$$\tilde{J}(j) = \sum_{y \in \mathcal{A}} \phi_{jy} r_y^*, \qquad j = 1, \ldots, n,$$

and is used in a one-step or multistep lookahead scheme.

This process involves approximation but constructs a transition mechanism for an *aggregate problem* whose states are just the representative ones. The transition probabilities between representative states x, y under control $u \in U(x)$ and the corresponding expected transition costs are

$$\hat{p}_{xy}(u) = \sum_{j=1}^{n} p_{xj}(u) \phi_{jy}, \qquad \hat{g}(x, u) = \sum_{j=1}^{n} p_{xj}(u) g(x, u, j). \tag{6.1}$$

We can solve the aggregate problem by any suitable exact DP method. Let r_x^* denote the corresponding optimal cost of representative state $x \in \mathcal{A}$. We can then approximate the optimal cost function of the original problem with the interpolation formula

$$\tilde{J}(j) = \sum_{y \in \mathcal{A}} \phi_{jy} r_y^*, \qquad j = 1, \ldots, n. \tag{6.2}$$

This function may in turn be used in a one-step or multistep lookahead scheme for approximation in value space of the original problem.

Note that there is a lot of freedom in selecting the aggregation probabilities ϕ_{jy}. Intuitively, ϕ_{jy} should express a measure of proximity between j and y, e.g., ϕ_{jy} should be relatively large when y is geometrically close to j. For example, we could set $\phi_{jy_j} = 1$ for the representative state y_j that is "closest" to j, and $\phi_{jy_j} = 0$ for all other representative states $y \neq y_j$. In this case, Eq. (6.2) yields a piecewise constant cost function approximation $\tilde{J}$ (the constant values are the scalars r_y^* of the representative states y).

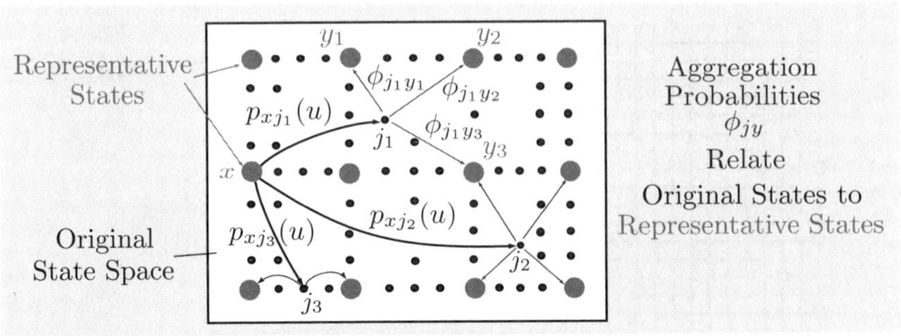

Figure 6.1.2 Illustration of the use of aggregation probabilities ϕ_{jy} from nonrepresentative states j to representative states y in Example 6.1.1. A transition from a state x to a nonrepresentative state j is followed by a transition to aggregate state y with probability ϕ_{jy}. In this figure, from representative state x, there are three possible transitions, to states j_1, j_2, and j_3, according to $p_{xj_1}(u), p_{xj_2}(u), p_{xj_3}(u)$, and each of these states is associated with a convex combination of representative states using the aggregation probabilities. For example in the figure the state j_1 is associated with the aggregate states y_1, y_2, y_3 with aggregation probabilities $\phi_{j_1y_1}, \phi_{j_1y_2}, \phi_{j_1y_3}$, respectively.

We will now formalize our framework for aggregation with representative states by straightforward generalization of the preceding example; see Fig. 6.1.3. We first consider the n-state α-discounted problem of Chapter 4, which we refer to as the "original problem," to distinguish from the "aggregate problem," which we define next.

Aggregation Framework with Representative States

We introduce a finite subset $\mathcal{A}$ of the original system states, which we call *representative states*, and we denote them by symbols such as x and y. We construct an *aggregate problem*, with state space $\mathcal{A}$, and transition probabilities and transition costs defined as follows:

(a) We relate the original system states j to representative states $y \in \mathcal{A}$ with aggregation probabilities ϕ_{jy}; these are scalar "weights" satisfying $\phi_{jy} \geq 0$ for all $y \in \mathcal{A}$, and $\sum_{y\in\mathcal{A}} \phi_{jy} = 1$.

(b) We define the transition probabilities between representative states x and y under control $u \in U(x)$ by

$$\hat{p}_{xy}(u) = \sum_{j=1}^{n} p_{xj}(u)\phi_{jy}. \tag{6.3}$$

(c) We define the expected transition costs at representative states x under control $u \in U(x)$ by

$$\hat{g}(x,u) = \sum_{j=1}^{n} p_{xj}(u)g(x,u,j). \tag{6.4}$$

The optimal costs of the representative states $y \in \mathcal{A}$ in the aggregate problem are denoted by r_y^*, and they define approximate costs for the original problem through the interpolation formula

$$\tilde{J}(j) = \sum_{y\in\mathcal{A}} \phi_{jy} r_y^*, \qquad j = 1,\ldots,n. \tag{6.5}$$

Aside from the selection of representative states, an important consideration is the choice of the aggregation probabilities. These probabilities express "similarity" or "proximity" of original to representative states (as in the case of the coarse grid Example 6.1.1), but in principle they can be arbitrary. Intuitively, ϕ_{jy} may be interpreted as some measure of "strength of relation" of j to y. The vectors $\{\phi_{jy} \mid j = 1,\ldots,n\}$ may also be viewed as basis functions for a linear cost function approximation via Eq. (6.5).

Hard Aggregation and Error Bound

A special case of interest, called *hard aggregation*, is when for every state j, we have $\phi_{jy} = 0$ for all representative states y, except a single one, denoted y_j, for which we have $\phi_{jy_j} = 1$. In this case, the one-step lookahead approximation

$$\tilde{J}(j) = \sum_{y\in\mathcal{A}} \phi_{jy} r_y^*, \qquad j = 1,\ldots,n,$$

is *piecewise constant*; it is constant and equal to r_y^* for all j in the set

$$S_y = \{j \mid \phi_{jy} = 1\}, \qquad y \in \mathcal{A},$$

called the *footprint* of representative state y; see Fig. 6.1.4. Moreover the footprints of all the representative states are disjoint and form a partition of the state space, i.e.,

$$\cup_{x\in\mathcal{A}} S_x = \{1,\ldots,n\}.$$

The footprint sets can be used to define a bound for the error $(J^* - \tilde{J})$. In particular, it can be shown that

$$\big|J^*(j) - \tilde{J}(j)\big| \le \frac{\epsilon}{1-\alpha}, \qquad j = 1,\ldots,n,$$

where

$$\epsilon = \max_{y\in\mathcal{A}} \max_{i,j\in S_y} \big|J^*(i) - J^*(j)\big|$$

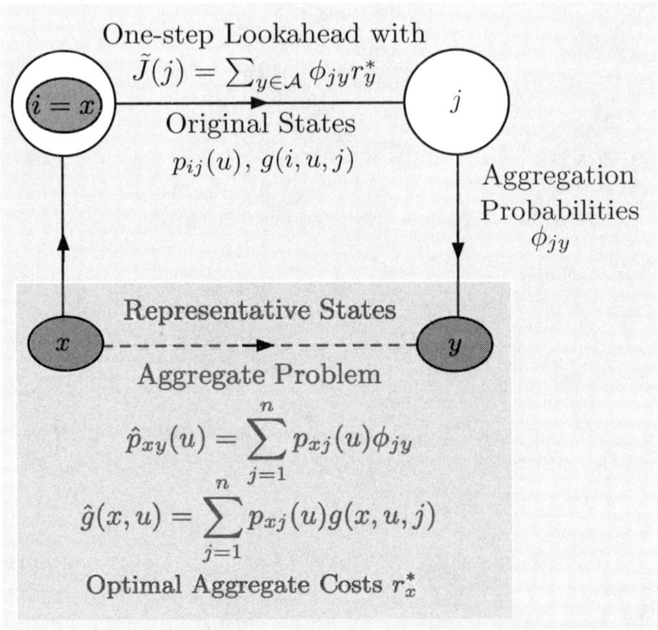

Figure 6.1.3 Illustration of the aggregate problem in the representative states framework. The transition probabilities $\hat{p}_{xy}(u)$ and transition costs $\hat{g}(x,u)$ are shown in the bottom part of the figure. Once the aggregate problem is solved (exactly) for its optimal costs r_y^*, we define approximate costs

$$\tilde{J}(j) = \sum_{y\in\mathcal{A}} \phi_{jy} r_y^*, \qquad j = 1,\ldots,n,$$

which are used for one-step lookahead approximation of the original problem.

is the *maximum variation of* J^* *within the footprint sets* S_y. We will show this fact in the context of an error bound result for the more general aggregation framework, which will be given in the next section (see Prop. 6.2.1). Note the primary intuition derived from this bound: *the error due to hard aggregation is small if* J^* *varies little within each* S_y.

For a special hard aggregation case of interest, consider the geometrical context of Example 6.1.1. There, aggregation probabilities are often based on a nearest neighbor approximation scheme, whereby each nonrepresentative state j takes the cost value of the "closest" representative state y, i.e.,

$$\phi_{jy_j} = 1 \qquad \text{if } y_j \text{ is the closest representative state to } j.$$

Then all states j for which a given representative state y is the closest to j (the footprint of y) are assigned equal approximate cost $\tilde{J}(j) = r_y^*$.

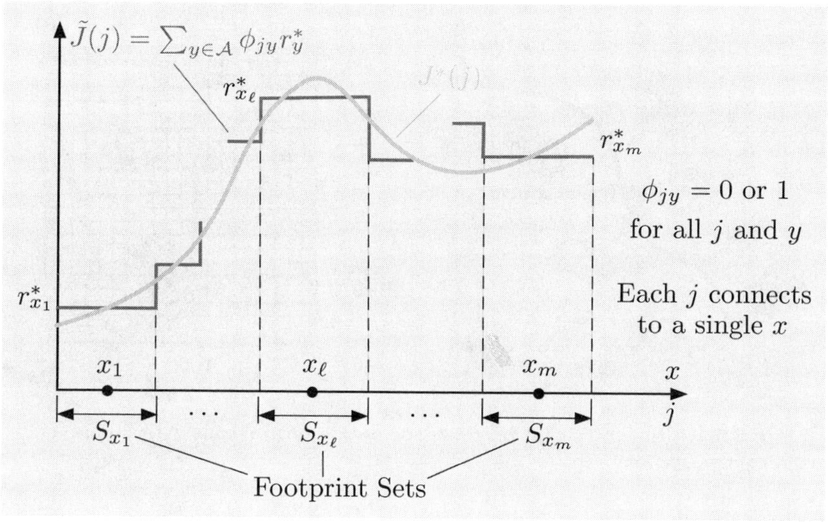

Figure 6.1.4 Illustration of the piecewise constant cost approximation

$$\tilde{J}(j) = \sum_{y\in\mathcal{A}} \phi_{jy} r_y^*, \qquad j = 1, \ldots, n,$$

in the hard aggregation case where we have $\phi_{jy} = 0$ for all representative states y, except a single one. Here $\tilde{J}$ is constant and equal to r_y^* for all j in the footprint set

$$S_y = \{j \mid \phi_{jy} = 1\}, \qquad y \in \mathcal{A}.$$

Methods for Solving the Aggregate Problem

The most straightforward way to solve the aggregate problem is to compute the aggregate problem transition probabilities $\hat{p}_{xy}(u)$ [cf. Eq. (6.3)] and transition costs $\hat{g}(x,u)$ [cf. Eq. (6.4)] by either an algebraic calculation or by simulation. The aggregate problem may then be solved by any one of the standard methods, such as VI, PI, or linear programming (cf. Chapter 4). This exact calculation is plausible if the number of representative states is relatively small.

Another possibility is to use a simulation-based VI or PI method. We postpone the discussion of these methods to Section 6.3, where we will consider them in the context of a more general aggregation framework.

An important observation is that if the original problem is deterministic and hard aggregation is used, the aggregate problem is also deterministic, and can be solved by shortest-path like methods. This is true for both discounted problems and for undiscounted shortest path-type problems. In the latter case, the termination state of the original problem must

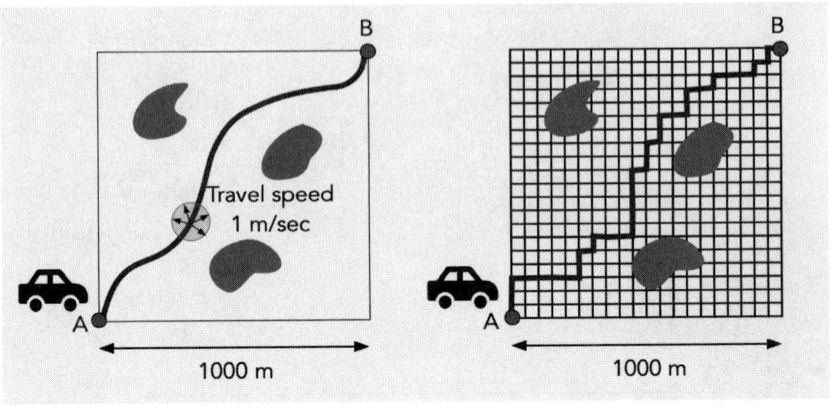

Figure 6.1.5 Illustration of discretization issues for problems with infinite state and control spaces.

be included as a representative state in the aggregate problem. However, if hard aggregation is not used, the aggregate problem will always be stochastic, because of the introduction of the aggregation probabilities. Of course, once the aggregate problem is solved and the lookahead approximation $\tilde{J}$ is obtained, a deterministic structure in the original problem can be exploited to facilitate the lookahead minimizations.

6.1.1 Continuous State and Control Space Discretization

Aggregation with representative states extends without difficulty to problems with a continuous state space, as long as the control space is finite. Then once the representative states and the aggregation probabilities have been defined, the corresponding aggregate problem is a finite state and control spaces discounted problem, which can be solved with the standard methods. The only potential difficulty arises when the disturbance space is also infinite, in which case the calculation of the transition probabilities and expected stage costs of the aggregate problem must be obtained by some form of integration process.

The case where both the state and the control spaces are continuous is somewhat more complicated, because both of these spaces must be discretized using representative state-control pairs, instead of just representative states. The following example illustrates what may happen if we use representative state discretization only.

Example 6.1.2 (Continuous Shortest Path Discretization)

Suppose that we want to find the fastest route for a car to travel between two points A and B located at the opposite ends of a square with side 1000 meters, while avoiding some known obstacles. We assume a constant car speed of 1 meter per second and that the car can drive in any direction; cf. Fig. 6.1.5.

Let us consider discretizing the space with a square grid (a set of representative states), and restrict the directions of motion to horizontal and vertical, so that at each stage the car moves from a grid point to one of the four closest grid points. Thus in the discretized version of the problem the car travels with a sequence of horizontal and vertical moves as indicated in the right side of Fig. 6.1.5. Is it possible to approximate the fastest route arbitrarily closely with the optimal solution of the discretized problem, assuming a sufficiently fine grid?

The answer is no! To see this note that in the discretized problem the optimal travel time is 2000 secs, regardless of how fine the discretization is. On the other hand, in the continuous space/nondiscretized problem the optimal travel time can be as little as $\sqrt{2} \cdot 1000$ secs (this corresponds to the favorable case where the straight line from A to B does not meet an obstacle).

The difficulty in the preceding example is that *the state space is discretized finely but the control space is not*. What is needed is to introduce a fine discretization of the control space as well, through some set of "representative controls." We can deal with this situation with a suitable form of discretized aggregate problem, which when solved provides an appropriate form of cost function approximation for use with one-step lookahead. The discretized problem is a stochastic infinite horizon problem, even if the original problem is deterministic. Further discussion of this approach is outside our scope in this book, and we refer to the sources cited at the end of the chapter. Under reasonable assumptions it is possible to show consistency, i.e., that the optimal cost function of the discretized problem converges to the optimal cost function of the original continuous spaces problem as the discretization of both the state and the control spaces becomes increasingly fine.

The type of difficulty illustrated in Example 6.1.2 does not arise if the state space is continuous but the control space is finite. In particular, this is true in partially observed finite spaces Markov decision problems (POMDP), which are defined over their belief space (the space of probability distributions over their states). We briefly discuss this case next.

6.1.2 Continuous State Space - POMDP Discretization

Let us consider any α-discounted DP problem, where the state space is a bounded convex subset B of a Euclidean space, such as the unit simplex, but the control space U is finite. We use b to denote the original problem states, to emphasize the connection with belief states in POMDP and to distinguish them from x, which we will use to denote representative states. Bellman's equation is $J = TJ$ with the Bellman operator T defined by

$$(TJ)(b) = \min_{u \in U} \mathop{E}_{w} \bigl\{ g(b,u,w) + \alpha J\bigl(f(b,u,w)\bigr) \bigr\}, \qquad b \in B.$$

We introduce a set of representative states $\{x_1, \ldots, x_m\} \subset B$. We assume that the convex hull of $\{x_1, \ldots, x_m\}$ is equal to B, so each state

$b \in B$ can be expressed as

$$b = \sum_{i=1}^{m} \phi_{bx_i} x_i,$$

where $\{\phi_{bx_i} \mid i = 1, \ldots, m\}$ is a probability distribution:

$$\phi_{bx_i} \geq 0, \; i = 1, \ldots, m, \quad \sum_{i=1}^{m} \phi_{bx_i} = 1, \qquad \text{for all } b \in B.$$

We view ϕ_{bx_i} as aggregation probabilities.

Consider the operator H that transforms a vector

$$r = (r_{x_1}, \ldots, r_{x_m})$$

into the vector Hr with components

$$(Hr)(x_1), \ldots, (Hr)(x_m)$$

defined by

$$(Hr)(x_i) = \min_{u \in U} \mathop{E}_{w} \left\{ g(x_i, u, w) + \alpha \sum_{j=1}^{m} \phi_{f(x_i,u,w)\, x_j} r_{x_j} \right\}, \qquad i = 1, \ldots, m,$$

where $\phi_{f(x_i,u,w)\, x_j}$ are the aggregation probabilities of the state $f(x_i, u, w)$. It can then be shown that H is a contraction mapping with respect to the maximum norm (we give a similar and more general result in the next section, and prove it in the appendix). Bellman's equation for an aggregate finite-state discounted DP problem whose states are $x_1, \ldots, x_m$ has the form

$$r_{x_i} = (Hr)(x_i), \qquad i = 1, \ldots, m,$$

and has a unique solution.

The transitions in this problem occur as follows: from state x_i under control u, we first move to $f(x_i, u, w)$ at cost $g(x_i, u, w)$, and then we move to a state x_j, $j = 1, \ldots, m$, according to the probabilities $\phi_{f(x_i,u,w)\, x_j}$. The optimal costs $r^*_{x_i}$, $i = 1, \ldots, m$, of this problem can often be obtained by standard VI and PI methods that may or may not use simulation. We may then approximate the optimal cost function of the original problem by

$$\tilde{J}(b) = \sum_{i=1}^{m} \phi_{bx_i} r^*_{x_i}, \qquad \text{for all } b \in B,$$

and reasonably expect that the optimal discretized solution converges to the optimal as the number of representative states increases.

In the case where B is the belief space of an α-discounted POMDP, the representative states/beliefs and the aggregation probabilities define an aggregate problem, which is a finite-state α-discounted problem with a perfect state information structure. This problem can be solved with the exact methods of Chapter 4 if either the aggregate transition probabilities and transition costs can be obtained analytically (in favorable cases) or if the number of representative states is small enough to allow their calculation by simulation. The aggregate problem can also be addressed with the approximate methods of Chapter 5, such as problem approximation/certainty equivalence approaches, or with a rollout method, which is well-suited for an on-line implementation.

6.2 AGGREGATION WITH REPRESENTATIVE FEATURES

We will now construct a more general aggregation framework for the infinite horizon n-state α-discounted problem, by using representative features in place of representative states. It is helpful to introduce this framework in a general form, and introduce features in the aggregate problem formulation later. To this end, we essentially replace the representative states x with *subsets* $I_x \subset \{1, \ldots, n\}$ of the original state space as follows.

General Aggregation Framework

We introduce a finite subset $\mathcal{A}$ of aggregate states, which we denote by symbols such as x and y. We define:

(a) A collection of disjoint subsets $I_x \subset \{1, \ldots, n\}$, $x \in \mathcal{A}$.

(b) A probability distribution over $\{1, \ldots, n\}$ for each $x \in \mathcal{A}$, denoted by $\{d_{xi} \mid i = 1, \ldots, n\}$, and referred to as the *disaggregation probabilities of* x. We require that the distribution corresponding to x is concentrated on the subset I_x:

$$d_{xi} = 0, \qquad \text{for all } i \notin I_x,\ x \in \mathcal{A}. \tag{6.6}$$

(c) For each original system state $j \in \{1, \ldots, n\}$, a probability distribution over $\mathcal{A}$, denoted by $\{\phi_{jy} \mid y \in \mathcal{A}\}$, and referred to as the *aggregation probabilities of* j. We require that

$$\phi_{jy} = 1, \qquad \text{for all } j \in I_y,\ y \in \mathcal{A}. \tag{6.7}$$

The aggregation and disaggregation probabilities specify a dynamic system involving both aggregate and original system states; cf. Fig. 6.2.1. In this system:

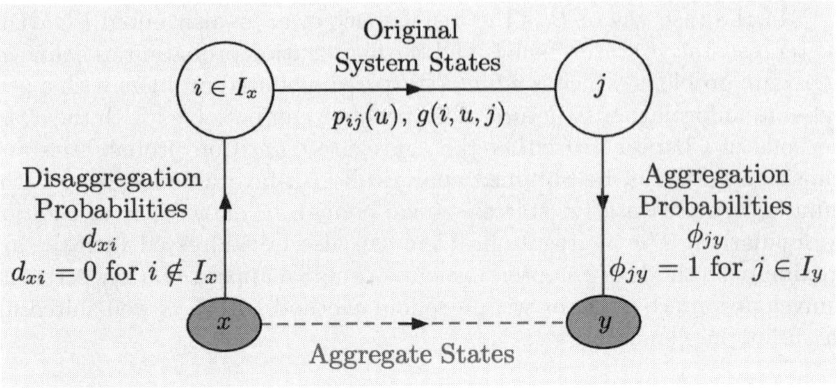

Figure 6.2.1 Illustration of the aggregate system, and the transition mechanism and the costs per stage of the aggregate problem.

(i) From aggregate state x, we generate an original system state $i \in I_x$ according to d_{xi}.

(ii) We generate transitions between original system states i and j according to $p_{ij}(u)$, with cost $g(i,u,j)$.

(iii) From original system state j, we generate aggregate state y according to ϕ_{jy}. [Note that in view of Eq. (6.7), all states j within a set I_y, are aggregated onto y with $\phi_{jy} = 1$.]

The optimal costs of the aggregate states $y \in \mathcal{A}$ in the aggregate problem are denoted by r_y^*, and they define approximate costs for the original problem through the interpolation formula

$$\tilde{J}(j) = \sum_{y \in \mathcal{A}} \phi_{jy} r_y^*, \qquad j = 1, \ldots, n. \tag{6.8}$$

The general aggregation framework is illustrated in Fig. 6.2.1. Note that if each set I_x consists of a single state, we obtain the representative states framework of the preceding section. In this case the disaggregation distribution $\{d_{xi} \mid i \in I_x\}$ is just the atomic distribution that assigns probability 1 to the unique state in I_x. Consistent with the special case of representative states, the disaggregation probability d_{xi} may be interpreted as a "measure of the relation of x and i."

The aggregate problem is a DP problem with an enlarged state space that consists of two copies of the original state space $\{1, \ldots, n\}$ plus the set of aggregate states $\mathcal{A}$. We introduce the corresponding optimal vectors $\tilde{J}_0$, $\tilde{J}_1$, and $r^* = \{r_x^* \mid x \in \mathcal{A}\}$ where:

r_x^* is the optimal cost-to-go from aggregate state x.

$\tilde{J}_0(i)$ is the optimal cost-to-go from original system state i that has just been generated from an aggregate state (left side of Fig. 6.2.1).

$\tilde{J}_1(j)$ is the optimal cost-to-go from original system state j that has just been generated from an original system state (right side of Fig. 6.2.1).

Note that because of the intermediate transitions to aggregate states, $\tilde{J}_0$ and $\tilde{J}_1$ are different.

These three vectors satisfy the following three Bellman equations:

$$r_x^* = \sum_{i \in I_x} d_{xi} \tilde{J}_0(i), \qquad x \in \mathcal{A}, \tag{6.9}$$

$$\tilde{J}_0(i) = \min_{u \in U(i)} \sum_{j=1}^{n} p_{ij}(u) \big(g(i,u,j) + \alpha \tilde{J}_1(j)\big), \qquad i = 1, \ldots, n, \tag{6.10}$$

$$\tilde{J}_1(j) = \sum_{y \in \mathcal{A}} \phi_{jy} r_y^*, \qquad j = 1, \ldots, n. \tag{6.11}$$

The objective is to solve for the optimal costs r_x^* of the aggregate states in order to obtain approximate costs for the original problem through the interpolation formula

$$\tilde{J}(j) = \sum_{y \in \mathcal{A}} \phi_{jy} r_y^*, \qquad j = 1, \ldots, n;$$

cf. Eq. (6.8).

By combining the three Bellman equations (6.9)-(6.11), we see that r^* satisfies

$$r_x^* = \sum_{i \in I_x} d_{xi} \min_{u \in U(i)} \sum_{j=1}^{n} p_{ij}(u) \left(g(i,u,j) + \alpha \sum_{y \in \mathcal{A}} \phi_{jy} \, r_y^* \right), \qquad x \in \mathcal{A}, \tag{6.12}$$

or equivalently $r^* = Hr^*$, where H is the operator that maps the vector r to the vector Hr with components

$$(Hr)(x) = \sum_{i \in I_x} d_{xi} \min_{u \in U(i)} \sum_{j=1}^{n} p_{ij}(u) \left(g(i,u,j) + \alpha \sum_{y \in \mathcal{A}} \phi_{jy} \, r_y \right), \qquad x \in \mathcal{A}. \tag{6.13}$$

It can be shown that H *is a contraction mapping with respect to the maximum norm*, and thus the composite Bellman equation (6.12) has r^*

as its unique solution. To see this, we note for any vectors r and r', we have

$$\begin{aligned}(Hr)(x) &= \sum_{i\in I_x} d_{xi} \min_{u\in U(i)} \sum_{j=1}^{n} p_{ij}(u)\left(g(i,u,j)+\alpha\sum_{y\in\mathcal{A}}\phi_{jy}\, r_y\right)\\ &\le \sum_{i\in I_x} d_{xi} \min_{u\in U(i)} \sum_{j=1}^{n} p_{ij}(u)\left(g(i,u,j)+\alpha\sum_{y\in\mathcal{A}}\phi_{jy}\left(r'_y+\|r-r'\|\right)\right)\\ &= (Hr')(x)+\alpha\|r-r'\|,\end{aligned}$$

where $\|\cdot\|$ is the maximum norm, and the equality follows from the definition of $(Hr')(x)$, and the fact that d_{xi}, $p_{ij}(u)$, and ϕ_{jy} are probabilities. Thus we have

$$(Hr)(x)-(Hr')(x)\le\alpha\|r-r'\|,\qquad x\in\mathcal{A}.$$

By reversing the roles of r and r', we also have

$$(Hr')(x)-(Hr)(x)\le\alpha\|r-r'\|,\qquad x\in\mathcal{A},$$

so that

$$\big|(Hr')(x)-(Hr)(x)\big|\le\alpha\|r-r'\|,\qquad x\in\mathcal{A}.$$

By taking the maximum over $x\in\mathcal{A}$, it follows that $\|Hr-Hr'\|\le\alpha\|r-r'\|$, and that H is a maximum norm contraction.

Note that the composite Bellman equation (6.12) has dimension equal to the number of aggregate states, which is potentially much smaller than n. To apply the aggregation framework of this section, we may solve exactly this equation for the optimal aggregate costs r^*_x, $x\in\mathcal{A}$, by simulation-based analogs of the VI and PI methods, and obtain a cost function approximation for the original problem through the interpolation formula (6.8). We will develop these methods later, but before doing so, we discuss various ways to formulate the aggregation framework, and in particular, how features can be used for this purpose.

6.2.1 Hard Aggregation and Error Bounds

Let us consider the special case of *hard aggregation*, where for every state j, we have $\phi_{jy}=0$ for all aggregate states y, except a single one, denoted y_j, for which we have $\phi_{jy_j}=1$. In this case, the one-step lookahead approximation

$$\tilde{J}(j)=\sum_{y\in\mathcal{A}}\phi_{jy}r^*_y,\qquad j=1,\ldots,n,$$

is piecewise constant; it is constant and equal to r^*_y for all j in the set

$$S_y=\{j\mid\phi_{jy}=1\},\qquad y\in\mathcal{A},\tag{6.14}$$

called the *footprint* of aggregate state y; see Fig. 6.1.4. Note that the footprints of all the aggregate states are disjoint and form a partition of the state space, i.e.,

$$\cup_{x\in\mathcal{A}} S_x = \{1,\ldots,n\}.$$

Note also that because of the requirement (6.7), we have $I_x \subset S_x$, and in particular if the sets I_x, $x \in \mathcal{A}$, form a partition of the state space, we must have $I_x = S_x$ for all $x \in \mathcal{A}$.

Similar to the representative states case of Section 6.1, in hard aggregation the cost function approximation

$$\tilde{J}(j) = \sum_{y\in\mathcal{A}} \phi_{jy} r_y^*, \qquad j = 1,\ldots,n,$$

is piecewise constant: it is constant on each of the footprint sets S_y, in view of the definition (6.14); cf. Fig. 6.1.4. Examples of this situation include cases where the aggregation probabilities are generated by a "nearest neighbor" scheme, where the cost $\tilde{J}_1(j)$ of a state $j \notin \cup_{y\in\mathcal{A}} S_y$ is taken to be equal to the cost of the "nearest" state within $\cup_{y\in\mathcal{A}} S_y$.

We can show the following error bound, due to Tsitsiklis and Van Roy [TsV96] (the proof is given in the appendix to this chapter).

Proposition 6.2.1: (Error Bound for Hard Aggregation) In the case of hard aggregation, we have

$$\big|J^*(j) - \tilde{J}(j)\big| \le \frac{\epsilon}{1-\alpha}, \qquad \text{for all } j \text{ such that } j \in S_y,\ y \in \mathcal{A},$$

where ϵ is the maximum variation of the optimal cost function J^* over the footprint sets S_y, $y \in \mathcal{A}$:

$$\epsilon = \max_{y\in\mathcal{A}} \max_{i,j\in S_y} \big|J^*(i) - J^*(j)\big|.$$

The meaning of the preceding proposition is that if the optimal cost function J^* varies by at most ϵ within each set S_y, the hard aggregation scheme yields a piecewise constant approximation to the optimal cost function that is within $\epsilon/(1-\alpha)$ of the optimal.

Aside from its intuitive nature and error bound properties, hard aggregation provides a connection with the projected equation approach for policy evaluation that we discussed in Section 5.5. In particular, it can be shown that for a given policy, the corresponding composite Bellman equation (6.12) for approximate evaluation of μ can be viewed as a projected equation, where a projection seminorm is used that is defined by the disaggregation probabilities. This was shown in the paper by Yu and Bertsekas [YuB12] (Section 5.5); see also the book [Ber12] (Exercise 6.10).

Selecting the Aggregate States

Generally, the method to select the aggregate states is an important issue, for which there is no mathematical theory at present. However, in practical problems, based on intuition and problem-specific knowledge, there are usually evident choices, which may be fine-tuned by experimentation. For example, suppose that the optimal cost function J^* is piecewise constant over a partition $\{S_y \mid y \in \mathcal{A}\}$ of the state space $\{1,\ldots,n\}$. By this we mean that for some vector

$$r^* = \{r_y^* \mid y \in \mathcal{A}\},$$

we have

$$J^*(j) = r_y^* \qquad \text{for all } j \in S_y,\ y \in \mathcal{A}.$$

Then from Prop. 6.2.1 it follows that the hard aggregation scheme is exact, so r_x^* are the optimal costs of the aggregate states x in the aggregate problem. This suggests that *in hard aggregation, the states in the footprint set S_y of aggregate state y should have roughly equal optimal cost*, consistently with the error bound of Prop. 6.2.1.

As an extension of the preceding argument, suppose that through some special insight into the problem's structure or some preliminary calculation, we know some features of the system's state that can "predict well" its optimal cost when combined through some approximation architecture, e.g., one that is linear. Then it seems reasonable to form the set aggregate states $\mathcal{A}$ of a hard aggregation scheme so that the sets I_y and S_y consist of states with "similar features" for every $y \in \mathcal{A}$. This is called *feature-based aggregation*, and was suggested in the book [BeT96], Section 3.1.2. The next section considers this possibility, and provides a way to introduce features and nonlinearities into the aggregation architecture, without compromising its other favorable aspects.

6.2.2 Aggregation Using Features

Let us consider the guideline for hard aggregation that we just discussed: *states i that belong to the same footprint set S_y should have nearly equal optimal costs*, i.e.,

$$\max_{i,j\in S_y} \big|J^*(i) - J^*(j)\big| \approx 0, \qquad \text{for all } y \in \mathcal{A}.$$

The question now is how to select the sets S_y according to this guideline.

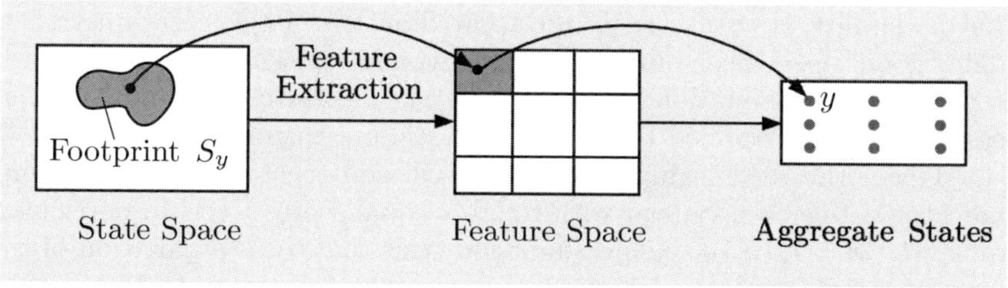

Figure 6.2.2 Feature-based hard aggregation using a partition of the space of features. Each aggregate state y has a footprint S_y that consists of states with "similar" features, i.e., states that map into the same subset of a partition in the space of features.

In this section we consider the possibility of using a *feature mapping*, i.e., a function F that maps a state i into an m-dimensional feature vector $F(i)$. In particular, suppose that F has the property that states i with nearly equal feature vector have nearly equal optimal cost $J^*(i)$. Then we can form the sets S_y by grouping together states with nearly equal feature vector. In particular, given F, we introduce a more or less regular partition of the feature space [the subset of $\Re^m$ that consists of all possible feature vectors $F(i)$]. The partition of the feature space induces a possibly irregular collection of subsets of the original state space. Each of these subsets is then used as the footprint of a distinct aggregate state; see Fig. 6.2.2.

Note that in the resulting aggregation scheme the number of aggregate states may become very large. On the other hand, there is a significant advantage over the linear feature-based architectures of Chapter 3, which assign a single weight to each feature: in feature-based hard aggregation we are assigning *a weight to each subset of the feature space partition* (possibly a weight to every possible feature value, in the extreme case where each feature value is viewed by itself as a distinct set of the partition). In effect we use aggregation to construct a *nonlinear* (piecewise constant) feature-based architecture, which may be much more powerful than the corresponding linear architecture.

The question now arises how to obtain a suitable feature vector when there is no obvious choice, based on problem-specific considerations. One possibility that we will consider later, in Section 6.4, is to obtain "good" features by using a neural network. In fact any method that automatically generates features from data may be used. We will first discuss a simpler possibility.

Using Scoring Functions

Suppose that we have obtained in some way a real-valued *scoring function* $V(i)$ of the state i, which serves as an index of undesirability of state i as a starting state (smaller values of V are assigned to more desirable states, consistent with the view of V as some form of "cost" function).

One possibility is to use as V an approximation of the cost function of some "good" (e.g., near-optimal) policy. Another possibility is to obtain V by problem approximation (cf. Section 2.3), i.e., as the cost function of some reasonable policy applied to an approximation of the original problem.

Given the scoring function V, we will construct a feature mapping that groups together states i with roughly equal scores $V(i)$. In particular, we let R_x, $x = 1, \ldots, q$, be q disjoint intervals that form a partition of the range of possible values of V [i.e., are such that for any state i, there is a unique interval R_x such that $V(i) \in R_x$]. We define a feature vector $F(i)$ of the state i according to

$$F(i) = x, \qquad \text{for all } i \text{ such that } V(i) \in R_x, \quad x = 1, \ldots, q. \tag{6.15}$$

This feature in turn defines a partition of the state space into the sets

$$I_x = \big\{i \mid F(i) = x\big\} = \big\{i \mid V(i) \in R_x\big\}, \qquad x = 1, \ldots, q. \tag{6.16}$$

Assuming that all the sets I_x are nonempty, we thus obtain a hard aggregation scheme, where the aggregate states are $x = 1, \ldots, q$, and the aggregation probabilities are defined by

$$\phi_{jx} = \begin{cases} 1 & \text{if } j \in I_x, \\ 0 & \text{otherwise,} \end{cases} \qquad j = 1, \ldots, n, \ x = 1, \ldots, q, \tag{6.17}$$

see Fig. 6.2.3. Note that the sets I_x coincide with the footprint sets S_x.

The following proposition illustrates the important role of the *quantization error*, defined as

$$\delta = \max_{x=1,\ldots,q} \max_{i,j \in I_x} \big|V(i) - V(j)\big|. \tag{6.18}$$

It represents the maximum error that can be incurred by approximating V within each set I_x with a single value from its range within the subset.

Proposition 6.2.2: Consider the hard aggregation scheme defined by a scoring function V as described above. Assume that the variations of J^* and V over the sets $I_1, \ldots, I_q$ are within a factor $\beta \geq 0$ of each other, i.e., that

$$\big|J^*(i) - J^*(j)\big| \leq \beta \big|V(i) - V(j)\big|, \qquad \text{for all } i, j \in I_x, \ x = 1, \ldots, q.$$

(a) We have

$$\big|J^*(i) - r_x^*\big| \leq \frac{\beta\delta}{1 - \alpha}, \qquad \text{for all } i \in I_x, \ x = 1, \ldots, q,$$

where δ is the quantization error of Eq. (6.18).

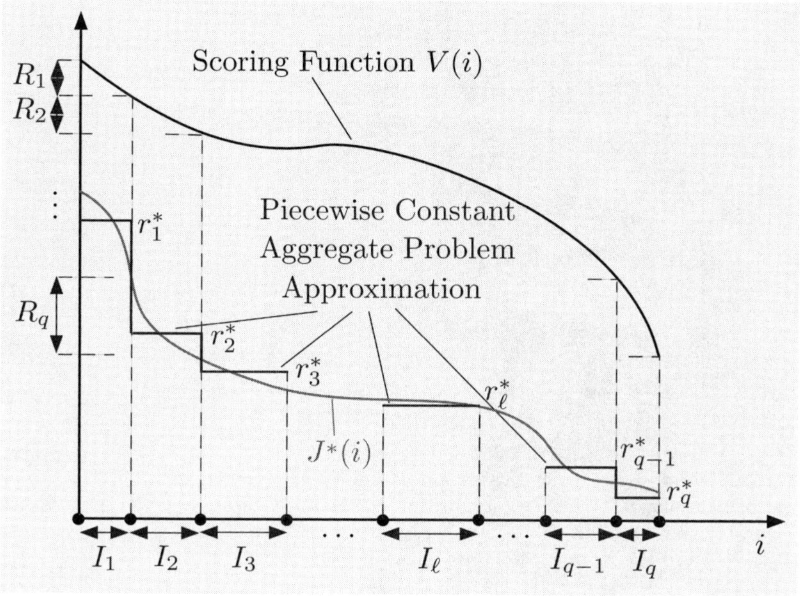

Figure 6.2.3. Hard aggregation scheme based on a single scoring function. We introduce q disjoint intervals $R_1, \ldots, R_q$ that form a partition of the set of possible values of V, and we define a feature vector $F(i)$ of the state i according to

$$F(i) = x, \qquad \text{for all } i \text{ such that } V(i) \in R_x,\ x = 1, \ldots, q.$$

This feature vector in turn defines a partition of the state space into the sets

$$I_x = \big\{i \mid F(i) = x\big\} = \big\{i \mid V(i) \in R_x\big\}, \qquad x = 1, \ldots, q.$$

The sets I_x coincide with the footprint sets S_x, and the solution of the aggregate problem yields a piecewise constant approximation of the optimal cost function of the original problem.

(b) Assume that there is no quantization error, i.e., V and J^* are constant within each set I_x. Then the aggregation scheme yields the optimal cost function J^* exactly, i.e.,

$$J^*(i) = r_x^*, \qquad \text{for all } i \in I_x,\ x = 1, \ldots, q.$$

The proposition essentially follows from Prop. 6.2.1. In particular, our assumptions imply that the maximum variation of J^* over the footprint sets I_x is bounded by $\epsilon = \beta\delta$, and the result of part (a) follows from Prop. 6.2.1. Part (b) is the special case of part (a) with $\delta = \epsilon = 0$.

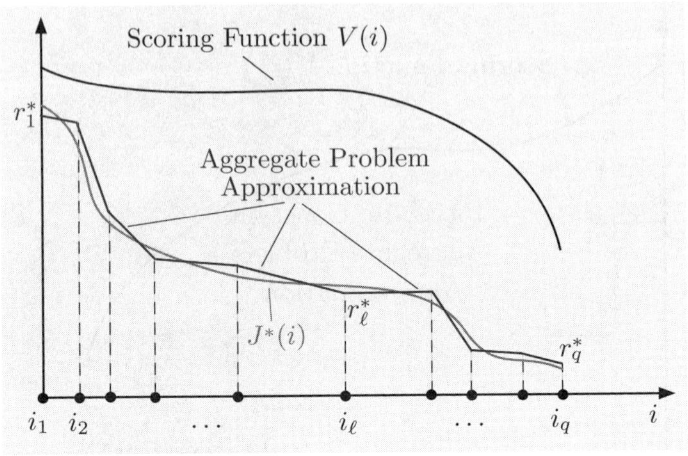

Figure 6.2.4. Schematic illustration of aggregation based on sampling states and using a scoring function V to form a representative set $i_1, \ldots, i_q$. A piecewise linear approximation of J^* is obtained by using the corresponding aggregate costs $r_1^*, \ldots, r_q^*$ and the aggregation probabilities.

Scoring Functions and Representative States

A related scoring function scheme may be based on representative states. Here the aggregate states and the disaggregation probabilities are obtained by forming a fairly large sample set of states $\{i_m \mid m = 1, \ldots, M\}$, by computing their corresponding scores $\big\{V(i_m) \mid m = 1, \ldots, M\big\}$, and by suitably dividing the range of these scores into disjoint intervals R_x, $x = 1, \ldots, q$, to form the aggregate states, similar to Eqs. (6.15)-(6.16). Simultaneously we obtain subsets of sampled states I_x to which we can assign positive disaggregation probabilities.

Figure 6.2.4 illustrates this idea for the case where each subset I_x consists of a single (representative) state. This is a form of "discretization" of the original state space based on the score values of the states. It is not a hard aggregation scheme, as the figure indicates. The role of the scoring function is to assist in forming a set of states that is small (to keep the aggregate DP problem computations manageable) but representative (to provide sufficient detail in the approximation of J^*, i.e., be dense in the parts of the state space where J^* varies a lot, and sparse in other parts).

Examples of scoring functions that may be useful in various settings are cost functions of nearly optimal policies, or approximations to such cost functions, provided for example by a neural network or other approximation schemes. Another example, suggested by an adaptive aggregation scheme proposed in the paper [BeC89], is to use as $V(i)$ the residual vector $(TJ)(i) - J(i)$, where J is some approximation to the optimal cost function J^*, or the residual vector $(T_\mu J)(i) - J(i)$, where J is some approximation to the cost function of a policy μ; see also the paper [KMP06]. Note that it is not essential that V approximates well J^* or J_μ. What is important is that states with similar values of J^* or J_μ also have similar values of V.

Scoring Function Scheme with a State Space Partition

Another useful scheme is based on a scoring function V, which is defined separately on each one of a collection of disjoint subsets $C_1,\ldots,C_m$ that form a partition of the state space. We define a feature vector $F(i)$ that depends not only on the value of $V(i)$ but also on the membership of i in the subsets of the partition.† In particular, for each $\theta = 1,\ldots,m$, let $R_{1\theta},\ldots,R_{q\theta}$ be q disjoint intervals that form a partition of the set of possible values of V over the set C_θ. We then define the feature vector

$$F(i) = (x,\theta), \qquad \text{for all } i \in C_\theta \text{ such that } V(i) \in R_{x\theta}.$$

This in turn defines a partition of the state space into the qm sets

$$I_{x\theta} = \big\{ i \in C_\theta \mid V(i) \in R_{x\theta} \big\}, \quad x = 1,\ldots,q,\ \theta = 1,\ldots,m,$$

which correspond to the resulting hard aggregation scheme.

Using Multiple Scoring Functions

The approach of forming features using a single scoring function can be extended to the case where we have a vector of scoring functions $V(i) = \big(V_1(i),\ldots,V_s(i)\big)$. Then we can partition the set of possible values of $V(i)$ into q disjoint subsets $R_1,\ldots,R_q$ of the s-dimensional space $\Re^s$, define a feature vector $F(i)$ according to

$$F(i) = x, \qquad \text{for all } i \text{ such that } V(i) \in R_x,\ x = 1,\ldots,q,$$

and proceed as in the case of a scalar scoring function, i.e., construct a hard aggregation scheme with disaggregation sets given by

$$I_x = \big\{ i \mid F(i) = x \big\} = \big\{ i \mid V(i) \in R_x \big\}, \qquad x = 1,\ldots,q.$$

One possibility to obtain multiple scoring functions is to start with a single fairly simple scoring function, obtain aggregate states as described earlier, solve the corresponding aggregate problem, and use the optimal cost function of that problem as an additional scoring function. A related and complementary possibility is to somehow construct multiple policies, evaluate each of these policies (perhaps approximately, using a neural network), and use the policy cost function evaluations as scoring functions.

---

† Partitioned architectures, based on problem-specific knowledge are commonly used, and they are often well-suited for distributed computation.

6.3 METHODS FOR SOLVING THE AGGREGATE PROBLEM

We will now focus on methods to compute the optimal costs r_x^* of the aggregate states $x \in \mathcal{A}$. This is the unique solution r^* of the composite Bellman Eq. (6.12). We first note that since r^* and the cost functions $\tilde{J}_0$ and $\tilde{J}_1$, form the solution of the Bellman equations (6.9)-(6.11), they can all be computed in principle with the standard PI and VI methods.

In this section, we will discuss specialized versions of PI and VI that compute just r^* (which has relatively low dimension), but not $\tilde{J}_0$ and $\tilde{J}_1$ (which may have large dimension). These methods are based on simulation as they involve the aggregate problem, which is stochastic because of the disaggregation and aggregation probabilities, even if the original problem is deterministic.

6.3.1 Simulation-Based Policy Iteration

One possible way to compute r^* is a PI-like algorithm, which generates sequences of policies $\{\mu^k\}$ for the original problem and vectors $\{r^k\}$ that converge to an optimal policy and r^*, respectively. The algorithm does not compute intermediate estimates of the high-dimensional vectors $\tilde{J}_0$ and $\tilde{J}_1$.

It starts with a stationary policy μ^0 for the original problem, and given μ^k, it performs a policy evaluation step by finding the unique fixed point r^k of the contraction mapping H_{μ^k} that maps the vector r to the vector $H_{\mu^k} r$ with components

$$(H_{\mu^k} r)(x) = \sum_{i=1}^{n} d_{xi} \sum_{j=1}^{n} p_{ij}\big(\mu^k(i)\big) \left(g\big(i, \mu^k(i), j\big) + \alpha \sum_{y \in \mathcal{A}} \phi_{jy}\, r_y \right), \quad x \in \mathcal{A},$$

cf. Eq. (6.13). Thus

$$r_x^k = \sum_{i=1}^{n} d_{xi} \sum_{j=1}^{n} p_{ij}\big(\mu^k(i)\big) \left(g\big(i, \mu^k(i), j\big) + \alpha \sum_{y \in \mathcal{A}} \phi_{jy}\, r_y^k \right), \quad x \in \mathcal{A}. \tag{6.19}$$

Following the policy evaluation step, the algorithm generates μ^{k+1} by

$$\mu^{k+1}(i) = \arg \min_{u \in U(i)} \sum_{j=1}^{n} p_{ij}(u) \left(g(i, u, j) + \alpha \sum_{y \in \mathcal{A}} \phi_{jy} r_y^k \right), \quad i = 1, \ldots, n; \tag{6.20}$$

this is the policy improvement step. In the preceding minimization we use one step lookahead, but a multistep lookahead or Monte Carlo tree search can also be used. Several other PI ideas that we have discussed in Chapters 4 and 5 also apply, e.g., optimistic PI schemes.

The following proposition shows that this PI algorithm (6.19)-(6.20) converges finitely to the unique solution of the composite Bellman Eq. (6.12) (the proof is given in the appendix).

Proposition 6.3.1: (PI Convergence) Let μ^0 be any policy and let $\{\mu^k, r^k\}$ be a sequence generated by the PI algorithm (6.19)-(6.20). Then the sequence $\{r^k\}$ is monotonically nonincreasing (i.e., we have $r_x^k \geq r_x^{k+1}$ for all x and k) and there exists an index $\bar{k}$ such that $r^{\bar{k}}$ is equal to r^*, the unique solution of the composite Bellman Eq. (6.12).

Let us now discuss how the preceding PI algorithm can be implemented by simulation. For notational convenience, let us introduce the matrices D and Φ whose rows are the disaggregation and aggregation distributions, respectively (i.e., row x of D has elements d_{xi}, $i = 1, \ldots, n$, while row j of Φ has elements ϕ_{jy}, $y \in \mathcal{A}$). Then the policy evaluation step (6.19) finds $r^k = \{r_x^k \mid x \in \mathcal{A}\}$ satisfying

$$r^k = H_{\mu^k} r^k = DT_{\mu^k}\Phi r^k = D\big(g_{\mu^k} + \alpha P_{\mu^k}\Phi r^k\big), \tag{6.21}$$

where P_{μ^k} is the transition probability matrix corresponding to μ^k, g_{μ^k} is the expected cost vector of μ^k, i.e., the vector whose ith component is

$$\sum_{j=1}^{n} p_{ij}\big(\mu^k(i)\big)g\big(i, \mu^k(i), j\big), \qquad i = 1, \ldots, n.$$

To avoid the n-dimensional calculations of the policy evaluation step (6.21), one may use simulation. In particular, the policy evaluation equation, $r = H_\mu r$, is linear of the form

$$r = Dg_\mu + \alpha DP_\mu \Phi r, \tag{6.22}$$

[cf. Eq. (6.21)]. Let us write this equation as $Cr = b$, where

$$C = I - \alpha DP_\mu\Phi, \qquad b = Dg_\mu,$$

and note that this is Bellman's equation for a policy with cost per stage vector equal to Dg_μ and transition probability matrix equal to $DP_\mu\Phi$. This is the transition matrix under policy μ for the Markov chain whose states are the aggregate states. The solution of the policy evaluation Eq. (6.22) is the cost vector corresponding to this Markov chain, and can be found by using simulation.

In particular, we may use simulation to approximate C and b, and then solve the system $Cr = b$ approximately. To this end, we obtain a sequence of sample transitions $\big\{(i_1, j_1), (i_2, j_2), \ldots\big\}$ by first generating a sequence of states $\{i_1, i_2, \ldots\}$ according to some distribution $\{\xi_i \mid i = 1, \ldots, n\}$ (with $\xi_i > 0$ for all i), and then generating for each $m \geq 1$ a sample transition (i_m, j_m) according to the distribution $\big\{p_{i_m j}\big(\mu^k(i)\big) \mid j = 1, \ldots, n\big\}$. Given the first M samples, we form the matrix $\widehat{C}_M$ and vector $\hat{b}_M$ given by

$$\widehat{C}_M = I - \frac{\alpha}{M} \sum_{m=1}^{M} \frac{1}{\xi_{i_m}} d(i_m)\phi(j_m)', \tag{6.23}$$

$$\hat{b}_M = \frac{1}{M} \sum_{m=1}^{M} \frac{1}{\xi_{i_m}} d(i_m) g\big(i_m, \mu(i_m), j_m\big), \tag{6.24}$$

where $d(i)$ is the ith column of D and $\phi(j)'$ is the jth row of Φ. We can then show that $\widehat{C}_M \to C$ and $\hat{b}_M \to d$ by using law of large numbers arguments, i.e., writing

$$C = I - \alpha \sum_{i=1}^{n} \sum_{j=1}^{n} p_{ij}\big(\mu(i)\big) d(i)\phi(j)',$$

$$b = \sum_{i=1}^{n} \sum_{j=1}^{n} p_{ij}\big(\mu(i)\big) d(i) g\big(i, \mu(i), j\big),$$

multiplying and dividing $p_{ij}\big(\mu(i)\big)$ by ξ_i in order to properly view these expressions as expected values, and using the relation

$$\lim_{M \to \infty} \frac{n_{ij}(M)}{M} = \xi_i \, p_{ij}\big(\mu(i)\big),$$

where $n_{ij}(M)$ is the number of occurrences of the i to j transition from time $m = 1$ to $m = M$.

Assuming $\widehat{C}_M$ is invertible, the corresponding estimates

$$\hat{r}_M = \widehat{C}_M^{-1} \hat{b}_M$$

converge to the unique solution of the policy evaluation Eq. (6.22) as $M \to \infty$, and provide the estimates $\Phi \hat{r}_M$ of the cost vector J_μ of μ:

$$\tilde{J}_\mu = \Phi \hat{r}_M.$$

This is the aggregation counterpart of the LSTD(0) method discussed in Section 5.5. One may also use an iterative simulation-based LSPE(0)-type method or a TD(0)-type method to solve the equation $Cr = b$; see Section 5.5 and [Ber12].

Note that instead of using the probabilities ξ_i to sample directly original system states, we may alternatively sample the aggregate states according to some distribution $\{\zeta_x \mid x \in \mathcal{A}\}$, generate a sequence of aggregate states $\{x_1, x_2, \ldots\}$, and then generate a state sequence $\{i_1, i_2, \ldots\}$ using the disaggregation probabilities, and corresponding transition sequence

$\{(i_1, j_1), (i_2, j_2), \ldots\}$. In this case Eqs. (6.23) and (6.24) should be modified as follows:

$$\widehat{C}_M = I - \frac{\alpha}{M} \sum_{m=1}^{M} \frac{1}{\zeta_{x_m} d_{x_m i_m}} d(i_m)\phi(j_m)',$$

$$\hat{b}_M = \frac{1}{M} \sum_{m=1}^{M} \frac{1}{\zeta_{x_m} d_{x_m i_m}} d(i_m) g\big(i_m, \mu(i_m), j_m\big).$$

The main difficulty with the policy improvement step at a given state i is the need to compute the expected value in the Q-factor expression

$$\sum_{j=1}^{n} p_{ij}(u) \left(g(i, u, j) + \alpha \sum_{y \in \mathcal{A}} \phi_{jy} r_y^k \right)$$

that is minimized over $u \in U(i)$. If the transition probabilities $p_{ij}(u)$ are available and the number of successor states [the states j with $p_{ij}(u) > 0$] is small, this expected value may be easily calculated (an important case where this is so is when the system is deterministic).

6.3.2 Simulation-Based Value Iteration

An exact VI algorithm for obtaining r^* is the fixed point iteration

$$r^{k+1} = Hr^k,$$

starting from some initial guess r^0, where H is the contraction mapping of Eq. (6.13). A stochastic approximation-type algorithm based on this fixed point iteration generates a sequence of aggregate states $\{x_0, x_1, \ldots\}$ by some probabilistic mechanism, which ensures that all aggregate states are adequately represented in the sequence. Given r^k and x_k, it independently generates an original system state i_k according to the probabilities $d_{x_k i}$, and updates the component r_{x_k} according to

$$r_{x_k}^{k+1} = (1 - \gamma^k) r_{x_k}^k + \gamma^k \min_{u \in U(i)} \sum_{j=1}^{n} p_{i_k j}(u) \left(g(i_k, u, j) + \alpha \sum_{y \in \mathcal{A}} \phi_{jy} r_y^k \right),$$

where γ^k is a diminishing positive stepsize, and leaves all the other components unchanged:

$$r_x^{k+1} = r_x^k, \qquad \text{if } x \neq x_k.$$

This algorithm can be viewed as an asynchronous stochastic approximation version of VI. The stepsize γ^k should be diminishing (typically at the rate of $1/k$), and its justification and convergence mechanism are very similar to the ones for the Q-learning algorithm. We refer to the paper by Tsitsiklis and Van Roy [TsV96] for further discussion and analysis (see also [BeT96], Section 6.7).

6.4 FEATURE-BASED AGGREGATION WITH A NEURAL NETWORK

We noted in Chapter 3 that neural networks can be used to construct features at the output of the last nonlinear layer. The neural network training process also yields linear weighting parameters for the feature vector $F(i)$ at the output of the last layer, thus obtaining an approximation $\hat{J}_\mu\big(F(i)\big)$ to the cost function of a given policy μ. Thus given the current policy μ, the typical PI produces the new policy $\hat{\mu}$ using an approximate policy improvement operation or a multistep variant, as illustrated in Fig. 6.4.1.

A similar PI scheme can be constructed based on feature-based aggregation with features supplied by the same neural network; see Fig. 6.4.2. The main idea is to replace the (approximate) policy improvement operation with the solution of an aggregate problem, which provides the (approximately) improved policy $\hat{\mu}$.

This is a more complicated policy improvement operation, but computes the new policy $\hat{\mu}$ based on a more accurate cost function approximation: one that is a nonlinear function of the features rather than linear. Moreover, $\hat{\mu}$ *not only aspires to be an improved policy relative to* μ, *but also to be an optimal policy based on the aggregate problem*, an approximation itself of the original DP problem. In particular, suppose that the neural network approximates J_μ perfectly. Then the scheme of Fig. 6.4.1 will replicate a single step of the PI algorithm starting from μ, while the aggregation scheme of Fig. 6.4.2, with sufficient number of aggregate states, will produce a policy that is arbitrarily close to optimal.

Let us now explain each of the steps of the aggregation-based PI procedure of Fig. 6.4.2, starting with the current policy μ.

(a) *Feature mapping construction*: We train the neural network using a training set of state-cost pairs that are generated using the current policy μ. This provides a feature vector $F(i)$ as described in Section 3.3.

(b) *Sampling to obtain the set of aggregate states x and the sets I_x*: We sample the state space, generating a subset of states $I \subset \{1, \ldots, n\}$. We partition this set into a collection of subsets I_x, $x = 1, \ldots, q$, with "similar" feature vector. We then consider the aggregation framework with $x = 1, \ldots, q$ as the aggregate states, and the corresponding aggregate problem. The sampling to obtain the set of states I may be combined with exploration to ensure that a sufficiently representative set of states is included.

(c) *Aggregate problem solution*: The aggregate DP problem is solved by using a simulation-based method to yield (perhaps approximately) the aggregate state optimal costs r_y^*, $y = 1, \ldots, q$ (cf. Section 6.3).

(d) *Definition of the improved policy*: The "improved" policy is simply the optimal policy of the aggregate problem (or an approximation

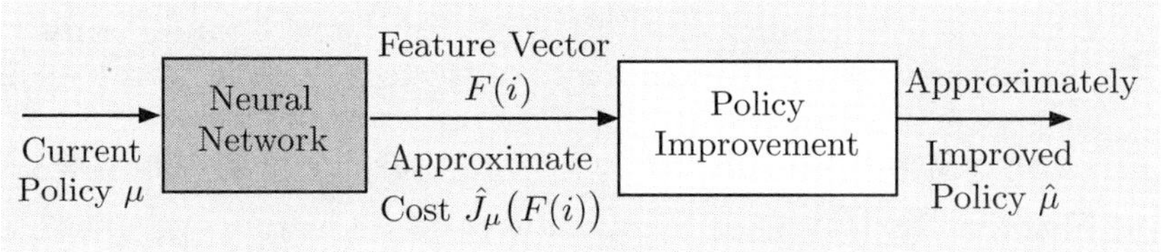

Figure 6.4.1. Schematic illustration of PI using a neural network-based cost approximation. Starting with a training set of state-cost pairs generated using the current policy μ, the neural network yields a set of features and an approximate cost evaluation $\hat{J}_\mu$ using a linear combination of the features. This is followed by policy improvement using $\hat{J}_\mu$ to generate the new policy $\hat{\mu}$.

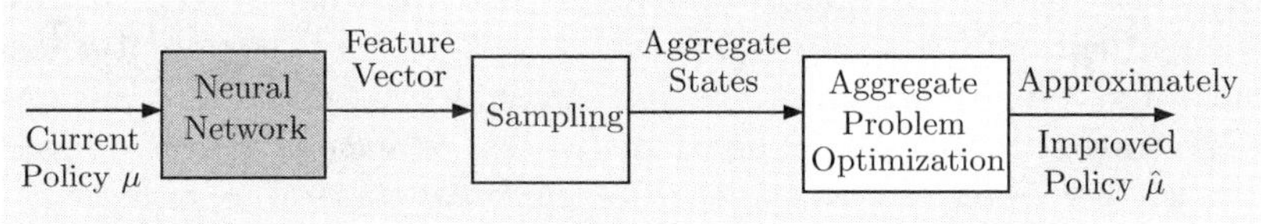

Figure 6.4.2. Illustration of PI using feature-based aggregation with features supplied by a neural network. Starting with a training set of state-cost pairs generated using the current policy μ, the neural network yields a set of features, which are used to construct a feature-based aggregation framework. The optimal policy of the corresponding aggregate problem is used as the new policy $\hat{\mu}$.

thereof, obtained for example after a few iterations of approximate simulation-based PI). This policy is defined implicitly by the aggregate costs r_y^*, $y = 1, \ldots, q$, using the one-step lookahead minimization

$$\hat{\mu}(i) \in \arg \min_{u \in U(i)} \sum_{j=1}^{n} p_{ij}(u) \left(g(i,u,j) + \alpha \sum_{y=1}^{q} \phi_{jy} \, r_y^* \right), \qquad i = 1, \ldots, n,$$

[cf. Eq. (6.10)] or a multistep lookahead variant. Alternatively, the "improved" policy can be implemented in model-free fashion using a Q-factor architecture.

Let us also note that there are several options for implementing the algorithmic ideas of this section.

(1) The neural network-based feature construction process may be performed any number of times, each time followed by an aggregate problem solution that constructs a new policy, which is then used to generate new training data for the neural network. Alternatively, the neural network training and feature construction process may be done only once, followed by the solution of the corresponding feature-based aggregate problem.

(2) Several neural network-based PI cycles may be performed, a subset of the features thus generated may be selected, and the corresponding aggregate problem is solved just once, as a way of improving the final policy generated by the neural network learning process.

(3) Following each cycle of neural network-based feature evaluation, the generated features may be supplemented with additional problem-specific handcrafted features, and/or features from previous cycles.

Finally, let us mention a potential weakness of using the features obtained at the output of the last nonlinear layer of the neural network in the context of aggregation: the sheer number of these features may be so large that the resulting number of aggregate states may become excessive. To address this situation one may consider pruning some of the features, or reducing their number using some form of regression, at the potential loss of some approximation accuracy. In this connection let us note an advantage of deep (rather than shallow) neural networks: *because with each additional layer, the generated features tend to be more complex, their number at the output of the final nonlinear layer of the network can be made smaller as the number of layers increases*. An extreme case is to use the cost function approximation obtained at the output of the neural network as a single feature/scoring function, in the spirit of Section 6.2.2.

6.5 BIASED AGGREGATION

In this section we will introduce an extension of the aggregation framework of Section 6.2. This extension involves a vector

$$V = \big(V(1), \ldots, V(n)\big)$$

called the *bias vector* or *bias function*, which affects the cost structure of the aggregate problem, and biases the values of its optimal cost function towards their correct levels. When $V = 0$, we will obtain the aggregation scheme of Section 6.2. When $V \neq 0$, we will obtain a different aggregation scheme, which yields an approximation to J^* that is equal to V plus a local correction; see Fig. 6.5.1. In this case the aggregate DP problem aims to provide a correction/improvement to V, which may itself be a reasonably good estimate of J^*.

An obvious context where biased aggregation can be used is to improve on an approximation to J^* obtained by using a different method, such as for example by neural network-based approximate PI, by rollout, or by problem approximation. Generally, we may speculate that if V captures a fair amount of the nonlinearity of J^*, we may reduce the number of aggregate states needed for adequate performance.

Let us now formulate the aggregate problem in biased aggregation. It is a discounted infinite horizon problem that is similar to the (unbiased)

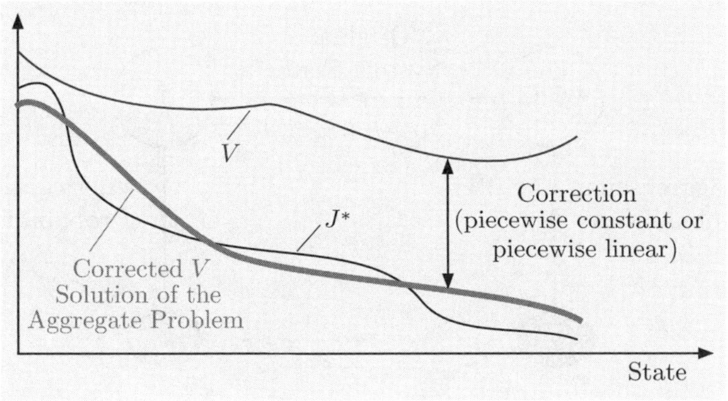

Figure 6.5.1 Schematic illustration of biased aggregation. It provides an approximation to J^* that is equal to the bias function V plus a local correction. When $V = 0$, we obtain the classical aggregation framework.

aggregate problem of Section 6.2. It involves three sets of states: two copies of the original state space, denoted I_0 and I_1, as well as a finite set $\mathcal{A}$ of aggregate states, as depicted in Fig. 6.5.2. The state transitions in the aggregate problem go from a state in $\mathcal{A}$ to a state in I_0, according to disaggregation probabilities, then to a state in I_1, and then back to a state in $\mathcal{A}$, according to aggregation probabilities, and the process is repeated. At state $i \in I_0$ we must choose a control $u \in U(i)$, and then transition to a state $j \in I_1$ at a cost $g(i, u, j)$ according to the original system transition probabilities $p_{ij}(u)$.

The salient new characteristic of the biased aggregation scheme is a (possibly nonzero) cost $-V(i)$ for transition from any aggregate state to a state $i \in I_0$, and of a cost $V(j)$ from a state $j \in I_1$ to any aggregate state; cf. Fig. 6.5.2. The function V is the bias function, and we will argue that V *should be chosen as close as possible to* J^*. Moreover, for practical purposes its values at various states should be easily computable.

A key insight is that *biased aggregation can be viewed as unbiased aggregation applied to a modified DP problem*, which is equivalent to the original DP problem in the sense that it has the same optimal policies. The modified DP problem is obtained from the original by changing its cost per stage from $g(i, u, j)$ to

$$g(i, u, j) - V(i) + \alpha V(j), \qquad i, j = 1, \ldots, n, \ u \in U(i). \tag{6.25}$$

In particular, by comparing Figs. 6.2.1 and 6.5.2 it can be seen that unbiased aggregation applied to the modified DP problem gives the same state-control trajectories as biased aggregation applied to the original DP problem, while the incurred transition costs (from aggregate state to aggregate state) are equal.

Moreover, there is a close connection between the optimal cost functions of the modified DP problem with cost per stage given by Eq. (6.25),

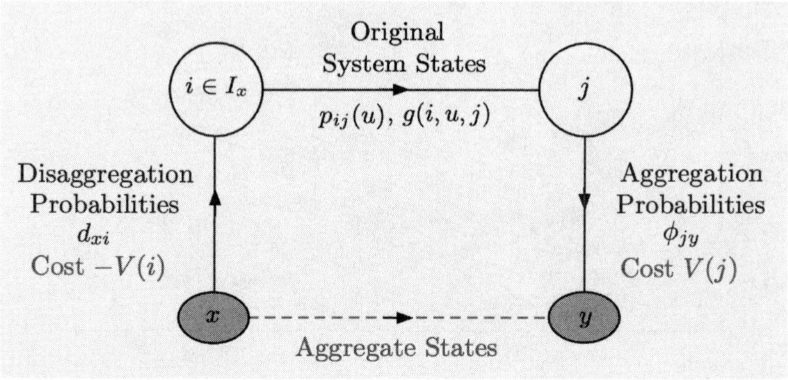

Figure 6.5.2 Illustration of the transition mechanism and the costs per stage of the aggregate problem in biased aggregation. When the bias function V is identically zero, we obtain the aggregation framework of Section 6.2.

and the original DP problem. In particular, the optimal cost function of the modified problem, call it $\tilde{J}$, satisfies the corresponding Bellman equation:

$$\tilde{J}(i) = \min_{u \in U(i)} \sum_{j=1}^{n} p_{ij}(u)\big(g(i,u,j) - V(i) + \alpha V(j) + \alpha \tilde{J}(j)\big), \quad i = 1, \ldots, n,$$

or equivalently

$$\tilde{J}(i) + V(i) = \min_{u \in U(i)} \sum_{j=1}^{n} p_{ij}(u)\Big(g(i,u,j) + \alpha\big(\tilde{J}(j) + V(j)\big)\Big), \quad i = 1, \ldots, n.$$

By comparing this equation with the Bellman equation for the original problem, we see that the optimal cost functions of the modified and the original problems are related by

$$J^*(i) = \tilde{J}(i) + V(i), \qquad i = 1, \ldots, n,$$

and that the two problems have the same optimal policies. This of course assumes that the original and modified problems are solved exactly. If instead they are solved approximately using aggregation or another approximation architecture, such as a neural network, the policies obtained may be substantially different. In particular, the choice of V and the approximation architecture may affect substantially the quality of suboptimal policies obtained. This is similar to the related idea of cost shaping that we discussed in Chapter 4. An interesting choice is to use as V an approximation to the cost function of a suitable policy; see [Ber18c].

To summarize, any unbiased aggregation scheme and algorithm, when applied to the modified DP problem with cost per stage given by Eq. (6.25), yields a biased aggregation scheme and algorithm for the original DP problem. Thus, we can straightforwardly transfer results, algorithms, and intuition from our earlier unbiased aggregation analysis to the biased aggregation framework, by applying them to the unbiased aggregation framework that corresponds to the modified stage cost (6.25). Moreover, we may use simulation-based algorithms for policy evaluation, policy improvement, and Q-learning for the aggregate problem (cf. Section 6.3), with the only requirement that the value $V(i)$ for any state i is available when needed.

6.6 NOTES AND SOURCES

Section 6.1: The aggregation approach has a long history in scientific computation and operations research (see for example Bean, Birge, and Smith [BBS87], Chatelin and Miranker [ChM82], Douglas and Douglas [DoD93], and Rogers et al. [RPW91]). It was introduced in the simulation-based approximate DP context, mostly in the form of VI; see Singh, Jaakkola, and Jordan [SJJ95], Gordon [Gor95], and Tsitsiklis and Van Roy [TsV96]. It was further discussed in the neuro-dynamic programming book [BeT96], Sections 3.1.2 and 6.7. More recently, aggregation was considered in RL contexts involving the notion of "options" by Ciosek and Silver [CiS15], and the notion of "bottleneck simulator" by Serban et. al. [SSP18].

For a discussion of discretization issues of optimal control problems involving continuous state and control spaces (cf. Section 6.1.1) we refer to the papers by Gonzalez and Rofman [GoR85], and Falcone [Fal87], and to the monograph by Kushner and Dupuis [KuD92] for a mathematically oriented analysis. For recent work that relates to aggregation and continuous spaces optimal control, see Zhong and Todorov [ZhT11]. An important special case of a deterministic optimal control problem is the continuous-time/continuous-space analog of a shortest path problem. This problem involves an origin and a destination, and aims to find a path from origin to destination, such that an integral of a cost per unit distance along the path is minimized; Example 6.1.2 is a special case. The terminal time is free and subject to optimization. By applying time discretization and then space discretization through aggregation with representative states, we obtain a special type of stochastic shortest path problem. Once this problem is solved, it yields an approximation to the original deterministic problem.

We note that for discrete-time continuous spaces optimal control problems, a finitely terminating adaptation of the Dijkstra shortest path algorithm has been developed by Tsitsiklis [Tsi95] (this is unexpected because the discretized shortest path problem is stochastic rather than deterministic). This Dijkstra-like algorithm and other related methods, together with many applications, are given by Sethian [Set99a], [Set99b], and by Helmsen et al. [HPC96]. Also, efficient analogs of label correcting algorithms for the same problem were developed by Bertsekas, Guerriero, and Musmanno [BGM95], and Polymenakos, Bertsekas, and Tsitsiklis [PBT98]. For more recent work, see Vladimirsky [Vla08], Chacon and Vladimirsky [ChV12], [ChV13], [ChV15], and Clawson [Cla17].

The material on POMDP discretization (cf. Section 6.1.2) is inspired by the paper by Yu and Bertsekas [YuB04]. This reference provides a scheme for the average cost case where the Bellman equation need not have a solution, but the scheme nevertheless provides lower bounds to optimal average cost functions which are difficult to compute. Here we focus on the simpler discounted case.

Our discussion has assumed that the set of aggregate states remains constant in the process of cost function approximation. An alternative is to modify this set adaptively, based on the results of interim computations; related ideas have been discussed by Bertsekas and Castanon [BeC89], and Singh, Jaakkola, and Jordan [SJJ95]. Multigrid methods, which have been applied to DP by Chow and Tsitsiklis [ChT91], involve ideas of multiple levels of aggregation detail, with the current solution at a given level used as a starting point for refinement of the solution at other levels.

Section 6.2: The aggregation framework with representative features was discussed in the author's book [Ber12], and was further developed in the author's survey paper [Ber18b], which provides an expanded view of the methodology. The book [Ber12] discusses also soft aggregation, which is similar to hard aggregation but allows the footprint sets to partially overlap. The error bounds of Section 6.2.1 are due to Tsitsiklis and Van Roy [TsV96]. The paper by Yu and Bertsekas [YuB12] describes a connection between feature-based aggregation and approximation using projected equations (see Section 5.5 of that paper).

Section 6.3: The book [Ber12] and the survey [Ber18b] contain a broad discussion of computational methods for aggregation, including variants and extensions of VI and PI that we have not covered here. The convergence of the VI method of Section 6.3.2 was established by Tsitsiklis and Van Roy [TsV96].

Section 6.4: The connections of PI methods for feature-based aggregation, using neural networks to extract features have been discussed in the author's paper [Ber18b].

Section 6.5: The material on biased aggregation was given recently in the author's paper [Ber18c], which contains further discussion, connections with rollout algorithms, and additional methods.

In this chapter we have focused on aggregation that aims at reduction of the size of the state space, including the classical case of continuous space discretization. A complementary framework is *temporal aggregation*, i.e., grouping multiple time periods (or continuous time intervals) and the corresponding state transitions and controls into single units. A simple example is time discretization of a continuous-time optimal control problem. Another example is semi-Markov problems, which are converted to

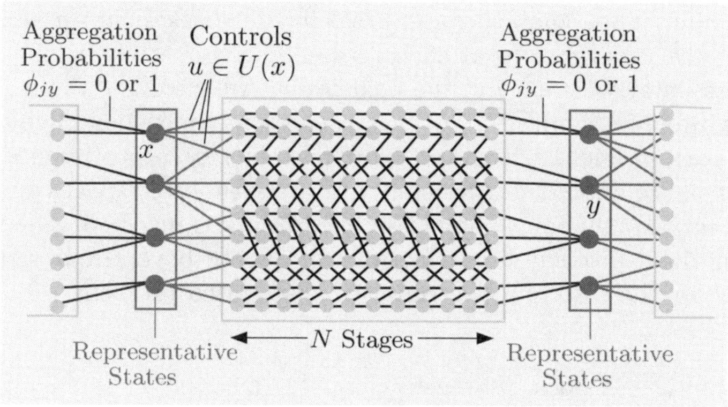

Figure 6.6.1 Schematic illustration of hard aggregation in space and time for a deterministic shortest path. The time horizon is divided into N-stage intervals with representative states inserted at the beginning of each interval.

discrete-time problems as described in Section 4.4. In fact in the RL literature, semi-Markov formulations have been used as a framework for temporal aggregation under the name "options," see e.g., Precup, Sutton, and Singh [PSS98], Ciosek and Silver [CiS15]. Temporal aggregation is also related to *multistep aggregation*, which is similar to the framework of Section 6.2, except that multiple transitions are allowed between original system states before return to an aggregate state; see [Ber12], Section 6.5.3.

Alternative frameworks for *simultaneous aggregation in space and time* have been developed in a variety of deterministic optimization contexts in transportation systems research, among others (see also the author's lecture notes and videos [Ber19a]). To illustrate one idea with an example, consider a 5-day auto travel from Boston to San Francisco. To plan this travel on a minute-by-minute and road-by-road basis would not make sense. Instead, the way most people would do it is to select major candidate stopping points (e.g. New York, Buffalo, Columbus, Atlanta, Chicago, St Louis, Salt Lake City, Dallas, Phoenix, Los Angeles, etc), then select major stopping times (times to stop for sleep, rest, etc), and finally decide on space and time schedules at a coarse level, while leaving the detailed optimization for later. We may view this process as an example of reduction of a very large-scale shortest path problem to a manageable shortest path problem, by aggregation in both space and time.

As an example of formalization of this type of aggregation process, consider a deterministic shortest path problem with an acyclic graph (all shortest path problems can be formulated in this way; cf. Sections 1.3.1-1.3.3). Let us introduce a "layer" of representative states after each block of a number of stages, and hard aggregation (probabilities $\phi_{jy} = 0$ or 1); see Fig. 6.6.1 (the representative states should include the destination, and the blocks may have a variable number of stages). We require that the shortest path passes through one of the representative states of each of the layers. Then *the aggregate problem decomposes into a sequence of shortest path problems*, from each representative state x in one layer to each representative state y in the next layer. After solving these problems with a suitable method, we may construct a low-dimensional deterministic shortest path problem, whose states are just the representative states. This approach can be extended to schemes where the time and space compression are less regular, and some stochastic uncertainty is involved, as well as for infinite horizon problems, but further details are beyond our scope. See Fig. 6.6.2 for the case of a deterministic discounted problem.

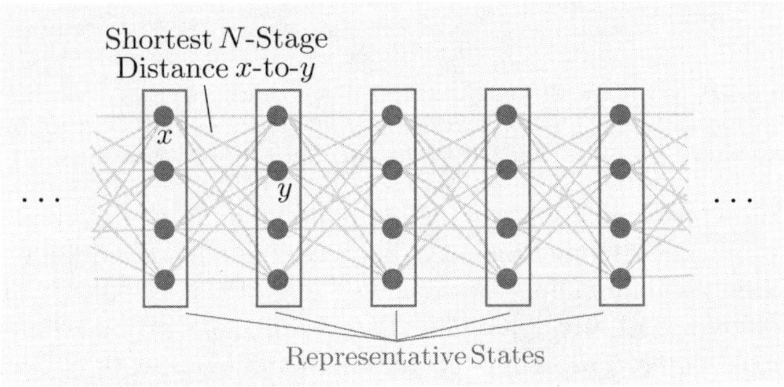

Figure 6.6.2 Schematic illustration of a decomposition process for solution of a discounted problem with aggregation in space and time. The transition cost between representative states x and y is the shortest distance between x and y, and can be obtained by a shortest path calculation. The aggregate problem is a deterministic discounted problem that involves just the representative states.

6.7 APPENDIX: MATHEMATICAL ANALYSIS

In this appendix we provide proofs of the mathematical results stated earlier. The following proposition shows the error bound of Section 6.2.1.

Proposition 6.2.1: (Error Bound for Hard Aggregation) In the case of hard aggregation, we have

$$\left|J^*(j) - \tilde{J}(j)\right| \leq \frac{\epsilon}{1-\alpha}, \qquad \text{for all } j \text{ such that } j \in S_y,\ y \in \mathcal{A},$$

where ϵ is the maximum variation of the optimal cost function J^* over the footprint sets S_y, $y \in \mathcal{A}$:

$$\epsilon = \max_{y \in \mathcal{A}} \max_{i,j \in S_y} \left|J^*(i) - J^*(j)\right|. \tag{6.26}$$

Proof: Consider the operator H defined by Eq. (6.13), and the vector $\overline{r}$ with components defined by

$$\overline{r}_x = \min_{i \in S_x} J^*(i) + \frac{\epsilon}{1-\alpha}, \qquad x \in \mathcal{A}.$$

Denote by y_j the aggregate state that corresponds to j, i.e., the index of the footprint set to which j belongs ($j \in S_{y_j}$). Then we have for all $x \in \mathcal{A}$,

$$\begin{aligned}
(H\overline{r})(x) &= \sum_{i=1}^{n} d_{xi} \min_{u \in U(i)} \sum_{j=1}^{n} p_{ij}(u)\Big(g(i,u,j) + \alpha \overline{r}_{y_j}\Big) \\
&\le \sum_{i=1}^{n} d_{xi} \min_{u \in U(i)} \sum_{j=1}^{n} p_{ij}(u)\left(g(i,u,j) + \alpha J^*(j) + \frac{\alpha\epsilon}{1-\alpha}\right) \\
&= \sum_{i=1}^{n} d_{xi}\left(J^*(i) + \frac{\alpha\epsilon}{1-\alpha}\right) \\
&\le \min_{i \in S_x}\big(J^*(i) + \epsilon\big) + \frac{\alpha\epsilon}{1-\alpha} \\
&= \min_{i \in S_x} J^*(i) + \frac{\epsilon}{1-\alpha} \\
&= \overline{r}_x,
\end{aligned}$$

where for the second equality we used the Bellman equation for J^*, and for the second inequality we used Eq. (6.26). Thus we have $H\overline{r} \le \overline{r}$, from which it follows that $r^* \le \overline{r}$ (since H is monotone, which implies that the sequence $\{H^k\overline{r}\}$ is monotonically nonincreasing, and we have

$$r^* = \lim_{k \to \infty} H^k \overline{r}$$

since H is a contraction). This proves one side of the desired error bound. The other side follows similarly. **Q.E.D.**

The following proposition shows the finite convergence property of the PI algorithm of Section 6.3.1.

> **Proposition 6.3.1: (PI Convergence)** Let μ^0 be any policy and let $\{\mu^k, r^k\}$ be a sequence generated by the PI algorithm (6.19)-(6.20). Then the sequence $\{r^k\}$ is monotonically nonincreasing (i.e., we have $r_x^k \ge r_x^{k+1}$ for all x and k) and there exists an index $\bar{k}$ such that $r^{\bar{k}}$ is equal to r^*, the unique solution of the composite Bellman Eq. (6.12).

Proof: For notational convenience, we introduce the matrices D and Φ whose rows are the disaggregation and aggregation distributions, respectively. For each policy μ, we consider the linear operator $\Phi D T_\mu : \Re^n \mapsto \Re^n$

given by

$$\Phi D T_\mu J = \Phi D(g_\mu + \alpha P_\mu J), \qquad J \in \Re^n,$$

where P_μ is the transition probability matrix corresponding to μ, and g_μ is the expected cost vector of μ. This operator is monotone in the sense that for all vectors J and J' with $J \geq J'$, we have

$$\Phi D T_\mu J \geq \Phi D T_\mu J',$$

since the matrix $\alpha \Phi D P_\mu$ of the operator has nonnegative components. Moreover, the operator is a contraction of modulus α with respect to the maximum norm. The reason is that the matrix $\Phi D P_\mu$ is a transition probability matrix, i.e., it has nonnegative components and its row sums are all equal to 1. This can be verified by a straightforward calculation, by writing its ith row sum as†

$$\sum_{j=1}^{n} (\Phi D P_\mu)_{ij} = \sum_{j=1}^{n} \sum_{\ell} \phi_{i\ell} \sum_{m} d_{\ell m} (P_\mu)_{mj},$$

rearranging the order of summations (making summation over j the innermost summation), using the fact that the rows of Φ, D, and P_μ are probability distributions, and concluding that

$$\sum_{j=1}^{n} (\Phi D P_\mu)_{ij} = 1.$$

Having established the monotonicity and contraction properties of the operators $\Phi D T_{\mu^k}$ and $D T_{\mu^k} \Phi$, we will next show that $\{r^k\}$ is nonincreasing. Since r^k is the unique fixed point of $D T_{\mu^k} \Phi$ [cf. Eq. (6.21)], we have $r^k = D T_{\mu^k} \Phi r^k$, so that the vector

$$\tilde{J}_{\mu^k} = \Phi r^k$$

satisfies

$$\tilde{J}_{\mu^k} = \Phi D T_{\mu^k} \tilde{J}_{\mu^k}.$$

It follows that $\tilde{J}_{\mu^k}$ is the unique fixed point of the contraction mapping $\Phi D T_{\mu^k}$. By using the definition

$$T_{\mu^{k+1}} \tilde{J}_{\mu^k} = T \tilde{J}_{\mu^k}$$

---

† It can also be intuitively verified from the structure of the aggregate problem: $\Phi D P_\mu$ is the matrix of transition probabilities under policy μ for the Markov chain whose n states are the states depicted in the top right part of the transition diagram of Fig. 6.2.1.

of μ^{k+1} [cf. Eq. (6.20)], we have

$$\tilde{J}_{\mu^k} = \Phi D T_{\mu^k} \tilde{J}_{\mu^k} \geq \Phi D T \tilde{J}_{\mu^k} = \Phi D T_{\mu^{k+1}} \tilde{J}_{\mu^k}. \tag{6.27}$$

Applying repeatedly the monotone operator $\Phi D T_{\mu^{k+1}}$ to this relation, we have for all $m \geq 1$,

$$\tilde{J}_{\mu^k} \geq (\Phi D T_{\mu^{k+1}})^m \tilde{J}_{\mu^k} \geq \lim_{m \to \infty} (\Phi D T_{\mu^{k+1}})^m \tilde{J}_{\mu^k} = \tilde{J}_{\mu^{k+1}}, \tag{6.28}$$

where the equality follows from the fact that $\tilde{J}_{\mu^{k+1}}$ is the fixed point of the contraction mapping $\Phi D T_{\mu^{k+1}}$. It follows that $\tilde{J}_{\mu^k} \geq \tilde{J}_{\mu^{k+1}}$, or equivalently

$$\Phi r^k \geq \Phi r^{k+1}, \qquad k = 0, 1, \ldots. \tag{6.29}$$

By the definition of the feature-based aggregation architecture [cf. Eq. (6.7)], each column of Φ has at least one component that is equal to 1, and in the row of that component, all other components are zero. Therefore Φ has full rank, while using also Eq. (6.29), we obtain for all k

$$r^k \geq r^{k+1}.$$

Moreover, since Φ has full rank, the equality $r^k = r^{k+1}$ holds if and only if $\Phi r^k = \Phi r^{k+1}$, or $\tilde{J}_{\mu^k} = \tilde{J}_{\mu^{k+1}}$.

We finally show that the sequence $\{r^k\}$ terminates and yields r^* upon termination. Indeed, the inequality $\tilde{J}_{\mu^k} \geq \tilde{J}_{\mu^{k+1}}$ implies that as long as $\tilde{J}_{\mu^k} \neq \tilde{J}_{\mu^{k+1}}$, a policy μ^k cannot be repeated. Since there are only a finite number of policies, it follows that we must eventually have $\tilde{J}_{\mu^k} = \tilde{J}_{\mu^{k+1}}$ and hence also $r^k = r^{k+1}$. Once this happens, from Eqs. (6.27) and (6.28), we see that

$$\tilde{J}_{\mu^k} = \Phi D T \tilde{J}_{\mu^k}$$

or

$$\Phi r^k = \Phi D T \Phi r^k.$$

Since each column of Φ has at least one component that is equal to 1, and in the row of that component, all other components are zero, it follows that $r^k = D T \Phi r^k$. Thus r^k is the fixed point of the contraction mapping $H = D T \Phi$ of Eq. (6.13), which is r^* by Eq. (6.12). Thus eventually we have $r^k = r^*$. **Q.E.D.**

References

[ACF02] Auer, P., Cesa-Bianchi, N., and Fischer, P., 2002. "Finite Time Analysis of the Multiarmed Bandit Problem," Machine Learning, Vol. 47, pp. 235-256.

[ACV09] Argall, B. D., Chernova, S., Veloso, M., and Browning, B., 2009. "A Survey of Robot Learning from Demonstration," Robotics and Autonomous Systems, Vol. 57, pp. 469-483.

[ADB17] Arulkumaran, K., Deisenroth, M. P., Brundage, M., and Bharath, A. A., 2017. "A Brief Survey of Deep Reinforcement Learning," arXiv preprint arXiv:1708.05866.

[ALZ08] Asmuth, J., Littman, M. L., and Zinkov, R., 2008. "Potential-Based Shaping in Model-Based Reinforcement Learning," Proc. of 23rd AAAI Conference, pp. 604-609.

[AMS07] Antos, A., Munos, R., and Szepesvari, C., 2007. "Fitted Q-Iteration in Continuous Action-Space MDPs," Proc. of NIPS, pp. 9-16.

[AMS09] Audibert, J.Y., Munos, R., and Szepesvari, C., 2009. "Exploration-Exploitation Tradeoff Using Variance Estimates in Multi-Armed Bandits," Theoretical Computer Science, Vol. 410, pp. 1876-1902.

[ASS68] Aleksandrov, V. M., Sysoyev, V. I, and Shemeneva, V. V., 1968. "Stochastic Optimization of Systems," Engineering Cybernetics, Vol. 5, pp.11-16.

[AbN04] Abbeel, P., and Ng, A. Y., 2004. "Apprenticeship Learning via Inverse Reinforcement Learning," in Proceedings of the 21st ICML.

[Abr90] Abramson, B., 1990. "Expected-Outcome: A General Model of Static Evaluation," IEEE Trans. on Pattern Analysis and Machine Intelligence, Vol. 12, pp. 182-193.

[Agr95] Agrawal, R., 1995. "Sample Mean Based Index Policies with $O(\log n)$ Regret for the Multiarmed Bandit Problem," Advances in Applied Probability, Vol. 27, pp. 1054-1078.

[AnH14] Antunes, D., and Heemels, W.P.M.H., 2014. "Rollout Event-Triggered Control: Beyond Periodic Control Performance," IEEE Transactions on Automatic Control, Vol. 59, pp. 3296-3311.

[AnM79] Anderson, B. D. O., and Moore, J. B., 1979. Optimal Filtering, Prentice-Hall, Englewood Cliffs, N. J.

[ArK03] Ariyur, K. B., and Krstic, M., 2003. Real-Time Optimization by Extremum-Seeking Control, John Wiley and Sons, N. Y.

[AsG10] Asmussen, S., and Glynn, P. W., 2010. Stochastic Simulation: Algorithms and Analysis, Springer, N. Y.

[AsH95] Aström, K. J., and Hagglund, T., 1995. PID Controllers: Theory, Design, and Tuning, Instrument Society of America, Research Triangle Park, N. C.

[AsH06] Aström, K. J., and Hagglund, T., 2006. Advanced PID Control, Instrument Society of America, Research Triangle Park, N. C.

[AsW94] Aström, K. J., and Wittenmark, B., 1994. Adaptive Control, 2nd Edition, Prentice-Hall, Englewood Cliffs, N. J.

[AtF66] Athans, M., and Falb, P., 1966. Optimal Control, McGraw-Hill, N. Y.

[BBD08] Bu, L., Babu, R., and De Schutter, B., 2008. "A Comprehensive Survey of Multiagent Reinforcement Learning," IEEE Transactions on Systems, Man, and Cybernetics, Part C, Vol. 38, pp. 156-172.

[BBD10] Busoniu, L., Babuska, R., De Schutter, B., and Ernst, D., 2010. Reinforcement Learning and Dynamic Programming Using Function Approximators, CRC Press, N. Y.

[BBG13] Bertazzi, L., Bosco, A., Guerriero, F., and Lagana, D., 2013. "A Stochastic Inventory Routing Problem with Stock-Out," Transportation Research, Part C, Vol. 27, pp. 89-107.

[BBM17] Borelli, F., Bemporad, A., and Morari, M., 2017. Predictive Control for Linear and Hybrid Systems, Cambridge Univ. Press, Cambridge, UK.

[BBN04] Bertsekas, D. P., Borkar, V., and Nedić, A., 2004. "Improved Temporal Difference Methods with Linear Function Approximation," in Learning and Approximate Dynamic Programming, by J. Si, A. Barto, W. Powell, and D. Wunsch, (Eds.), IEEE Press, N. Y.

[BBP13] Bhatnagar, S., Borkar, V. S., and Prashanth, L. A., 2013. "Adaptive Feature Pursuit: Online Adaptation of Features in Reinforcement Learning," in *Reinforcement Learning and Approximate Dynamic Programming for Feedback Control*, by F. Lewis and D. Liu (eds.), IEEE Press, Piscataway, N. J., pp. 517-534.

[BBS87] Bean, J. C., Birge, J. R., and Smith, R. L., 1987. "Aggregation in Dynamic Programming," Operations Research, Vol. 35, pp. 215-220.

[BBS95] Barto, A. G., Bradtke, S. J., and Singh, S. P., 1995. "Real-Time Learning and Control Using Asynchronous Dynamic Programming," Artificial Intelligence, Vol. 72, pp. 81-138.

[BCN18] Bottou, L., Curtis, F. E., and Nocedal, J., 2018. "Optimization Methods for Large-Scale Machine Learning," SIAM Review, Vol. 60, pp. 223-311.

[BDT18] Busoniu, L., de Bruin, T., Tolic, D., Kober, J., and Palunko, I., 2018. "Reinforcement Learning for Control: Performance, Stability, and Deep Approximators," Annual Reviews in Control, Vol. 46, pp. 8-28.

[BGM95] Bertsekas, D. P., Guerriero, F., and Musmanno, R., 1995. "Parallel Shortest Path Methods for Globally Optimal Trajectories," High Performance Computing: Technology, Methods, and Applications, (J. Dongarra et al., Eds.), Elsevier.

[BKM05] de Boer, P. T., Kroese, D. P., Mannor, S., and Rubinstein, R. Y. 2005. "A Tutorial on the Cross-Entropy Method," Annals of Operations Research, Vol. 134, pp. 19-67.

[BLL19] Bartlett, P. L., Long, P. M., Lugosi, G., and Tsigler, A., 2019. "Benign Overfitting in Linear Regression," arXiv preprint arXiv:1906.11300.

[BMM18] Belkin, M., Ma, S., and Mandal, S., 2018. "To Understand Deep Learning we Need to Understand Kernel Learning," arXiv preprint arXiv:1802.01396.

[BNO03] Bertsekas, D. P., Nedić, A., and Ozdaglar, A. E., 2003. Convex Analysis and Optimization, Athena Scientific, Belmont, MA.

[BPP13] Bhatnagar, S., Prasad, H., and Prashanth, L. A., 2013. Stochastic Recursive Algorithms for Optimization, Lecture Notes in Control and Information Sciences, Springer, N. Y.

[BPW12] Browne, C., Powley, E., Whitehouse, D., Lucas, L., Cowling, P. I., Rohlfshagen, P., Tavener, S., Perez, D., Samothrakis, S., and Colton, S., 2012. "A Survey of Monte Carlo Tree Search Methods," IEEE Trans. on Computational Intelligence and AI in Games, Vol. 4, pp. 1-43.

[BRT18] Belkin, M., Rakhlin, A., and Tsybakov, A. B., 2018. "Does Data Interpolation Contradict Statistical Optimality?," arXiv preprint arXiv:1806.09471.

[BSA83] Barto, A. G., Sutton, R. S., and Anderson, C. W., 1983. "Neuronlike Elements that Can Solve Difficult Learning Control Problems," IEEE Trans. on Systems, Man, and Cybernetics, Vol. 13, pp. 835-846.

[BTW97] Bertsekas, D. P., Tsitsiklis, J. N., and Wu, C., 1997. "Rollout Algorithms for Combinatorial Optimization," Heuristics, Vol. 3, pp. 245-262.

[BVE13] Ben Amor, H, Vogt, D., Ewerton, M., Berger, E., Jung, B., and Peters, J., 2013. "Learning Responsive Robot Behavior by Imitation," in 2013 IEEE/RSJ International Conference on Intelligent Robots and Systems, pp. 3257-3264.

[BWL19] Beuchat, P.N., Warrington, J., and Lygeros, J., 2019. "Accelerated Point-Wise Maximum Approach to Approximate Dynamic Programming," arXiv preprint arXiv:1901.03619.

[BYB94] Bradtke, S. J., Ydstie, B. E., and Barto, A. G., 1994. "Adaptive Linear Quadratic Control Using Policy Iteration," Proc. IEEE American Control Conference, Vol. 3, pp. 3475-3479.

[BaB01] Baxter, J., and Bartlett, P. L., 2001. "Infinite-Horizon Policy-Gradient Estimation," Journal of Artificial Intelligence Research, Vol. 15, pp. 319-350.

[Bac96] Back, T., 1996. Evolutionary Algorithms in Theory and Practice: Evolution Strategies, Evolutionary Programming, Genetic Algorithms, Oxford University Press.

[Bai93] Baird, L. C., 1993. "Advantage Updating," Report WL-TR-93-1146, Wright Patterson AFB, OH.

[Bai94] Baird, L. C., 1994. "Reinforcement Learning in Continuous Time: Advantage Updating," International Conf. on Neural Networks, Orlando, Fla.

[BeC89] Bertsekas, D. P., and Castanon, D. A., 1989. "Adaptive Aggregation Methods for Infinite Horizon Dynamic Programming," IEEE Trans. on Aut. Control, Vol. AC-34, pp. 589-598.

[BeC99] Bertsekas, D. P., and Castanon, D. A., 1999. "Rollout Algorithms for Stochastic Scheduling Problems," Heuristics, Vol. 5, pp. 89-108.

[BeC08] Besse, C., and Chaib-draa, B., 2008. "Parallel Rollout for Online Solution of DEC-POMDPs," Proc. of 21st International FLAIRS Conference, pp. 619-624.

[BeL14] Beyme, S., and Leung, C., 2014. "Rollout Algorithm for Target Search in a Wireless Sensor Network," 80th Vehicular Technology Conference (VTC2014), IEEE, pp. 1-5.

[BeI96] Bertsekas, D. P., and Ioffe, S., 1996. "Temporal Differences-Based Policy Iteration and Applications in Neuro-Dynamic Programming," Lab. for Info. and Decision

Systems Report LIDS-P-2349, Massachusetts Institute of Technology.

[BeP03] Bertsimas, D., and Popescu, I., 2003. "Revenue Management in a Dynamic Network Environment," Transportation Science, Vol. 37, pp. 257-277.

[BeR71] Bertsekas, D. P., and Rhodes, I. B., 1971. "On the Minimax Reachability of Target Sets and Target Tubes," Automatica, Vol. 7, pp. 233-247.

[BeR73] Bertsekas, D. P., and Rhodes, I. B., 1973. "Sufficiently Informative Functions and the Minimax Feedback Control of Uncertain Dynamic Systems," IEEE Trans. Automatic Control, Vol. AC-18, pp. 117-124.

[BeS78] Bertsekas, D. P., and Shreve, S. E., 1978. Stochastic Optimal Control: The Discrete Time Case, Academic Press, N. Y.; republished by Athena Scientific, Belmont, MA, 1996 (can be downloaded from the author's website).

[BeS18] Bertazzi, L., and Secomandi, N., 2018. "Faster Rollout Search for the Vehicle Routing Problem with Stochastic Demands and Restocking," European J. of Operational Research, Vol. 270, pp.487-497.

[BeT89] Bertsekas, D. P., and Tsitsiklis, J. N., 1989. Parallel and Distributed Computation: Numerical Methods, Prentice-Hall, Englewood Cliffs, N. J.; republished by Athena Scientific, Belmont, MA, 1997 (can be downloaded from the author's website).

[BeT91] Bertsekas, D. P., and Tsitsiklis, J. N., 1991. "An Analysis of Stochastic Shortest Path Problems," Math. Operations Res., Vol. 16, pp. 580-595.

[BeT96] Bertsekas, D. P., and Tsitsiklis, J. N., 1996. Neuro-Dynamic Programming, Athena Scientific, Belmont, MA.

[BeT97] Bertsimas, D., and Tsitsiklis, J. N., 1997. Introduction to Linear Optimization, Athena Scientific, Belmont, MA.

[BeT00] Bertsekas, D. P., and Tsitsiklis, J. N., 2000. "Gradient Convergence of Gradient Methods with Errors," SIAM J. on Optimization, Vol. 36, pp. 627-642.

[BeT08] Bertsekas, D. P., and Tsitsiklis, J. N., 2008. Introduction to Probability, 2nd Edition, Athena Scientific, Belmont, MA.

[BeY09] Bertsekas, D. P., and Yu, H., 2009. "Projected Equation Methods for Approximate Solution of Large Linear Systems," J. of Computational and Applied Mathematics, Vol. 227, pp. 27-50.

[BeY10] Bertsekas, D. P., and Yu, H., 2010. "Asynchronous Distributed Policy Iteration in Dynamic Programming," Proc. of Allerton Conf. on Communication, Control and Computing, Allerton Park, Ill, pp. 1368-1374.

[BeY12] Bertsekas, D. P., and Yu, H., 2012. "Q-Learning and Enhanced Policy Iteration in Discounted Dynamic Programming," Math. of Operations Research, Vol. 37, pp. 66-94.

[BeY16] Bertsekas, D. P., and Yu, H., 2016. "Stochastic Shortest Path Problems Under Weak Conditions," Lab. for Information and Decision Systems Report LIDS-2909, MIT.

[Bel57] Bellman, R., 1957. Dynamic Programming, Princeton University Press, Princeton, N. J.

[Bel67] Bellman, R., 1967. Introduction to the Mathematical Theory of Control Processes, Academic Press, Vols. I and II, New York, N. Y.

[Bel84] Bellman, R., 1984. Eye of the Hurricane, World Scientific Publishing, Singapore.

[Ben09] Bengio, Y., 2009. "Learning Deep Architectures for AI," Foundations and Trends in Machine Learning, Vol. 2, pp. 1-127.

[Ber71] Bertsekas, D. P., 1971. "Control of Uncertain Systems With a Set-Membership Description of the Uncertainty," Ph.D. Dissertation, Massachusetts Institute of Technology, Cambridge, MA (can be downloaded from the author's website).

[Ber72] Bertsekas, D. P., 1972. "Infinite Time Reachability of State Space Regions by Using Feedback Control," IEEE Trans. Automatic Control, Vol. AC-17, pp. 604-613.

[Ber73] Bertsekas, D. P., 1973. "Stochastic Optimization Problems with Nondifferentiable Cost Functionals," J. of Optimization Theory and Applications, Vol. 12, pp. 218-231.

[Ber82] Bertsekas, D. P., 1982. "Distributed Dynamic Programming," IEEE Trans. Automatic Control, Vol. AC-27, pp. 610-616.

[Ber83] Bertsekas, D. P., 1983. "Asynchronous Distributed Computation of Fixed Points," Math. Programming, Vol. 27, pp. 107-120.

[Ber91] Bertsekas, D. P., 1991. Linear Network Optimization: Algorithms and Codes, M.I.T. Press, Cambridge, MA (can be downloaded from the author's website).

[Ber95] Bertsekas, D. P., 1995. "A Counterexample to Temporal Differences Learning," Neural Computation, Vol. 7, pp. 270-279.

[Ber96a] Bertsekas, D. P., 1996. "Incremental Least Squares Methods and the Extended Kalman Filter," SIAM J. on Optimization, Vol. 6, pp. 807-822.

[Ber96b] Bertsekas, D. P., 1996. Lecture at NSF Workshop on Reinforcement Learning, Hilltop House, Harper's Ferry, N. Y.

[Ber97a] Bertsekas, D. P., 1997. "A New Class of Incremental Gradient Methods for Least Squares Problems," SIAM J. on Optimization, Vol. 7, pp. 913-926.

[Ber97b] Bertsekas, D. P., 1997. "Differential Training of Rollout Policies," Proc. of the 35th Allerton Conference on Communication, Control, and Computing, Allerton Park, Ill.

[Ber98] Bertsekas, D. P., 1998. Network Optimization: Continuous and Discrete Models, Athena Scientific, Belmont, MA (can be downloaded from the author's website).

[Ber05a] Bertsekas, D. P., 2005. "Dynamic Programming and Suboptimal Control: A Survey from ADP to MPC," European J. of Control, Vol. 11, pp. 310-334.

[Ber05b] Bertsekas, D. P., 2005. "Rollout Algorithms for Constrained Dynamic Programming," LIDS Report 2646, MIT.

[Ber07] Bertsekas, D. P., 2007. "Separable Dynamic Programming and Approximate Decomposition Methods," IEEE Trans. on Aut. Control, Vol. 52, pp. 911-916.

[Ber10] Bertsekas, D. P., 2010. "Incremental Gradient, Subgradient, and Proximal Methods for Convex Optimization: A Survey," Lab. for Information and Decision Systems Report LIDS-P-2848, MIT; a condensed version with the same title appears in Optimization for Machine Learning, by S. Sra, S. Nowozin, and S. J. Wright, (eds.), MIT Press, Cambridge, MA, 2012, pp. 85-119.

[Ber11a] Bertsekas, D. P., 2011. "Approximate Policy Iteration: A Survey and Some New Methods," J. of Control Theory and Applications, Vol. 9, pp. 310-335.

[Ber11b] Bertsekas, D. P., 2011. "Temporal Difference Methods for General Projected Equations," IEEE Trans. on Aut. Control, Vol. 56, pp. 2128-2139.

[Ber11c] Bertsekas, D. P., 2011. "Incremental Proximal Methods for Large Scale Convex Optimization," Math. Programming, Vol. 129, pp. 163-195.

[Ber12] Bertsekas, D. P., 2012. Dynamic Programming and Optimal Control, Vol. II, 4th Edition, Athena Scientific, Belmont, MA.

[Ber13a] Bertsekas, D. P., 2013. "Rollout Algorithms for Discrete Optimization: A Survey," Handbook of Combinatorial Optimization, Springer.

[Ber13b] Bertsekas, D. P., 2013. "λ-Policy Iteration: A Review and a New Implementation," in Reinforcement Learning and Approximate Dynamic Programming for Feedback Control, by F. Lewis and D. Liu (eds.), IEEE Press, Piscataway, N. J., pp. 381-409.

[Ber15a] Bertsekas, D. P., 2015. Convex Optimization Algorithms, Athena Scientific, Belmont, MA.

[Ber15b] Bertsekas, D. P., 2015. "Incremental Aggregated Proximal and Augmented Lagrangian Algorithms," Lab. for Information and Decision Systems Report LIDS-P-3176, MIT; arXiv preprint arXiv:1507.1365936.

[Ber16a] Bertsekas, D. P., 2016. Nonlinear Programming, 3rd Edition, Athena Scientific, Belmont, MA.

[Ber16b] Bertsekas, D. P., 2016. "Affine Monotonic and Risk-Sensitive Models in Dynamic Programming," arXiv preprint arXiv:1608.01393; IEEE Trans. on Automatic Control, Vol. 64, 2019, pp. 3117-3128.

[Ber16c] Bertsekas, D. P., 2016. "Proximal Algorithms and Temporal Differences for Large Linear Systems: Extrapolation, Approximation, and Simulation," arXiv preprint arXiv:1610.05427; Computational Optimization and Applications J., Vol. 70, 2018, pp. 709-736.

[Ber17] Bertsekas, D. P., 2017. Dynamic Programming and Optimal Control, Vol. I, 4th Edition, Athena Scientific, Belmont, MA.

[Ber18a] Bertsekas, D. P., 2018. Abstract Dynamic Programming, 2nd Edition, Athena Scientific, Belmont, MA (can be downloaded from the author's website).

[Ber18b] Bertsekas, D. P., 2018. "Feature-Based Aggregation and Deep Reinforcement Learning: A Survey and Some New Implementations," Lab. for Information and Decision Systems Report, MIT; arXiv preprint arXiv:1804.04577; IEEE/CAA Journal of Automatica Sinica, Vol. 6, 2019, pp. 1-31.

[Ber18c] Bertsekas, D. P., 2018. "Biased Aggregation, Rollout, and Enhanced Policy Improvement for Reinforcement Learning," Lab. for Information and Decision Systems Report, MIT; arXiv preprint arXiv:1910.02426.

[Ber18d] Bertsekas, D. P., 2018. "Proximal Algorithms and Temporal Difference Methods for Solving Fixed Point Problems," Computational Optim. Appl., Vol. 70, pp. 709-736.

[Ber18e] Bertsekas, D.P., 2018. "Proper Policies in Infinite-State Stochastic Shortest Path Problems," IEEE Trans. on Automatic Control, Vol. 63, pp. 3787-3792.

[Ber19a] Bertsekas, D. P., 2019. Lecture Slides and Videolectures on Reinforcement Learning and Optimal Control, ASU, at http://web.mit.edu/dimitrib/www/RLbook.html.

[Ber19b] Bertsekas, D. P., 2019. "Robust Shortest Path Planning and Semicontractive Dynamic Programming," Naval Research Logistics, Vol. 66, pp. 15-37.

[Ber19c] Bertsekas, D. P., 2019. "Multiagent Rollout Algorithms and Reinforcement Learning," arXiv preprint arXiv:1910.00120.

[Bet10] Bethke, B. M., 2010. Kernel-Based Approximate Dynamic Programming Using Bellman Residual Elimination, Ph.D. Thesis, MIT.

[Bia16] Bianchi, P., 2016. "Ergodic Convergence of a Stochastic Proximal Point Algorithm," SIAM J. on Optimization, Vol. 26, pp. 2235-2260.

[Bis95] Bishop, C. M, 1995. Neural Networks for Pattern Recognition, Oxford University Press, N. Y.

[Bis06] Bishop, C. M, 2006. Pattern Recognition and Machine Learning, Springer, N. Y.

[Bla99] Blanchini, F., 1999. "Set Invariance in Control – A Survey," Automatica, Vol. 35, pp. 1747-1768.

[BoV79] Borkar, V., and Varaiya, P. P., 1979. "Adaptive Control of Markov Chains, I: Finite Parameter Set," IEEE Trans. Automatic Control, Vol. AC-24, pp. 953-958.

[Bor08] Borkar, V. S., 2008. Stochastic Approximation: A Dynamical Systems Viewpoint, Cambridge Univ. Press.

[Bor09] Borkar, V. S., 2009. "Reinforcement Learning: A Bridge Between Numerical Methods and Monte Carlo," in World Scientific Review, Vol. 9, Ch. 4.

[Boy02] Boyan, J. A., 2002. "Technical Update: Least-Squares Temporal Difference Learning," Machine Learning, Vol. 49, pp. 1-15.

[BrB96] Bradtke, S. J., and Barto, A. G., 1996. "Linear Least-Squares Algorithms for Temporal Difference Learning," Machine Learning, Vol. 22, pp. 33-57.

[BrH75] Bryson, A., and Ho, Y. C., 1975. Applied Optimal Control: Optimization, Estimation, and Control, (revised edition), Taylor and Francis, Levittown, Penn.

[BuK97] Burnetas, A. N., and Katehakis, M. N., 1997. "Optimal Adaptive Policies for Markov Decision Processes," Math. of Operations Research, Vol. 22, pp. 222-255.

[CFH05] Chang, H. S., Hu, J., Fu, M. C., and Marcus, S. I., 2005. "An Adaptive Sampling Algorithm for Solving Markov Decision Processes," Operations Research, Vol. 53, pp. 126-139.

[CFH13] Chang, H. S., Hu, J., Fu, M. C., and Marcus, S. I., 2013. Simulation-Based Algorithms for Markov Decision Processes, 2nd Edition, Springer, N. Y.

[CFH16] Chang, H. S., Hu, J., Fu, M. C., and Marcus, S. I., 2016. "Google DeepMind's AlphaGo," ORMS Today, INFORMS, Vol. 43.

[CGC00] Chong, E. K., Givan, R. L., and Chang, H. S., 2000. "A Framework for Simulation-Based Network Control via Hindsight Optimization," Proc. 39th IEEE CDC, pp. 1433-1438.

[CGC04] Chang, H. S., Givan, R. L., and Chong, E. K. P., 2004. "Parallel Rollout for Online Solution of Partially Observable Markov Decision Processes," Discrete Event Dynamic Systems, Vol. 14, pp. 309-341.

[CLT19] Chapman, M. P., Lacotte, J., Tamar, A., Lee, D., Smith, K. M., Cheng, V., Fisac, J. F., Jha, S., Pavone, M., and Tomlin, C. J., 2019. "A Risk-Sensitive Finite-Time Reachability Approach for Safety of Stochastic Dynamic Systems," arXiv preprint arXiv:1902.11277.

[CRV06] Cogill, R., Rotkowitz, M., Van Roy, B., and Lall, S., 2006. "An Approximate Dynamic Programming Approach to Decentralized Control of Stochastic Systems," in Control of Uncertain Systems: Modelling, Approximation, and Design, Springer, Berlin, pp. 243-256.

[CXL19] Chu, Z., Xu, Z., and Li, H., 2019. "New Heuristics for the RCPSP with Multiple Overlapping Modes," Computers and Industrial Engineering, Vol. 131, pp. 146-156.

[CaB04] Camacho, E. F., and Bordons, C., 2004. Model Predictive Control, 2nd Edition, Springer, New York, N. Y.

[CaC97] Cao, X. R., and Chen, H. F., 1997. "Perturbation Realization Potentials and Sensitivity Analysis of Markov Processes," IEEE Trans. on Aut. Control, Vol. 32, pp. 1382-1393.

[CaC18] Caterini, A. L., and Chang, D. E., 2018. Deep Neural Networks in a Mathematical Framework, Springer, Oxford, UK.

[CaW98] Cao, X. R., and Wan, Y. W., 1998. "Algorithms for Sensitivity Analysis of Markov Systems Through Potentials and Perturbation Realization," IEEE Trans. Control Systems Technology, Vol. 6, pp. 482-494.

[Can16] Candy, J. V., 2016. Bayesian Signal Processing: Classical, Modern, and Particle Filtering Methods, Wiley-IEEE Press.

[Cao07] Cao, X. R., 2007. Stochastic Learning and Optimization: A Sensitivity-Based Approach, Springer, N. Y.

[ChC17] Chui, C. K., and Chen, G., 2017. Kalman Filtering, Springer International Publishing.

[ChM82] Chatelin, F., and Miranker, W. L., 1982. "Acceleration by Aggregation of Successive Approximation Methods," Linear Algebra and its Applications, Vol. 43, pp. 17-47.

[ChS00] Christianini, N., and Shawe-Taylor, J., 2000. Support Vector Machines and Other Kernel-Based Learning Methods, Cambridge Univ. Press.

[ChT91] Chow, C.-S., and Tsitsiklis, J. N., 1991. "An Optimal One–Way Multigrid Algorithm for Discrete–Time Stochastic Control," IEEE Trans. on Aut. Control, Vol. AC-36, pp. 898-914.

[ChV12] Chacon, A., and Vladimirsky, A., 2012. "Fast Two-Scale Methods for Eikonal Equations," SIAM J. on Scientific Computing, Vol. 34, pp. A547-A578.

[ChV13] Chacon, A., and Vladimirsky, A., 2013. "A Parallel Heap-Cell Method for Eikonal Equations," arXiv preprint arXiv:1306.4743.

[ChV15] Chacon, A., and Vladimirsky, A., 2015. "A Parallel Two-Scale Method for Eikonal Equations," SIAM J. on Scientific Computing, Vol. 37, pp. A156-A180.

[CiS15] Ciosek, K., and Silver, D., 2015. "Value Iteration with Options and State Aggregation," Report, Centre for Computational Statistics and Machine Learning University College London.

[Cla17] Clawson, Z., 2017. Shortest path problems: Domain restriction, anytime planning, and multi-objective optimization. Ph.D. Thesis, Cornell University.

[Cou06] Coulom, R., 2006. "Efficient Selectivity and Backup Operators in Monte-Carlo Tree Search," International Conference on Computers and Games, Springer, pp. 72-83.

[CrS00] Cristianini, N., and Shawe-Taylor, J., 2000. An Introduction to Support Vector Machines and Other Kernel-Based Learning Methods, Cambridge Univ. Press.

[Cyb89] Cybenko, 1989. "Approximation by Superpositions of a Sigmoidal Function," Math. of Control, Signals, and Systems, Vol. 2, pp. 303-314.

[DDF19] Daubechies, I., DeVore, R., Foucart, S., Hanin, B., and Petrova, G., 2019. "Nonlinear Approximation and (Deep) ReLU Networks," arXiv preprint arXiv:1905.02199.

[D'Ep60] D'Epenoux, F., 1960. "Sur un Probleme de Production et de Stockage Dans l'Aleatoire," Rev. Francaise Aut. Infor. Recherche Operationnelle, Vol. 14, (English Transl.: Management Sci., Vol. 10, 1963, pp. 98-108).

[DFM12] Desai, V. V., Farias, V. F., and Moallemi, C. C., 2012. "Aproximate Dynamic Programming via a Smoothed Approximate Linear Program," Operations Research, Vol. 60, pp. 655-674.

[DFM13] Desai, V. V., Farias, V. F., and Moallemi, C. C., 2013. "Bounds for Markov Decision Processes," in Reinforcement Learning and Approximate Dynamic Programming for Feedback Control, by F. Lewis and D. Liu (eds.), IEEE Press, Piscataway, N. J., pp. 452-473.

[DFV03] de Farias, D. P., and Van Roy, B., 2003. "The Linear Programming Approach to Approximate Dynamic Programming," Operations Research, Vol. 51, pp. 850-865.

[DFV04] de Farias, D. P., and Van Roy, B., 2004. "On Constraint Sampling in the Linear Programming Approach to Approximate Dynamic Programming," Mathematics of Operations Research, Vol. 29, pp. 462-478.

[DHS12] Duda, R. O., Hart, P. E., and Stork, D. G., 2012. Pattern Classification, J. Wiley, N. Y.

[DJW12] Duchi, J., Jordan, M. I., Wainwright, M. J., and Wibisono, A., 2012. "Finite Sample Convergence Rate of Zero-Order Stochastic Optimization Methods," NIPS, pp. 1448-1456.

[DJW15] Duchi, J., Jordan, M. I., Wainwright, M. J., and Wibisono, A., 2015. "Optimal Rates for Zero-Order Convex Optimization: The Power of Two Function Evaluations," IEEE Trans. on Information Theory, Vol. 61, pp. 2788-2806.

[DNP11] Deisenroth, M. P., Neumann, G., and Peters, J., 2011. "A Survey on Policy Search for Robotics," Foundations and Trends in Robotics, Vol. 2, pp. 1-142.

[DNW16] David, O. E., Netanyahu, N. S., and Wolf, L., 2016. "Deepchess: End-to-End Deep Neural Network for Automatic Learning in Chess," in International Conference on Artificial Neural Networks, pp. 88-96.

[DeF04] De Farias, D. P., 2004. "The Linear Programming Approach to Approximate Dynamic Programming," in Learning and Approximate Dynamic Programming, by J. Si, A. Barto, W. Powell, and D. Wunsch, (Eds.), IEEE Press, N. Y.

[DeJ06] De Jong, K. A., 2006. Evolutionary Computation: A Unified Approach, MIT Press, Cambridge, MA.

[DeK11] Devlin, S., and Kudenko, D., 2011. "Theoretical Considerations of Potential-Based Reward Shaping for Multi-Agent Systems," in Proceedings of AAMAS.

[DeR79] Denardo, E. V., and Rothblum, U. G., 1979. "Optimal Stopping, Exponential Utility, and Linear Programming," Math. Programming, Vol. 16, pp. 228-244.

[DiL08] Dimitrakakis, C., and Lagoudakis, M. G., 2008. "Rollout Sampling Approximate Policy Iteration," Machine Learning, Vol. 72, pp. 157-171.

[DiM10] Di Castro, D., and Mannor, S., 2010. "Adaptive Bases for Reinforcement Learning," Machine Learning and Knowledge Discovery in Databases, Vol. 6321, pp. 312-327.

[DiW02] Dietterich, T. G., and Wang, X., 2002. "Batch Value Function Approximation via Support Vectors," in Advances in Neural Information Processing Systems, pp. 1491-

1498.

[Die00] Dietterich, T., 2000. "Hierarchical Reinforcement Learning with the MAXQ Value Function Decomposition," J. of Artificial Intelligence Research, Vol. 13, pp. 227-303.

[DoD93] Douglas, C. C., and Douglas, J., 1993. "A Unified Convergence Theory for Abstract Multigrid or Multilevel Algorithms, Serial and Parallel," SIAM J. Num. Anal., Vol. 30, pp. 136-158.

[DoJ09] Doucet, A., and Johansen, A. M., 2009. "A Tutorial on Particle Filtering and Smoothing: Fifteen Years Later," Handbook of Nonlinear Filtering, Oxford University Press, Vol. 12, p. 3.

[Dre02] Dreyfus, S. D., 2002. "Richard Bellman on the Birth of Dynamic Programming," Operations Research, Vol. 50, pp. 48-51.

[EGW06] Ernst, D., Geurts, P., and Wehenkel, L., 2006. "Tree-Based Batch Mode Reinforcement Learning," J. of Machine Learning Research, Vol. 6, pp. 503-556.

[ELP12] Estanjini, R. M., Li, K., and Paschalidis, I. C., 2012. "A Least Squares Temporal Difference Actor-Critic Algorithm with Applications to Warehouse Management," Naval Research Logistics, Vol. 59, pp. 197-211.

[EMM05] Engel, Y., Mannor, S., and Meir, R., 2005. "Reinforcement Learning with Gaussian Processes," in Proc. of the 22nd ICML, pp. 201-208.

[FHS14] Feinberg, E. A., Huang, J., and Scherrer, B., 2014. "Modified Policy Iteration Algorithms are not Strongly Polynomial for Discounted Dynamic Programming," Operations Research Letters, Vol. 42, pp. 429-431.

[FKB13] Frihauf, P., Krstic, M., and Basar, T., 2013. "Finite-Horizon LQ Control for Unknown Discrete-Time Linear Systems via Extremum Seeking," European Journal of Control, Vol. 19, pp. 399-407.

[FPB15] Farahmand, A. M., Precup, D., Barreto, A. M., and Ghavamzadeh, M., 2015. "Classification-Based Approximate Policy Iteration," IEEE Trans. on Automatic Control, Vol. 60, pp. 2989-2993.

[FYG06] Fern, A., Yoon, S., and Givan, R., 2006. "Approximate Policy Iteration with a Policy Language Bias: Solving Relational Markov Decision Processes," J. of Artificial Intelligence Research, Vol. 25, pp. 75-118.

[Fal87] Falcone, M., 1987. "A Numerical Approach to the Infinite Horizon Problem of Deterministic Control Theory," Appl. Math. Opt., Vol. 15, pp. 1-13.

[FeS04] Ferrari, S., and Stengel, R. F., 2004. "Model-Based Adaptive Critic Designs," in Learning and Approximate Dynamic Programming, by J. Si, A. Barto, W. Powell, and D. Wunsch, (Eds.), IEEE Press, N. Y.

[FeV02] Ferris, M. C., and Voelker, M. M., 2002. "Neuro-Dynamic Programming for Radiation Treatment Planning," Numerical Analysis Group Research Report NA-02/06, Oxford University Computing Laboratory, Oxford University.

[FeV04] Ferris, M. C., and Voelker, M. M., 2004. "Fractionation in Radiation Treatment Planning," Mathematical Programming B, Vol. 102, pp. 387-413.

[Fer10] Fernau, H., 2010. "Minimum Dominating Set of Queens: A Trivial Programming Exercise?" Discrete Applied Mathematics, Vol. 158, pp. 308-318.

[Fu17] Fu, M. C., 2017. "Markov Decision Processes, AlphaGo, and Monte Carlo Tree Search: Back to the Future," Leading Developments from INFORMS Communities,

INFORMS, pp. 68-88.

[FuH94] Fu, M. C., and Hu, J.-Q., 1994. "Smoothed Perturbation Analysis Derivative Estimation for Markov Chains," Oper. Res. Letters, Vol. 41, pp. 241-251.

[Fun89] Funahashi, K., 1989. "On the Approximate Realization of Continuous Mappings by Neural Networks," Neural Networks, Vol. 2, pp. 183-192.

[GBB04] Greensmith, E., Bartlett, P. L., and Baxter, J., 2004. "Variance Reduction Techniques for Gradient Estimates in Reinforcement Learning," Journal of Machine Learning Research, Vol. 5, pp. 1471-1530.

[GBC16] Goodfellow, I., Bengio, J., and Courville, A., Deep Learning, MIT Press, Cambridge, MA.

[GBL12] Grondman, I., Busoniu, L., Lopes, G. A. D., and Babuska, R., 2012. "A Survey of Actor-Critic Reinforcement Learning: Standard and Natural Policy Gradients," IEEE Trans. on Systems, Man, and Cybernetics, Part C, Vol. 42, pp. 1291-1307.

[GDP18] Guerriero, F., Di Puglia Pugliese, L., and Macrina, G., 2018. "A Rollout Algorithm for the Resource Constrained Elementary Shortest Path Problem," Optimization Methods and Software, pp. 1-19.

[GGS13] Gabillon, V., Ghavamzadeh, M., and Scherrer, B., 2013. "Approximate Dynamic Programming Finally Performs Well in the Game of Tetris," in NIPS, pp. 1754-1762.

[GLG11] Gabillon, V., Lazaric, A., Ghavamzadeh, M., and Scherrer, B., 2011. "Classification-Based Policy Iteration with a Critic," in Proc. of ICML.

[GSS93] Gordon, N. J., Salmond, D. J., and Smith, A. F., 1993. "Novel Approach to Nonlinear/Non-Gaussian Bayesian State Estimation," in IEE Proceedings, Vol. 140, pp. 107-113.

[GTO15] Goodson, J. C., Thomas, B. W., and Ohlmann, J. W., 2015. "Restocking-Based Rollout Policies for the Vehicle Routing Problem with Stochastic Demand and Duration Limits," Transportation Science, Vol. 50, pp. 591-607.

[Gla13] Glasserman, P., 2013. Monte Carlo Methods in Financial Engineering, Springer, N. Y.

[Gly87] Glynn, P. W., 1987. "Likelihood Ratio Gradient Estimation: An Overview," Proc. of the 1987 Winter Simulation Conference, pp. 366-375.

[Gly90] Glynn, P. W., 1990. "Likelihood Ratio Gradient Estimation for Stochastic Systems," Communications of the ACM, Vol. 33, pp. 75-84.

[GoR85] Gonzalez, R., and Rofman, E., 1985. "On Deterministic Control Problems: An Approximation Procedure for the Optimal Cost, Parts I, II," SIAM J. Control Optimization, Vol. 23, pp. 242-285.

[GoS84] Goodwin, G. C., and Sin, K. S. S., 1984. Adaptive Filtering, Prediction, and Control, Prentice-Hall, Englewood Cliffs, N. J.

[Gor95] Gordon, G. J., 1995. "Stable Function Approximation in Dynamic Programming," in Machine Learning: Proceedings of the Twelfth International Conference, Morgan Kaufmann, San Francisco, CA.

[Gos15] Gosavi, A., 2015. Simulation-Based Optimization: Parametric Optimization Techniques and Reinforcement Learning, 2nd Edition, Springer, N. Y.

[Gre05] Greensmith, E., 2005. Policy Gradient Methods: Variance Reduction and Stochastic Convergence, Ph.D. Thesis, The Australian National University.

[Grz17] Grzes, M., 2017. "Reward Shaping in Episodic Reinforcement Learning," in Proc. of the 16th Conference on Autonomous Agents and MultiAgent Systems, pp. 565-573.

[GuM03] Guerriero, F., and Mancini, M., 2003. "A Cooperative Parallel Rollout Algorithm for the Sequential Ordering Problem," Parallel Computing, Vol. 29, pp. 663-677.

[GuS18] Guillot, M., and Stauffer, G., 2018. The Stochastic Shortest Path Problem: A Polyhedral Combinatorics Perspective, European J. of Operational Research.

[HJG16] Huang, Q., Jia, Q. S., and Guan, X., 2016. "Robust Scheduling of EV Charging Load with Uncertain Wind Power Integration," IEEE Trans. on Smart Grid, Vol. 9, pp. 1043-1054.

[HLZ18] Hanawal, M. K., Liu, H., Zhu, H., and Paschalidis, I. C., 2018. "Learning Policies for Markov Decision Processes from Data," IEEE Trans. on Automatic Control.

[HMR19] Hastie, T., Montanari, A., Rosset, S., and Tibshirani, R. J., 2019. "Surprises in High-Dimensional Ridgeless Least Squares Interpolation," arXiv preprint arXiv:1903-.08560.

[HPC96] Helmsen, J., Puckett, E. G., Colella, P., and Dorr, M., 1996. "Two New Methods for Simulating Photolithography Development," SPIE's 1996 International Symposium on Microlithography, pp. 253-261.

[HSS08] Hofmann, T., Scholkopf, B., and Smola, A. J., 2008. "Kernel Methods in Machine Learning," The Annals of Statistics, Vol. 36, pp. 1171-1220.

[HSW89] Hornick, K., Stinchcombe, M., and White, H., 1989. "Multilayer Feedforward Networks are Universal Approximators," Neural Networks, Vol. 2, pp. 359-159.

[Han98] Hansen, E. A., 1998. "Solving POMDPs by Searching in Policy Space," in Proc. of the 14th Conf. on Uncertainty in Artificial Intelligence, pp. 211-219.

[Hay08] Haykin, S., 2008. Neural Networks and Learning Machines, 3rd Edition, Prentice-Hall, Englewood-Cliffs, N. J.

[IJT18] Iusem, Jofre, A., and Thompson, P., 2018. "Incremental Constraint Projection Methods for Monotone Stochastic Variational Inequalities," Math. of Operations Research, Vol. 44, pp. 236-263.

[IoS96] Ioannou, P. A., and Sun, J., 1996. Robust Adaptive Control, Prentice-Hall, Englewood Cliffs, N. J.

[Iva68] Ivakhnenko, A. G., 1968. "The Group Method of Data Handling: A Rival of the Method of Stochastic Approximation," Soviet Automatic Control, Vol. 13, pp. 43-55.

[Iva71] Ivakhnenko, A. G., 1971. "Polynomial Theory of Complex Systems," IEEE Trans. on Systems, Man and Cybernetics, Vol. 4, pp. 364-378.

[JSJ95] Jaakkola, T., Singh, S. P., and Jordan, M. I., 1995. "Reinforcement Learning Algorithm for Partially Observable Markov Decision Problems," NIPS, Vol. 7, pp. 345-352.

[JiJ17] Jiang, Y., and Jiang, Z. P., 2017. Robust Adaptive Dynamic Programming, J. Wiley, N. Y.

[JoB16] Joseph, A. G., and Bhatnagar, S., 2016. "Revisiting the Cross Entropy Method with Applications in Stochastic Global Optimization and Reinforcement Learning," in Proc. of the 22nd European Conference on Artificial Intelligence, pp. 1026-1034.

[JoB18] Joseph, A. G., and Bhatnagar, S., 2018. "A Cross Entropy Based Optimization Algorithm with Global Convergence Guarantees," arXiv preprint arXiv:1801.10291.

[Jon90] Jones, L. K., 1990. "Constructive Approximations for Neural Networks by Sigmoidal Functions," Proceedings of the IEEE, Vol. 78, pp. 1586-1589.

[JuP07] Jung, T., and Polani, D., 2007. "Kernelizing LSPE(λ)," Proc. 2007 IEEE Symposium on Approximate Dynamic Programming and Reinforcement Learning, Honolulu, Ha., pp. 338-345.

[KAC15] Kochenderfer, M. J., with Amato, C., Chowdhary, G., How, J. P., Davison Reynolds, H. J., Thornton, J. R., Torres-Carrasquillo, P. A., Ore, N. K., Vian, J., 2015. Decision Making under Uncertainty: Theory and Application, MIT Press, Cambridge, MA.

[KAH15] Khashooei, B. A., Antunes, D. J. and Heemels, W.P.M.H., 2015. "Rollout Strategies for Output-Based Event-Triggered Control," in Proc. 2015 European Control Conference, pp. 2168-2173.

[KGB82] Kimemia, J., Gershwin, S. B., and Bertsekas, D. P., 1982. "Computation of Production Control Policies by a Dynamic Programming Technique," in Analysis and Optimization of Systems, A. Bensoussan and J. L. Lions (eds.), Springer, N. Y., pp. 243-269.

[KKK95] Krstic, M., Kanellakopoulos, I., Kokotovic, P., 1995. Nonlinear and Adaptive Control Design, J. Wiley, N. Y.

[KLC98] Kaelbling, L. P., Littman, M. L., and Cassandra, A. R., 1998. "Planning and Acting in Partially Observable Stochastic Domains," Artificial Intelligence, Vol. 101, pp. 99-134.

[KLM96] Kaelbling, L. P., Littman, M. L., and Moore, A. W., 1996. "Reinforcement Learning: A Survey," J. of Artificial Intelligence Res., Vol. 4, pp. 237-285.

[KMP06] Keller, P. W., Mannor, S., and Precup, D., 2006. "Automatic Basis Function Construction for Approximate Dynamic Programming and Reinforcement Learning," Proc. of the 23rd ICML, Pittsburgh, Penn.

[KMP13] Kar, S., Moura, J. M., and Poor, H. V., 2013. "Qd-Learning: A Collaborative Distributed Strategy for Multi-Agent Reinforcement Learning Through Consensus and Innovations," IEEE Trans. on Signal Processing, Vol. 61, pp. 1848?1862.

[KRC13] Kroese, D. P., Rubinstein, R. Y., Cohen, I., Porotsky, S., and Taimre, T., 2013. "Cross-Entropy Method," in Encyclopedia of Operations Research and Management Science, Springer, Boston, MA, pp. 326-333.

[Kak02] Kakade, S. A., 2002. "Natural Policy Gradient," NIPS, Vol. 14, pp. 1531-1538.

[KeG88] Keerthi, S. S., and Gilbert, E. G., 1988. "Optimal, Infinite Horizon Feedback Laws for a General Class of Constrained Discrete Time Systems: Stability and Moving-Horizon Approximations," J. Optimization Theory Appl., Vo. 57, pp. 265-293.

[KiK06] Killingsworth, N. J., and Krstic, M., 2006. "PID Tuning Using Extremum Seeking," IEEE Control Systems Magazine, pp. 70-79.

[Kim82] Kimemia, J., 1982. "Hierarchical Control of Production in Flexible Manufacturing Systems," Ph.D. Thesis, Dep. of Electrical Engineering and Computer Science, Massachusetts Institute of Technology.

[KoC15] Kouvaritakis, B., and Cannon, M., 2015. Model Predictive Control: Classical, Robust and Stochastic, Springer, N. Y.

[KoS06] Kocsis, L., and Szepesvari, C., 2006. "Bandit Based Monte-Carlo Planning," Proc. of 17th European Conference on Machine Learning, Berlin, pp. 282-293.

[KoT99] Konda, V. R., and Tsitsiklis, J. N., 1999. "Actor-Critic Algorithms," NIPS, Denver, Colorado, pp. 1008-1014.

[KoT03] Konda, V. R., and Tsitsiklis, J. N., 2003. "Actor-Critic Algorithms," SIAM J. on Control and Optimization, Vol. 42, pp. 1143-1166.

[Kor90] Korf, R. E., 1990. "Real-Time Heuristic Search," Artificial Intelligence, Vol. 42, pp. 189-211.

[Kre19] Krener, A. J., 2019. "Adaptive Horizon Model Predictive Control and Al'brekht's Method," arXiv preprint arXiv:1904.00053.

[Kri16] Krishnamurthy, V., 2016. Partially Observed Markov Decision Processes, Cambridge Univ. Press.

[KuD92] Kushner, H. J., and Dupuis, P. G., 1992. Numerical Methods for Stochastic Control Problems in Continuous Time, Springer, N. Y.

[KuV86] Kumar, P. R., and Varaiya, P. P., 1986. Stochastic Systems: Estimation, Identification, and Adaptive Control, Prentice-Hall, Englewood Cliffs, N. J.

[KuY03] Kushner, H. J., and Yin, G., 2003. Stochastic Approximation and Recursive Algorithms and Applications, (2nd Ed.), Springer-Verlag, New York.

[Kum83] Kumar, P. R., 1983. "Optimal Adaptive Control of Linear-Quadratic-Gaussian Systems," SIAM J. on Control and Optimization, Vol. 21, pp. 163-178.

[Kun14] Kung, S. Y., 2014. Kernel Methods and Machine Learning, Cambridge Univ. Press.

[L'Ec91] L'Ecuyer, P., 1991. "An Overview of Derivative Estimation," Proceedings of the 1991 Winter Simulation Conference, pp. 207-217.

[LGM03] Lequin, O., Gevers, M., Mossberg, M., Bosmans, E., and Triest, L., 2003. "Iterative Feedback Tuning of PID Parameters: Comparison with Classical Tuning Rules," Control Engineering Practice, Vol. 11, pp. 1023-1033.

[LGM10] Lazaric, A., Ghavamzadeh, M., and Munos, R., 2010. "Analysis of a Classification-Based Policy Iteration Algorithm," INRIA Report.

[LGW16] Lan, Y., Guan, X., and Wu, J., 2016. "Rollout Strategies for Real-Time Multi-Energy Scheduling in Microgrid with Storage System," IET Generation, Transmission and Distribution, Vol. 10, pp. 688-696.

[LLL08] Lewis, F. L., Liu, D., and Lendaris, G. G., 2008. Special Issue on Adaptive Dynamic Programming and Reinforcement Learning in Feedback Control, IEEE Trans. on Systems, Man, and Cybernetics, Part B, Vol. 38, No. 4.

[LLP93] Leshno, M., Lin, V. Y., Pinkus, A., and Schocken, S., 1993. "Multilayer Feedforward Networks with a Nonpolynomial Activation Function can Approximate any Function," Neural Networks, Vol. 6, pp. 861-867.

[LWW17] Liu, D., Wei, Q., Wang, D., Yang, X., and Li, H., 2017. Adaptive Dynamic Programming with Applications in Optimal Control, Springer, Berlin.

[LaP03] Lagoudakis, M. G., and Parr, R., 2003. "Reinforcement Learning as Classification: Leveraging Modern Classifiers," in Proc. of ICML, pp. 424-431.

[LaR85] Lai, T., and Robbins, H., 1985. "Asymptotically Efficient Adaptive Allocation Rules," Advances in Applied Mathematics, Vol. 6, pp. 4-22.

[LeL13] Lewis, F. L., and Liu, D., (Eds), 2013. Reinforcement Learning and Approximate Dynamic Programming for Feedback Control, Wiley, Hoboken, N. J.

[LeV09] Lewis, F. L., and Vrabie, D., 2009. "Reinforcement Learning and Adaptive Dynamic Programming for Feedback Control," IEEE Circuits and Systems Magazine, 3rd Q. Issue.

[Lee17] Lee, J., 2017. "A Survey of Robot Learning from Demonstrations for Human-Robot Collaboration," arXiv preprint arXiv:1710.08789.

[LiW14] Liu, D., and Wei, Q., 2014. "Policy Iteration Adaptive Dynamic Programming Algorithm for Discrete-Time Nonlinear Systems," IEEE Trans. on Neural Networks and Learning Systems, Vol. 25, pp. 621-634.

[Li17] Li, Y., 2017. "Deep Reinforcement Learning: An Overview," arXiv preprint ArXiv: 1701.07274v5.

[LiR06] Lincoln, B., and Rantzer, A., 2006. "Relaxing Dynamic Programming," IEEE Trans. Automatic Control, Vol. 51, pp. 1249-1260.

[LiS16] Liang, S., and Srikant, R., 2016. "Why Deep Neural Networks for Function Approximation?" arXiv preprint arXiv:1610.04161.

[LiW15] Li, H., and Womer, N. K., 2015. "Solving Stochastic Resource-Constrained Project Scheduling Problems by Closed-Loop Approximate Dynamic Programming," European J. of Operational Research, Vol. 246, pp. 20-33.

[Lib11] Liberzon, D., 2011. Calculus of Variations and Optimal Control Theory: A Concise Introduction, Princeton Univ. Press.

[Liu01] Liu, J. S., 2001. Monte Carlo Strategies in Scientific Computing, Springer, N. Y.

[LoS01] Longstaff, F. A., and Schwartz, E. S., 2001. "Valuing American Options by Simulation: A Simple Least-Squares Approach," Review of Financial Studies, Vol. 14, pp. 113-147.

[MBT05] Mitchell, I. M., Bayen, A. M., and Tomlin, C. J., 2005. "A Time-Dependent Hamilton-Jacobi Formulation of Reachable Sets for Continuous Dynamic Games," IEEE Trans. on Automatic Control, Vol. 50, pp. 947-957.

[MKS15] Mnih, V., Kavukcuoglu, K., Silver, D., Rusu, A. A., Veness, J., Bellemare, M. G., Graves, A., Riedmiller, M., Fidjeland, A. K., Ostrovski, G., and Petersen, S., 2015. "Human-Level Control Through Deep Reinforcement Learning," Nature, Vol. 518, p. 529.

[MMB02] McGovern, A., Moss, E., and Barto, A., 2002. "Building a Basic Building Block Scheduler Using Reinforcement Learning and Rollouts," Machine Learning, Vol. 49, pp. 141-160.

[MMS05] Menache, I., Mannor, S., and Shimkin, N., 2005. "Basis Function Adaptation in Temporal Difference Reinforcement Learning," Ann. Oper. Res., Vol. 134, pp. 215-238.

[MPK99] Meuleau, N., Peshkin, L., Kim, K. E., and Kaelbling, L. P., 1999. "Learning Finite-State Controllers for Partially Observable Environments," in Proc. of the 15th Conference on Uncertainty in Artificial Intelligence, pp. 427-436.

[MPP04] Meloni, C., Pacciarelli, D., and Pranzo, M., 2004. "A Rollout Metaheuristic for Job Shop Scheduling Problems," Annals of Operations Research, Vol. 131, pp. 215-235.

[MRG03] Mannor, S., Rubinstein, R. Y., and Gat, Y., 2003. "The Cross Entropy Method for Fast Policy Search," in Proc. of the 20th International Conference on Machine Learning (ICML-03), pp. 512-519.

[MVS19] Muthukumar, V., Vodrahalli, K., and Sahai, A., 2019. "Harmless Interpolation of Noisy Data in Regression," arXiv preprint arXiv:1903.09139.

[MYF03] Moriyama, H., Yamashita, N., and Fukushima, M., 2003. "The Incremental Gauss-Newton Algorithm with Adaptive Stepsize Rule," Computational Optimization and Applications, Vol. 26, pp. 107-141.

[MaJ15] Mastin, A., and Jaillet, P., 2015. "Average-Case Performance of Rollout Algorithms for Knapsack Problems," J. of Optimization Theory and Applications, Vol. 165, pp. 964-984.

[MaK12] Mausam, and Kolobov, A., 2012. "Planning with Markov Decision Processes: An AI Perspective," Synthesis Lectures on Artificial Intelligence and Machine Learning, Vol. 6, pp. 1-210.

[MaT01] Marbach, P., and Tsitsiklis, J. N., 2001. "Simulation-Based Optimization of Markov Reward Processes," IEEE Trans. on Aut. Control, Vol. 46, pp. 191-209.

[MaT03] Marbach, P., and Tsitsiklis, J. N., 2003. "Approximate Gradient Methods in Policy-Space Optimization of Markov Reward Processes," J. Discrete Event Dynamic Systems, Vol. 13, pp. 111-148.

[Mac02] Maciejowski, J. M., 2002. Predictive Control with Constraints, Addison-Wesley, Reading, MA.

[Mat65] J. Matyas, J., 1965. "Random Optimization," Automation and Remote Control, Vol. 26, pp. 246-253.

[May14] Mayne, D. Q., 2014. "Model Predictive Control: Recent Developments and Future Promise," Automatica, Vol. 50, pp. 2967-2986.

[MeB99] Meuleau, N., and Bourgine, P., 1999. "Exploration of Multi-State Environments: Local Measures and Back-Propagation of Uncertainty," Machine Learning, Vol. 35, pp. 117-154.

[Mey07] Meyn, S., 2007. Control Techniques for Complex Networks, Cambridge Univ. Press, N. Y.

[MoL99] Morari, M., and Lee, J. H., 1999. "Model Predictive Control: Past, Present, and Future," Computers and Chemical Engineering, Vol. 23, pp. 667-682.

[MuS08] Munos, R., and Szepesvari, C, 2008. "Finite-Time Bounds for Fitted Value Iteration," J. of Machine Learning Research, Vol. 1, pp. 815-857.

[Mun14] Munos, R., 2014. "From Bandits to Monte-Carlo Tree Search: The Optimistic Principle Applied to Optimization and Planning," Foundations and Trends in Machine Learning, Vol. 7, pp. 1-129.

[NHR99] Ng, A. Y., Harada, D., and Russell, S. J., 1999. "Policy Invariance Under Reward Transformations: Theory and Application to Reward Shaping," in Proc. of the 16th International Conference on Machine Learning, pp. 278-287.

[NaA12] Narendra, K. S., and Annaswamy, A. M., 2012. Stable Adaptive Systems, Courier Corporation.

[NeB00] Nedić, A., and Bertsekas, D. P., 2000. "Convergence Rate of Incremental Subgradient Algorithms," in Stochastic Optimization: Algorithms and Applications, by S. Uryasev and P. M. Pardalos, Eds., Kluwer, pp. 263-304.

[NeB01] Nedić, A., and Bertsekas, D. P., 2001. "Incremental Subgradient Methods for Nondifferentiable Optimization," SIAM J. on Optimization, Vol. 12, pp. 109-138.

[NeB03] Nedić, A., and Bertsekas, D. P., 2003. "Least-Squares Policy Evaluation Algorithms with Linear Function Approximation," J. of Discrete Event Systems, Vol. 13, pp. 79-110.

[NeS12] Neu, G., and Szepesvari, C., 2012. "Apprenticeship Learning Using Inverse Reinforcement Learning and Gradient Methods," arXiv preprint arXiv:1206.5264.

[NeS17] Nesterov, Y., and Spokoiny, V., 2017. "Random Gradient-Free Minimization of Convex Functions," Foundations of Computational Mathematics, Vol. 17, pp. 527-566.

[Ned11] Nedić, A., 2011. "Random Algorithms for Convex Minimization Problems," Math. Programming, Ser. B, Vol. 129, pp. 225-253.

[OVR19] Osband, I., Van Roy, B., Russo, D. J., and Wen, Z., 2019. "Deep Exploration via Randomized Value Functions," arXiv preprint arXiv: 1703.07608v4.

[OrS02] Ormoneit, D., and Sen, S., 2002. "Kernel-Based Reinforcement Learning," Machine Learning, Vol. 49, pp. 161-178.

[PBT98] Polymenakos, L. C., Bertsekas, D. P., and Tsitsiklis, J. N., 1998. "Efficient Algorithms for Continuous-Space Shortest Path Problems," IEEE Trans. on Automatic Control, Vol. 43, pp. 278-283.

[PDC14] Pillonetto, G., Dinuzzo, F., Chen, T., De Nicolao, G., and Ljung, L., 2014. "Kernel Methods in System Identification, Machine Learning and Function Estimation: A Survey," Automatica, Vol. 50, pp. 657-682.

[PSS98] Precup, D., Sutton, R. S., and Singh, S., 1998. "Theoretical Results on Reinforcement Learning with Temporally Abstract Options," in European Conf. on Machine Learning, Springer, Berlin, pp. 382-393.

[PaR12] Papahristou, N., and Refanidis, I., 2012. "On the Design and Training of Bots to Play Backgammon Variants," in IFIP International Conference on Artificial Intelligence Applications and Innovations, pp. 78-87.

[PaT00] Paschalidis, I. C., and Tsitsiklis, J. N., 2000. "Congestion-Dependent Pricing of Network Services," IEEE/ACM Trans. on Networking, Vol. 8, pp. 171-184.

[Pat01] Patek, S. D., 2001. "On Terminating Markov Decision Processes with a Risk Averse Objective Function," Automatica, Vol. 37, pp. 1379-1386.

[Pat07] Patek, S. D., 2007. "Partially Observed Stochastic Shortest Path Problems with Approximate Solution by Neuro-Dynamic Programming," IEEE Trans. on Systems, Man, and Cybernetics Part A, Vol. 37, pp. 710-720.

[PeG04] Peret, L., and Garcia, F., 2004. "On-Line Search for Solving Markov Decision Processes via Heuristic Sampling," in Proc. of the 16th European Conference on Artificial Intelligence, pp. 530-534.

[PeS08] Peters, J., and Schaal, S., 2008. "Reinforcement Learning of Motor Skills with Policy Gradients," Neural Networks, Vol. 4, pp. 682-697.

[PeW96] Peng, J., and Williams, R., 1996. "Incremental Multi-Step Q-Learning," Machine Learning, Vol. 22, pp. 283-290.

[PoB04] Poupart, P., and Boutilier, C., 2004. "Bounded Finite State Controllers," in Advances in Neural Information Processing Systems, pp. 823-830.

[PoV04] Powell, W. B., and Van Roy, B., 2004. "Approximate Dynamic Programming for High-Dimensional Resource Allocation Problems," in Learning and Approximate Dynamic Programming, by J. Si, A. Barto, W. Powell, and D. Wunsch, (Eds.), IEEE Press, N. Y.

[Pow11] Powell, W. B., 2011. Approximate Dynamic Programming: Solving the Curses of Dimensionality, 2nd Edition, J. Wiley and Sons, Hoboken, N. J.

[Pre95] Prekopa, A., 1995. Stochastic Programming, Kluwer, Boston.

[PuS78] Puterman, M. L., and Shin, M. C., 1978. "Modified Policy Iteration Algorithms for Discounted Markov Decision Problems," Management Sci., Vol. 24, pp. 1127-1137.

[PuS82] Puterman, M. L., and Shin, M. C., 1982. "Action Elimination Procedures for Modified Policy Iteration Algorithms," Operations Research, Vol. 30, pp. 301-318.

[Put94] Puterman, M. L., 1994. Markovian Decision Problems, J. Wiley, N. Y.

[QuL19] Qu, G., Li, N., "Exploiting Fast Decaying and Locality in Multi-Agent MDP with Tree Dependence Structure," Proc. of 2019 CDC, Nice, France.

[RPW91] Rogers, D. F., Plante, R. D., Wong, R. T., and Evans, J. R., 1991. "Aggregation and Disaggregation Techniques and Methodology in Optimization," Operations Research, Vol. 39, pp. 553-582.

[RSM08] Reisinger, J., Stone, P., and Miikkulainen, R., 2008. "Online Kernel Selection for Bayesian Reinforcement Learning," in Proc. of the 25th International Conference on Machine Learning, pp. 816-823.

[RVK18] Russo, D. J., Van Roy, B., Kazerouni, A., Osband, I., and Wen, Z., 2018. A Tutorial on Thompson Sampling, Foundations and Trends in Machine Learning.

[RaK18] Radenkovic, S., and Krstic, M., 2018. "Extremum Seeking-Based Perfect Adaptive Tracking of Non-PE References Despite Nonvanishing Variance of Perturbations," Automatica, Vol. 93, pp. 189-196.

[Ras63] Rastrigin, R. A., 1963. "About Convergence of Random Search Method in Extremal Control of Multi-Parameter Systems," Avtomat. i Telemekh., Vol. 24, pp. 1467-1473.

[Rec18a] Recht, B., 2018. "A Tour of Reinforcement Learning: The View from Continuous Control," Annual Review of Control, Robotics, and Autonomous Systems.

[Rec18b] Recht, B., 2018. "An Outsider's Tour of Reinforcement Learning," at http://www.argmin.net/2018/06/25/outsider-rl/

[RoC10] Robert, C. P., and Casella, G., 2010. Monte Carlo Statistical Methods, Springer, N. Y.

[Ros70] Ross, S. M., 1970. Applied Probability Models with Optimization Applications, Holden-Day, San Francisco, CA.

[Ros12] Ross, S. M., 2012. Simulation, 5th Edition, Academic Press, Orlando, Fla.

[RuK04] Rubinstein, R. Y., and Kroese, D. P., 2004. The Cross-Entropy Method: A Unified Approach to Combinatorial Optimization, Springer, N. Y.

[RuK13] Rubinstein, R. Y., and Kroese, D. P., 2013. The Cross-Entropy Method: A Unified Approach to Combinatorial Optimization, Monte-Carlo Simulation and Machine Learning, Springer Science and Business Media.

[RuK16] Rubinstein, R. Y., and Kroese, D. P., 2016. Simulation and the Monte Carlo Method, 3rd Edition, J. Wiley, N. Y.

[RuN94] Rummery, G. A., and Niranjan, M., 1994. "On-Line Q-Learning Using Connectionist Systems," University of Cambridge, England, Department of Engineering, TR-166.

[RuN16] Russell, S. J., and Norvig, P., 2016. Artificial Intelligence: A Modern Approach, Pearson Education Limited, Malaysia.

[RuV16] Russo, D., and Van Roy, B., 2016. "An Information-Theoretic Analysis of Thompson Sampling," The Journal of Machine Learning Research, Vol. 17, pp. 2442-2471.

[Rub69] Rubinstein, R. Y., 1969. Some Problems in Monte Carlo Optimization, Ph.D. Thesis.

[SBP04] Si, J., Barto, A., Powell, W., and Wunsch, D., (Eds.) 2004. Learning and Approximate Dynamic Programming, IEEE Press, N. Y.

[SGC02] Savagaonkar, U., Givan, R., and Chong, E. K. P., 2002. "Sampling Techniques for Zero-Sum, Discounted Markov Games," in Proc. 40th Allerton Conference on Communication, Control and Computing, Monticello, Ill.

[SGG15] Scherrer, B., Ghavamzadeh, M., Gabillon, V., Lesner, B., and Geist, M., 2015. "Approximate Modified Policy Iteration and its Application to the Game of Tetris," J. of Machine Learning Research, Vol. 16, pp. 1629-1676.

[SHB15] Simroth, A., Holfeld, D., and Brunsch, R., 2015. "Job Shop Production Planning under Uncertainty: A Monte Carlo Rollout Approach," Proc. of the International Scientific and Practical Conference, Vol. 3, pp. 175-179.

[SHC17] Salimans, T., Ho, J., Chen, X., Sidor, S., and Sutskever, I., 2017. "Evolution Strategies as a Scalable Alternative to Reinforcement Learning," arXiv preprint arXiv:1703.03864.

[SHM16] Silver, D., Huang, A., Maddison, C. J., Guez, A., Sifre, L., Van Den Driessche, G., Schrittwieser, J., Antonoglou, I., Panneershelvam, V., Lanctot, M., and Dieleman, S., 2016. "Mastering the Game of Go with Deep Neural Networks and Tree Search," Nature, Vol. 529, pp. 484-489.

[SHS17] Silver, D., Hubert, T., Schrittwieser, J., Antonoglou, I., Lai, M., Guez, A., Lanctot, M., Sifre, L., Kumaran, D., Graepel, T., and Lillicrap, T., 2017. "Mastering Chess and Shogi by Self-Play with a General Reinforcement Learning Algorithm," arXiv preprint arXiv:1712.01815.

[SJJ95] Singh, S. P., Jaakkola, T., and Jordan, M. I., 1995. "Reinforcement Learning with Soft State Aggregation," in Advances in Neural Information Processing Systems 7, MIT Press, Cambridge, MA.

[SJL18] Soltanolkotabi, M., Javanmard, A., and Lee, J. D., 2018. "Theoretical Insights into the Optimization Landscape of Over-Parameterized Shallow Neural Networks," IEEE Trans. on Information Theory, Vol. 65, pp. 742-769.

[SLJ13] Sun, B., Luh, P. B., Jia, Q. S., Jiang, Z., Wang, F., and Song, C., 2013. "Building Energy Management: Integrated Control of Active and Passive Heating, Cooling, Lighting, Shading, and Ventilation Systems," IEEE Trans. on Automation Science and Engineering, Vol. 10, pp. 588-602.

[SMS99] Sutton, R. S., McAllester, D., Singh, S. P., and Mansour, Y., 1999. "Policy Gradient Methods for Reinforcement Learning with Function Approximation," NIPS, Denver, Colorado.

[SNC18] Sarkale, Y., Nozhati, S., Chong, E. K., Ellingwood, B. R., and Mahmoud, H., 2018. "Solving Markov Decision Processes for Network-Level Post-Hazard Recovery via Simulation Optimization and Rollout," in 2018 IEEE 14th International Conference on Automation Science and Engineering, pp. 906-912.

[SSP18] Serban, I. V., Sankar, C., Pieper, M., Pineau, J., Bengio, J., 2018. "The Bottleneck Simulator: A Model-Based Deep Reinforcement Learning Approach," arXiv preprint arXiv:1807.04723.v1.

[SYL04] Si, J., Yang, L., and Liu, D., 2004. "Direct Neural Dynamic Programming," in Learning and Approximate Dynamic Programming, by J. Si, A. Barto, W. Powell, and D. Wunsch, (Eds.), IEEE Press, N. Y.

[SYL17] Saldi, N., Yuksel, S., and Linder, T., 2017. "Finite Model Approximations for Partially Observed Markov Decision Processes with Discounted Cost," arXiv preprint arXiv:1710.07009.

[SZL08] Sun, T., Zhao, Q., Lun, P., and Tomastik, R., 2008. "Optimization of Joint Replacement Policies for Multipart Systems by a Rollout Framework," IEEE Trans. on Automation Science and Engineering, Vol. 5, pp. 609-619.

[SaB11] Sastry, S., and Bodson, M., 2011. Adaptive Control: Stability, Convergence and Robustness, Courier Corporation.

[Sam59] Samuel, A. L., 1959. "Some Studies in Machine Learning Using the Game of Checkers," IBM J. of Research and Development, pp. 210-229.

[Sam67] Samuel, A. L., 1967. "Some Studies in Machine Learning Using the Game of Checkers. II – Recent Progress," IBM J. of Research and Development, pp. 601-617.

[ScS85] Schweitzer, P. J., and Seidman, A., 1985. "Generalized Polynomial Approximations in Markovian Decision Problems," J. Math. Anal. and Appl., Vol. 110, pp. 568-582.

[ScS02] Scholkopf, B., and Smola, A. J., 2002. Learning with Kernels: Support Vector Machines, Regularization, Optimization, and Beyond, MIT Press, Cambridge, MA.

[Sch99] Schaal, S., 1999. "Is Imitation Learning the Route to Humanoid Robots?" Trends in Cognitive Sciences, Vol. 3, pp. 233-242.

[Sch13] Scherrer, B., 2013. "Performance Bounds for Lambda Policy Iteration and Application to the Game of Tetris," J. of Machine Learning Research, Vol. 14, pp. 1181-1227.

[Sch15] Schmidhuber, J., 2015. "Deep Learning in Neural Networks: An Overview," Neural Networks, pp. 85-117.

[Sec00] Secomandi, N., 2000. "Comparing Neuro-Dynamic Programming Algorithms for the Vehicle Routing Problem with Stochastic Demands," Computers and Operations Research, Vol. 27, pp. 1201-1225.

[Sec01] Secomandi, N., 2001. "A Rollout Policy for the Vehicle Routing Problem with Stochastic Demands," Operations Research, Vol. 49, pp. 796-802.

[Sec03] Secomandi, N., 2003. "Analysis of a Rollout Approach to Sequencing Problems with Stochastic Routing Applications," J. of Heuristics, Vol. 9, pp. 321-352.

[Set99a] Sethian, J. A., 1999. Level Set Methods and Fast Marching Methods Evolving Interfaces in Computational Geometry, Fluid Mechanics, Computer Vision, and Materials Science, Cambridge University Press, N. Y.

[Set99b] Sethian, J. A., 1999. "Fast Marching Methods," SIAM Review, Vol. 41, pp. 199-235.

[ShC04] Shawe-Taylor, J., and Cristianini, N., 2004. Kernel Methods for Pattern Analysis, Cambridge Univ. Press.

[Sha50] Shannon, C., 1950. "Programming a Digital Computer for Playing Chess," Phil. Mag., Vol. 41, pp. 356-375.

[SlL91] Slotine, J.-J. E., and Li, W., Applied Nonlinear Control, Prentice-Hall, Englewood Cliffs, N. J.

[SpV05] Spaan, M. T., and Vlassis, N., 2005. "Perseus: Randomized Point-Based Value Iteration for POMDPs," J. of Artificial Intelligence Research, Vol. 24, pp. 195-220.

[Spa92] Spall, J. C., "Multivariate Stochastic Approximation Using a Simultaneous Perturbation Gradient Approximation," IEEE Trans. on Automatic Control, Vol. pp. 332-341.

[Spa03] Spall, J. C., 2003. Introduction to Stochastic Search and Optimization: Estimation, Simulation, and Control, J. Wiley, Hoboken, N. J.

[Str18] Strogatz, S., 2018. "One Giant Step for a Chess-Playing Machine," NY Times article on Dec. 26, 2018.

[SuB18] Sutton, R., and Barto, A. G., 2018. Reinforcement Learning, 2nd Edition, MIT Press, Cambridge, MA.

[Sut88] Sutton, R. S., 1988. "Learning to Predict by the Methods of Temporal Differences," Machine Learning, Vol. 3, pp. 9-44.

[SzL06] Szita, I., and Lorinz, A., 2006. "Learning Tetris Using the Noisy Cross-Entropy Method," Neural Computation, Vol. 18, pp. 2936-2941.

[Sze10] Szepesvari, C., 2010. Algorithms for Reinforcement Learning, Morgan and Claypool Publishers, San Franscisco, CA.

[TCW19] Tseng, W. J., Chen, J. C., Wu, I. C., and Wei, T. H., 2019. "Comparison Training for Computer Chinese Chess," IEEE Trans. on Games.

[TGL13] Tesauro, G., Gondek, D. C., Lenchner, J., Fan, J., and Prager, J. M., 2013. "Analysis of Watson's Strategies for Playing Jeopardy!," J. of Artificial Intelligence Research, Vol. 47, pp. 205-251.

[TeG96] Tesauro, G., and Galperin, G. R., 1996. "On-Line Policy Improvement Using Monte Carlo Search," NIPS, Denver, CO.

[Tes89a] Tesauro, G. J., 1989. "Neurogammon Wins Computer Olympiad," Neural Computation, Vol. 1, pp. 321-323.

[Tes89b] Tesauro, G. J., 1989. "Connectionist Learning of Expert Preferences by Comparison Training," in Advances in Neural Information Processing Systems, pp. 99-106.

[Tes92] Tesauro, G. J., 1992. "Practical Issues in Temporal Difference Learning," Machine Learning, Vol. 8, pp. 257-277.

[Tes94] Tesauro, G. J., 1994. "TD-Gammon, a Self-Teaching Backgammon Program, Achieves Master-Level Play," Neural Computation, Vol. 6, pp. 215-219.

[Tes95] Tesauro, G. J., 1995. "Temporal Difference Learning and TD-Gammon," Communications of the ACM, Vol. 38, pp. 58-68.

[Tes01] Tesauro, G. J., 2001. "Comparison Training of Chess Evaluation Functions," in Machines that Learn to Play Games, Nova Science Publishers, pp. 117-130.

[Tes02] Tesauro, G. J., 2002. "Programming Backgammon Using Self-Teaching Neural Nets," Artificial Intelligence, Vol. 134, pp. 181-199.

[ThS09] Thiery, C., and Scherrer, B., 2009. "Improvements on Learning Tetris with Cross-Entropy," International Computer Games Association J., Vol. 32, pp. 23-33.

[ThS10] Thiery, C., and Scherrer, B., 2010. "Performane Bound for Approximate Optimistic Policy Iteration," Technical Report, INRIA, France.

[TsV96] Tsitsiklis, J. N., and Van Roy, B., 1996. "Feature-Based Methods for Large-Scale Dynamic Programming," Machine Learning, Vol. 22, pp. 59-94.

[TsV97] Tsitsiklis, J. N., and Van Roy, B., 1997. "An Analysis of Temporal-Difference Learning with Function Approximation," IEEE Trans. on Aut. Control, Vol. 42, pp. 674-690.

[TsV99a] Tsitsiklis, J. N., and Van Roy, B., 1999. "Average Cost Temporal-Difference Learning," Automatica, Vol. 35, pp. 1799-1808.

[TsV99b] Tsitsiklis, J. N., and Van Roy, B., 1999. "Optimal Stopping of Markov Processes: Hilbert Space Theory, Approximation Algorithms, and an Application to Pricing Financial Derivatives", IEEE Trans. on Aut. Control, Vol. 44, pp. 1840-1851.

[Tse98] Tseng, P., 1998. "Incremental Gradient(-Projection) Method with Momentum Term and Adaptive Stepsize Rule," SIAM J. on Optimization, Vol. 8, pp. 506-531.

[Tsi94] Tsitsiklis, J. N., 1994. "Asynchronous Stochastic Approximation and Q-Learning," Machine Learning, Vol. 16, pp. 185-202.

[Tsi95] Tsitsiklis, J. N., 1995. "Efficient Algorithms for Globally Optimal Trajectories," IEEE Trans. Automatic Control, Vol. AC-40, pp. 1528-1538.

[TuP03] Tu, F., and Pattipati, K. R., 2003. "Rollout Strategies for Sequential Fault Diagnosis," IEEE Trans. on Systems, Man and Cybernetics, Part A, pp. 86-99.

[UGM18] Ulmer, M. W., Goodson, J. C., Mattfeld, D. C., and Hennig, M., 2018. "Offline-Online Approximate Dynamic Programming for Dynamic Vehicle Routing with Stochastic Requests," Transportation Science, Vol. 53, pp. 185-202.

[Ulm17] Ulmer, M. W., 2017. Approximate Dynamic Programming for Dynamic Vehicle Routing, Springer, Berlin.

[VVL13] Vrabie, D., Vamvoudakis, K. G., and Lewis, F. L., 2013. Optimal Adaptive Control and Differential Games by Reinforcement Learning Principles, The Institution of Engineering and Technology, London.

[Vla08] Vladimirsky, A., 2008. "Label-Setting Methods for Multimode Stochastic Shortest Path Problems on Graphs," Math. of Operations Research, Vol. 33, pp. 821-838.

[WCG02] Wu, G., Chong, E. K. P., and Givan, R. L., 2002. "Burst-Level Congestion Control Using Hindsight Optimization," IEEE Transactions on Aut. Control, Vol. 47, pp. 979-991.

[WCG03] Wu, G., Chong, E. K. P., and Givan, R. L., 2003. "Congestion Control Using Policy Rollout," Proc. 2nd IEEE CDC, Maui, Hawaii, pp. 4825-4830.

[WOB15] Wang, Y., O'Donoghue, B., and Boyd, S., 2015. "Approximate Dynamic Programming via Iterated Bellman Inequalities," International J. of Robust and Nonlinear Control, Vol. 25, pp. 1472-1496.

[WaB13a] Wang, M., and Bertsekas, D. P., 2013. "Stabilization of Stochastic Iterative Methods for Singular and Nearly Singular Linear Systems," Mathematics of Operations Research, Vol. 39, pp. 1-30.

[WaB13b] Wang, M., and Bertsekas, D. P., 2013. "Convergence of Iterative Simulation-Based Methods for Singular Linear Systems," Stochastic Systems, Vol. 3, pp. 39-96.

[WaB14] Wang, M., and Bertsekas, D. P., 2014. "Incremental Constraint Projection Methods for Variational Inequalities," Mathematical Programming, pp. 1-43.

[WaB16] Wang, M., and Bertsekas, D. P., 2016. "Stochastic First-Order Methods with Random Constraint Projection," SIAM Journal on Optimization, Vol. 26, pp. 681-717.

[WaP17] Wang, J., and Paschalidis, I. C., 2017. "An Actor-Critic Algorithm with Second-Order Actor and Critic," IEEE Trans. on Automatic Control, Vol. 62, pp. 2689-2703.

[Van76] Van Nunen, J. A., 1976. Contracting Markov Decision Processes, Mathematical Centre Report, Amsterdam.

[Wat89] Watkins, C. J. C. H., Learning from Delayed Rewards, Ph.D. Thesis, Cambridge Univ., England.

[WeB99] Weaver, L., and Baxter, J., 1999. "Reinforcement Learning From State and Temporal Differences," Tech. Report, Department of Computer Science, Australian National University.

[WhS92] White, D., and Sofge, D., (Eds.), 1992. Handbook of Intelligent Control, Van Nostrand, N. Y.

[WhS94] White, C. C., and Scherer, W. T., 1994. "Finite-Memory Suboptimal Design for Partially Observed Markov Decision Processes," Operations Research, Vol. 42, pp. 439-455.

[Whi88] Whittle, P., 1988. "Restless Bandits: Activity Allocation in a Changing World," J. of Applied Probability, pp. 287-298.

[Whi91] White, C. C., 1991. "A Survey of Solution Techniques for the Partially Observed Markov Decision Process," Annals of Operations Research, Vol. 32, pp. 215-230.

[WiB93] Williams, R. J., and Baird, L. C., 1993. "Analysis of Some Incremental Variants of Policy Iteration: First Steps Toward Understanding Actor-Critic Learning Systems," Report NU-CCS-93-11, College of Computer Science, Northeastern Univ., Boston, MA.

[WiS98] Wiering, M., and Schmidhuber, J., 1998. "Fast Online Q(λ)," Machine Learning, Vol. 33, pp. 105-115.

[Wie03] Wiewiora, E., 2003. Potential-Based Shaping and Q-Value Initialization are Equivalent," J. of Artificial Intelligence Research, Vol. 19, pp. 205-208.

[Wil92] Williams, R. J., 1992. "Simple Statistical Gradient-Following Algorithms for Connectionist Reinforcement Learning," Machine Learning, Vol. 8, pp. 229-256.

[YDR04] Yan, X., Diaconis, P., Rusmevichientong, P., and Van Roy, B., 2004. "Solitaire: Man Versus Machine," Advances in Neural Information Processing Systems, Vol. 17, pp. 1553-1560.

[YFG08] Yoon, S. W., Fern, A., Givan, R., and Kambhampati, S., 2008. "Probabilistic Planning via Determinization in Hindsight," in Proc. AAAI, pp. 1010-1016.

[YSH17] Ye, N., Somani, A., Hsu, D., and Lee, W. S., 2017. "DESPOT: Online POMDP Planning with Regularization," Journal of Artificial Intelligence Research, Vol. 58, pp. 231-266.

[Yar17] Yarotsky, D., 2017. "Error Bounds for Approximations with Deep ReLU Networks," Neural Networks, Vol. 94, pp. 103-114.

[YuB04] Yu, H., and Bertsekas, D. P., 2004. "Discretized Approximations for POMDP with Average Cost," Proc. of the 20th Conference on Uncertainty in Artificial Intelligence, Banff, Canada.

[YuB06] Yu, H., and Bertsekas, D. P., 2006. "On Near-Optimality of the Set of Finite-State Controllers for Average Cost POMDP," Lab. for Information and Decision Systems Report LIDS-P-2689, MIT; Mathematics of Operations Research, Vol. 33, 2008, pp. 1-11.

[YuB07] Yu, H., and Bertsekas, D. P., 2007. "A Least Squares Q-Learning Algorithm for Optimal Stopping Problems," Proc. European Control Conference 2007, Kos, Greece, pp. 2368-2375; an extended version appears in report LIDS-P-2731, MIT.

[YuB09a] Yu, H., and Bertsekas, D. P., 2009. "Convergence Results for Some Temporal Difference Methods Based on Least Squares," IEEE Trans. on Aut. Control, Vol. 54, pp. 1515-1531.

[YuB09b] Yu, H., and Bertsekas, D. P., 2009. "Basis Function Adaptation Methods for Cost Approximation in MDP," Proceedings of 2009 IEEE Symposium on Approximate Dynamic Programming and Reinforcement Learning (ADPRL 2009), Nashville, Tenn.

[YuB10] Yu, H., and Bertsekas, D. P., 2010. "Error Bounds for Approximations from Projected Linear Equations," Mathematics of Operations Research, Vol. 35, pp. 306-329.

[YuB12] Yu, H., and Bertsekas, D. P., 2012. "Weighted Bellman Equations and their Applications in Dynamic Programming," Lab. for Information and Decision Systems Report LIDS-P-2876, MIT.

[YuB13a] Yu, H., and Bertsekas, D. P., 2013. "Q-Learning and Policy Iteration Algorithms for Stochastic Shortest Path Problems," Annals of Operations Research, Vol. 208, pp. 95-132.

[YuB13b] Yu, H., and Bertsekas, D. P., 2013. "On Boundedness of Q-Learning Iterates for Stochastic Shortest Path Problems," Math. of OR, Vol. 38, pp. 209-227.

[YuB15] Yu, H., and Bertsekas, D. P., 2015. "A Mixed Value and Policy Iteration Method for Stochastic Control with Universally Measurable Policies," Math. of OR, Vol. 40, pp. 926-968.

[Yu05] Yu, H., 2005. "A Function Approximation Approach to Estimation of Policy Gradient for POMDP with Structured Policies," Proc. of the 21st Conference on Uncertainty in Artificial Intelligence, Edinburgh, Scotland.

[Yu12] Yu, H., 2012. "Least Squares Temporal Difference Methods: An Analysis Under General Conditions," SIAM J. on Control and Optimization, Vol. 50, pp. 3310-3343.

[Yu15] Yu, H., 2015. "On Convergence of Value Iteration for a Class of Total Cost Markov Decision Processes," SIAM J. on Control and Optimization, Vol. 53, pp. 1982-2016.

[ZBH16] Zhang, C., Bengio, S., Hardt, M., Recht, B., and Vinyals, O., 2016. "Understanding Deep Learning Requires Rethinking Generalization," arXiv preprint arXiv: 1611.03530.

[ZYL18] Zhang, K., Yang, Z., Liu, H., Zhang, T., and Basar, T., 2018. "Fully Decentralized Multi-Agent Reinforcement Learning with Networked Agents," arXiv preprint arXiv:1802.08757.

[ZhO11] Zhang, C., and Ordonez, R., 2011. Extremum-Seeking Control and Applications: A Numerical Optimization-Based Approach, Springer.

[ZhT11] Zhong, M., and Todorov, E., 2011. "Aggregation Methods for Linearly-Solvable Markov Decision Process," IFAC Proc., Vol. 44, pp. 11220-11225.

INDEX

Neuro-Dynamic Programming
Dimitri P. Bertsekas and John N. Tsitsiklis

Athena Scientific, 1996, 512 pp., hardcover, ISBN 1-886529-10-8

This is the first textbook that provided a treatment of the reinforcement learning methodology, a breakthrough in the practical application of neural networks and dynamic programming to complex problems of planning, optimal decision making, and intelligent control.

From the review by George Cybenko for IEEE Computational Science and Engineering, May 1998:

"Neuro-Dynamic Programming is a remarkable monograph that integrates a sweeping mathematical and computational landscape into a coherent body of rigorous knowledge. The topics are current, the writing is clear and to the point, the examples are comprehensive and the historical notes and comments are scholarly."

"In this monograph, Bertsekas and Tsitsiklis have performed a Herculean task that will be studied and appreciated by generations to come. I strongly recommend it to scientists and engineers eager to seriously understand the mathematics and computations behind modern behavioral machine learning."

Among its special features, the book:

- Describes and unifies a large number of NDP methods, including several that are new
- Describes new approaches to formulation and solution of important problems in stochastic optimal control, sequential decision making, and discrete optimization
- Rigorously explains the mathematical principles behind NDP
- Illustrates through examples and case studies the practical application of NDP to complex problems from optimal resource allocation, optimal feedback control, data communications, game playing, and combinatorial optimization
- Presents extensive background and new research material on dynamic programming and neural network training

In 1997, Dimitri Bertsekas and John Tsitsiklis were awarded the 1997 INFORMS CSTS prize for research excellence in the interface between Operations Research and Computer Science for the book "Neuro-Dynamic Programming."

In 2018, they were awarded the INFORMS John von Neumann Theory Prize for the contributions of the "Neuro-Dynamic Programming" book as well as the book "Parallel and Distributed Computation."